T0230720

THE EMERGING ECONOMIC GEOGRAPHY IN EU ACCESSION COUNTRIES

The Emerging Economic Geography in EU Accession Countries

Edited by

IULIA TRAISTARU
Center for European Integration Studies (ZEI), University of Bonn

PETER NIJKAMP
Department of Spatial Economics, Free University of Amsterdam

LAURA RESMINI
Institute of Latin American Studies and Transition Countries (ISLA), University 'Luigi Bocconi', Milan

Routledge
Taylor & Francis Group

LONDON AND NEW YORK

Contents

PART I: ANALYTICAL FRAMEWORK

PART II: COUNTRY STUDIES

PART III: COMPARATIVE ANALYSIS AND LESSONS

List of Figures and Boxes

List of Tables

Editors

Iulia Traistaru is a Senior Researcher at the Center for European Integration Studies (ZEI) of the University of Bonn. Her key areas of expertise include the economics of European integration, international trade, transition economics, economic geography and the economics of international security. She is currently scientific coordinator of the research program at ZEI on 'European integration, regional development and policy', which includes the undertaking of research projects, policy advising and the coordination of a European network of researchers and policy makers. She has studied Economics, International Economics and International Relations at the Academy of Economic Studies, Bucharest, University 'Luigi Bocconi', Milan and the University of Warwick, and received her doctoral degree in Economics from the Academy of Economic Studies, Bucharest. Her previous appointments include research fellowships at the Centre for Transition Economics, Catholic University of Leuven, Erasmus University Rotterdam, Conservatoire National des Arts et Metiers, Paris, and positions as Assistant and Associate Professor of Economics at the University 'Politehnica' of Bucharest. She has been a Consultant to the World Bank, European Commission, Inter-American Development Bank, and World Economic Forum.

Peter Nijkamp is Professor of regional and urban economics and economic geography at the Free University, Amsterdam. His main research interests include plan evaluation, multicriteria analysis, regional and urban planning, transport systems analysis, mathematical modeling, technological innovation, and resource management. In the past years he has focused his research, in particular, on quantitative methods for policy analysis, as well as on behavioural analysis of economic agents. He has broad expertise in the areas of public policy, services planning, infrastructure management and environmental protection. In all of these fields he has published many books and numerous articles. He is a member of the editorial boards of more than 20 journals. He has been a visiting professor at many universities all over the world. He is also a past president of the European Regional Science Association and of the Regional Science Association International. He is also a fellow of the Royal Netherlands Academy of Sciences, and is the immediate past vice-president of this organization. Since June 2002 he has been serving as president of the governing board of the Netherlands Research Council (NWO).

Laura Resmini is Assistant Professor at the Institute of Economics, 'L. Bocconi' University, Milan, and Senior Researcher at ISLA, Institute of Latin American and Transition Countries Studies, 'L. Bocconi' University. She holds a doctoral degree in Economics from the University of Venice (Italy) and a Certificate of International Studies with a specialization in International Economics from the Graduate Institute of International Studies, Geneva. Her main research interests include the economics of foreign direct investments, international trade and economic geography, with particular emphasis on Central and Eastern Europe.

Contributors

Jože P. Damijan
University of Ljubljana
Faculty of Economics
Kardeljeva pl. 17
1000 Ljubljana
Slovenia

Grigory Fainshtein
Estonian Institute of Economics
7 Estonia Avenue
10143, Tallinn
Estonia

Anna Iara
University of Bonn
Center for European Integration
Studies (ZEI)
Walter-Flex-Straße, 3
53113 Bonn
Germany

Črt Kostevc
University of Ljubljana
Faculty of Economics
Kardeljeva pl. 17
1000 Ljubljana
Slovenia

Simonetta Longhi
Tinbergen Institute Amsterdam
Keizersgracht 482
1017 EG Amsterdam
The Netherlands

Natalie Lubenets
Tallinn Technical University
Department of Economics
101, Kopli str.
11712, Tallinn
Estonia

Alessandro Maffioli
University of Insubria (Varese) and
University 'Luigi Bocconi'
Institute of Latin American Studies
and Transition Countries (ISLA)
Via Sarfatti 25
20136 Milan
Italy

Peter Nijkamp
Free University Amsterdam
Department of Spatial Economics
De Boelelaan 1105
1081 HV Amsterdam
The Netherlands

Carmen Pauna
Institute for Economic Forecasting
Regional Development Department
Calea 13 Septembrie 13
Casa Academiei, Sector 5
76117 Bucharest
Romania

Laura Resmini
University 'Luigi Bocconi'
Institute of Latin American Studies
and Transition Countries (ISLA)
Via Sarfatti 25
20136 Milan
Italy

Iulia Traistaru
University of Bonn
Center for European Integration
Studies (ZEI)
Walter-Flex-Straße, 3
53113 Bonn
Germany

Julia Spiridonova
National Center for Regional
Development
50, Levski Blvd.
1142 Sofia
Bulgaria

Preface

Over the last two decades, academic and policy interest in the spatial implications of economic integration has grown. This interest has been stimulated, on the one hand, by the challenges posed by the widening and deepening of economic integration, particularly in Europe and North America. On the other hand, recently-developed location and trade theories have provided scientists with new ways to analyze these spatial implications which has heightened their interest. These theories permit the modeling of convergence and divergences forces in a common analytical framework and have helped us to understand the uneven spatial distribution of the benefits and costs associated with economic integration. According to these new economic theories, the structural change that accompanies economic integration is likely to increase the degree of regional specialization and geographic concentration of industrial activity, which may make regions vulnerable to asymmetric shocks. Industry demand shocks may become region-specific shocks and the short-run adjustment costs may be high in the case of the relocation of firms. Offsetting benefits may occur however, because higher specialization and greater concentration of industrial activity are expected to increase productivity via economies of scale.

Since 1990, Central and East European countries (CEECs) have experienced increased economic integration with the European Union (EU), which has led to a reallocation of resources across sectors and space. While sectoral shifts in CEECs have frequently been analysed, the spatial implications of increasing economic integration in the EU accession countries have not been investigated in-depth so far. Where is industrial activity located? Have patterns of regional specialization and industrial concentration changed during the 1990s? How does regional specialization relate to economic performance? What are the determinants of industrial location patterns? How has increased trade liberalization affected regional relative wages? What types of regions are winners and what types of regions are losers?

This book is the first to provide answers to these policy relevant questions by bringing together the results of a research project entitled 'European Integration, Regional Specialization and Location of Industrial Activity in Accession Countries', which was undertaken during the period 1 October 2000–30 March 2002. We would like to acknowledge the financial support of the European Community's PHARE ACE Programme 1998, which allowed us to conduct this research in a European comparative context and to include researchers on our team from Germany, the Netherlands, Italy, Bulgaria, Estonia, Hungary, Romania and Slovenia.

We have benefited from excellent cooperation among the team members in fulfilling our demanding research objectives. We wish to thank the contributing authors for their efforts in following our research agenda and editorial guidelines. We have also benefited from stimulating discussions with Donat Magyari from the Ministry of Economic Affairs in Hungary, Narciza-Adela Nica from the Ministry for Development and Forecasting in Romania, and Andrej Horvath, from the Ministry for Economics in Slovenia. We wish to acknowledge the support we have received from the Center for European Integration Studies at the University of Bonn, the Department of Spatial Economics at Free University of Amsterdam and the Institute of Latin American Studies and Transition Countries (ISLA) at the University 'Luigi Bocconi' Milan. In particular, we thank Jürgen von Hagen, Director of the Center for European Integration Studies at the University of Bonn, and Carlo Secchi, Director of the Institute of Latin American Studies and Transition Countries at the University 'Luigi Bocconi' Milan, for their constant encouragement to conduct this research and for inspiring discussions on European integration and the accession of Central and Eastern European countries to the European Union. Special thanks go to Manfred Fischer, Ronald Moomaw, and George Petrakos, for their sound and constructive comments and suggestions. There are also many friends and colleagues we would like to thank for very stimulating discussions we have had on the occasion of seminars, workshops and conferences and informal meetings. Our gratitude goes to all of them. Finally, we wish to thank Dawn Blizard for her excellent editorial assistance. Her careful and dedicated linguistic editing has contributed to the quality of this book.

Iulia Traistaru, Center for European Integration Studies,
University of Bonn
Peter Nijkamp, Department of Spatial Economics,
Free University of Amsterdam
Laura Resmini, Institute of Latin American Studies and Transition
Countries (ISLA), University 'Luigi Bocconi' Milan

PART I
ANALYTICAL FRAMEWORK

Chapter 1

Spatial Implications of Economic Integration in EU Accession Countries

Laura Resmini and Iulia Traistaru

1 Introduction

Over the last decade, the economic and political landscape of Europe has changed considerably in response to two important phenomena. On the one hand, economic integration has evolved through the Single European Market, the Economic and Monetary Union and the fourth enlargement of the European Union (EU), together with the implementation of policies aimed at increasing the efficiency and competitiveness of the European economy in light of the challenges posed by the global economy. On the other hand, Central and Eastern European countries (CEECs) have experienced the process of transition to a market economy and democracy with its tremendous pressures on national economies for economic and institutional reforms.

These phenomena are more complex than one would think at first, for at least two reasons, as underlined by Petrakos (2000). First of all, the structural changes they imply aim at gradually eliminating historical divides, thus giving the countries involved a real opportunity to remove barriers that for decades have restricted economic and social interactions, though in different ways and degrees, depending upon whether the countries concerned were located in Western or Eastern Europe. Secondly, transition is intertwined with integration, because it is deeply involved with the EU. Because of its proximity and economic weight, the EU quickly assumed political responsibility for helping CEECs reintegrate themselves into the world economy. The EU first upgraded the status of CEECs to that of the least developed countries by granting them the Generalized System of Preferences, and then by signing the Europe Agreements (EAs), which went well beyond the scope of market access by opening the path to deep policy-induced integration (Kaminski, 2001).

For the CEECs, integration into the European economy is a crucial pillar of the transition process. Free trade and Foreign Direct Investment (FDI) have fostered the restructuring of these economies, thus establishing a more solid basis for growth and economic development. Furthermore, the opening of CEECs' markets has also brought advan tages to Western economies, which besides geographical proximity, have close historical and cultural links with CEECs.

While research initially focused on the transformation of formerly centrally planned economies and its optimal speed (Aghion and Blanchard, 1994; Blanchard, 1997), as well as on its effects on Western European economies (Baldwin, 1997), the spatial implications of economic integration have not been investigated in-depth so far. A general concern, however, exists, that the structural changes accompanying economic integration are likely to result in increasing regional specialization and concentration of economic activity, which may make regions in the current EU and in the CEECs more vulnerable to asymmetric shocks.

In this context, the main purpose of this introductory chapter is to take a closer look at the process of economic integration of Western and Eastern Europe, emphasizing two critical issues: trade and FDI . First of all, we will provide a critical overview of trade patterns between the two areas. Trade is usually the simplest and most common link between economic units and is a very sensitive indicator of changes in the economic environment. Secondly, patterns of FDI un dertaken by EU firms in CEECs will be analyzed in order to provide a broader picture of the system of economic integration. FDI involves mergers and acquisitions, the purchase of privatized firms and investments in new production plants, as well as agreements on cooperation and technological transfers. Consequently, FDI can be considered a measure of industrial relationships between investing and receiving firms, countries and/or regions.

Finally, an attempt is made to understand the possible spatial implications of the aforementioned phenomena. Economic integration does not necessarily affect spatial patterns of economic activity, but the evidence shows that geographical proximity and strong economic ties are strictly correlated. A nalysis of in ter national trade and investment flows using gravity models provides an example of this evidence (Brainard, 1993 and 1997). Moreover, it is also true that economic integration, as well as political and social integration, reinforces the central dimension of spatial integration, i.e., accessibility, making it more likely that changes in the locations of economic activities within and across countries will occur.

The remaining chapters of the book are aimed at understanding whether and to what extend the åintegration of CEECs into European and world market systems has contributed to changes in the economic geography of CEECs.

2 Economic Integration with the EU

2.1 Trade Relations

All indicators derived from trade statistics point to the exceptional expansion of trade between CEECs and the EU during the 1990s.

In 1999, CEEC exports to the EU were on average about six times higher than they had been in 1989, while imports were seven times higher (Figure 1.1). In relative terms, the change in the importance of the EU as a trade partner for CEECs is even more evident: since 1989, exports to and imports from the EU have out-performed growth in total exports and imports by an average of 10 per centage points. As a result, the EU increased its share in CEEC imports from 27 to 62 per cent between 1989 and 1999 (Table 1.1). As the Community market for industrial products tended to become more open, the increase in the EU's share in CEEC exports was even more pronounced, rising from 28 to 69 per cent. From an EU perspective, CEECs are still marginal partners, both in absolute and relative terms. However, CEECs' shares (taken as a whole) in extra-EU imports and exports have steadily increased during the 1990s, from 3 per cent in 1990 up to 12 and 9 per cent in 1999, respectively. Given these impressive changes, at the end of the period the trade balance was in favour of the EU.

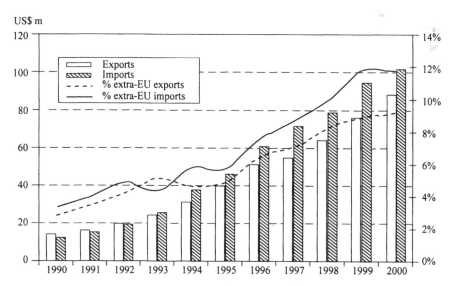

Figure 1.1   CEECs' trade with the EU and its share in extra-EU trade

Source: Authors' calculation from IMF, *Direction of Trade Statistics*, 2002.

These basic patterns are common to all countries, with a few differences among them, as is shown in Table 1.1. Trade dependence on the EU varies somewhat: the Czech Republic, Hungary, Poland and Slovenia trade significantly more with the EU than the CEECs, on average. In 1999 the simple average for the four countries was about 65 per cent on the import side and almost 71 per cent on the export side. It is worth noting that the EU's share in Slovenian external trade has remained more or less constant over the period, suggesting that Slovenia was already well integrated with the EU at the beginning of the transition.

Trade growth rates, however, vary substantially over countries and over time. Although Estonia experienced the largest increases in exports, the base was very low, in contrast, for instance, to Hungary. Nonetheless, both have had superior performance with re spect to other countries.[1] Bulgaria and Slovenia experienced the smallest increases, though this is weighted by a different base. On the imports side, the picture is not much different. The Baltic states perform better than the others, with Romania, Slovenia and Bulgaria at the bottom of the ranking. Hungary had impressive gains in the early phase of transition, mainly in exports into the EU (Table 1.2).

The redirection of trade flows from East to West has been accompanied by changes in the composition of trade flows, both on the import side and the export side. At the beginning of the transition process, the composition of trade flows was not favourable to CEECs: they imported consumer goods to satisfy particularly high internal demand and exported low-skilled labour intensive products (such as textiles, clothing and leather products) to the EU, as well as fuels, basic chemicals and metals, a specialization inherited from the central planning period. However, the situation changed rapidly. Exports soon came to include machinery, electronic products and motor vehicles as well, while imports from the EU shifted towards technology-intensive goods, increasing the share of intra-industry trade in total trade (Hoekman and Djankov 1996; Landesmann, 1995; Dohrn, 2001; Weise, Bachtler, Down et al., 2001,) and further increasing (or deepening) economic integration with the EU.

Sorting out which factors are responsible for this considerable trade integration is an arduous task, since integration-induced effects are impossible to separate from those stemming from the dismantling of planned economies or those that were policy-induced. However, a few attempts have been made to explain these changes. According to these studies, it is likely that these interrelated effects have played different roles during different phases of the transition. At the beginning, the shift in commercial relations towards the EU might have been driven by natural causes (Dohrn, 1998; Kaminski,

Table 1.1 **CEECs' exports to and imports from the EU (shares in total CEECs exports and imports)**

	1989	1990	1991	1992	1993	1994	1995	1996	1997	1998	1999
Bulgaria											
Exports	0.19	0.34	0.40	0.29	0.29	0.40	0.38	0.29	0.36	0.50	0.52
Imports	0.35	0.47	0.48	0.36	0.40	0.56	0.37	0.26	0.31	0.47	0.49
Czech R.*											
Exports	0.25	0.32	0.41	0.54	0.44	0.46	0.43	0.58	0.60	0.64	0.71
Imports	0.28	0.36	0.34	0.49	0.44	0.46	0.46	0.58	0.52	0.64	0.65
Estonia											
Exports	–	–	–	0.31	0.48	0.48	0.55	0.51	0.49	0.57	0.63
Imports	–	–	–	0.28	0.61	0.63	0.66	0.64	0.59	0.62	0.58
Hungary											
Exports	0.24	0.35	0.46	0.62	0.56	0.64	0.65	0.65	0.73	0.73	0.76
Imports	0.29	0.37	0.39	0.57	0.54	0.62	0.63	0.61	0.64	0.64	0.64
Latvia											
Exports	–	–	–	0.40	0.33	0.39	0.44	0.44	0.49	0.57	0.63
Imports	–	–	–	0.29	0.33	0.44	0.47	0.47	0.51	0.53	0.45
Lithuania											
Exports	–	–	–	0.89	0.38	0.30	0.36	0.33	0.33	0.38	0.49
Imports	–	–	–	0.83	0.30	0.32	0.37	0.41	0.44	0.47	0.46
Poland											
Exports	0.32	0.47	0.56	0.62	0.69	0.70	0.70	0.66	0.64	0.71	0.71
Imports	0.36	0.45	0.54	0.58	0.65	0.66	0.65	0.64	0.64	0.67	0.65
Romania											
Exports	0.31	0.32	0.34	0.35	0.41	0.48	0.55	0.53	0.56	0.63	0.65
Imports	0.06	0.20	0.26	0.40	0.45	0.48	0.51	0.45	0.51	0.56	0.59
Slovenia											
Exports	–	–	–	–	0.63	0.66	0.68	0.65	0.64	0.65	0.65
Imports	–	–	–	–	0.66	0.70	0.70	0.68	0.67	0.69	0.68
Slovakia											
Exports	–	–	–	–	0.30	0.35	0.37	0.41	0.55	0.56	0.60
Imports	–	–	–	–	0.28	0.33	0.35	0.36	0.45	0.46	0.47
CEECs											
Exports	*0.28*	*0.37*	*0.46*	*0.54*	*0.51*	*0.55*	*0.55*	*0.57*	*0.60*	*0.65*	*0.69*
Imports	*0.27*	*0.35*	*0.42*	*0.51*	*0.51*	*0.56*	*0.55*	*0.56*	*0.57*	*0.62*	*0.62*

* Czechoslovakia before 1993.

Source: Authors' calculations from IMF, *Direction of Trade Statistics*, 2002.

Table 1.2   Annual changes in CEECs' total exports and imports, 1990–99

		1990	1991	1992	1993	1994	1995	1996	1997	1998	1999
Bulgaria	Total	-0.26	0.01	0.87	-0.05	0.06	0.36	0.23	-0.19	-0.19	-0.09
	EU	0.31	0.19	0.35	-0.05	0.43	0.29	-0.05	0.02	0.10	-0.04
Czech R.*	Total	-0.18	-0.08	0.03	0.28	0.12	0.34	0.01	0.04	0.16	-0.01
	EU	0.03	0.19	0.37	0.04	0.18	0.24	0.38	0.06	0.25	0.09
Estonia	Total	–	–	–	1.26	0.64	0.40	0.13	0.41	0.07	-0.06
	EU	–	–	–	2.57	0.61	0.60	0.05	0.34	0.26	0.03
Hungary	Total	-0.03	0.05	0.05	-0.17	0.20	0.16	0.02	0.47	0.23	0.09
	EU	0.42	0.37	0.43	-0.25	0.37	0.18	0.02	0.65	0.23	0.13
Latvia	Total	–	–	–	0.28	-0.01	0.32	0.11	0.16	0.08	-0.05
	EU	–	–	–	0.08	0.16	0.46	0.11	0.30	0.26	0.05
Lithuania	Total	–	–	–	0.67	0.76	0.33	0.24	0.15	-0.04	-0.18
	EU	–	–	–	0.27	-0.21	0.61	0.11	0.15	0.12	0.07
Poland	Total	0.01	0.09	-0.11	0.06	0.20	0.34	0.07	0.05	0.06	0.01
	EU	0.47	0.30	-0.01	0.19	0.22	0.34	0.01	0.02	0.17	0.00
Romania	Total	-0.45	-0.26	0.02	0.12	0.26	0.29	0.02	0.04	-0.02	0.02
	EU	-0.42	-0.22	0.07	0.32	0.47	0.48	-0.03	0.11	0.11	0.05
Slovenia	Total	–	–	–	–	0.12	0.22	0.00	0.01	0.08	-0.05
	EU	–	–	–	–	0.18	0.24	-0.05	-0.01	0.11	-0.05
Slovakia	Total	–	–	–	–	0.23	0.28	0.03	-0.06	0.30	-0.05
	EU	–	–	–	–	0.45	0.37	0.14	0.25	0.31	0.02
CEECs	Total	-0.16	-0.02	0.07	0.33	0.19	0.29	0.05	0.08	0.11	0.00
	EU	0.14	0.22	0.25	0.27	0.26	0.30	0.08	0.15	0.20	0.05

Imports

		1990	1991	1992	1993	1994	1995	1996	1997	1998	1999
Bulgaria	Total	-0.33	-0.21	0.64	0.07	-0.12	0.35	-0.11	-0.02	0.01	0.09
	EU	-0.11	-0.18	0.22	0.19	0.25	-0.12	-0.15	-0.09	0.43	0.13
Czech R.*	Total	-0.09	-0.19	0.14	0.20	0.21	0.45	0.10	-0.02	0.06	0.00
	EU	0.16	-0.23	0.62	0.10	0.25	0.46	0.39	-0.14	0.31	0.02
Estonia	Total	–	–	–	0.67	0.87	0.53	0.27	0.37	0.04	-0.11
	EU	–	–	–	2.58	0.95	0.59	0.23	0.27	0.10	-0.18
Hungary	Total	-0.02	0.33	-0.03	0.13	0.15	0.05	0.05	0.30	0.24	0.09
	EU	0.25	0.41	0.41	0.07	0.31	0.07	0.02	0.37	0.25	0.09
Latvia	Total	–	–	–	0.07	0.44	0.53	0.27	0.17	0.17	0.05
	EU	–	–	–	0.19	0.95	0.63	0.25	0.27	0.22	-0.10
Lithuania	Total	–	–	–	1.26	0.03	0.55	0.25	0.24	0.03	-0.16
	EU	–	–	–	0.38	0.09	0.79	0.38	0.33	0.09	-0.18
Poland	Total	-0.21	0.87	0.00	0.20	0.14	0.36	0.28	0.14	0.10	-0.01
	EU	-0.01	1.23	0.06	0.35	0.15	0.33	0.26	0.14	0.15	-0.04

Table 1.2 cont'd

		1990	1991	1992	1993	1994	1995	1996	1997	1998	1999
Exports											
Romania	Total	0.09	-0.42	0.08	0.04	0.09	0.45	0.11	-0.01	0.05	-0.12
	EU	2.60	-0.23	0.65	0.15	0.16	0.54	-0.02	0.12	0.16	-0.08
Slovenia	Total	–	–	–	–	0.12	0.30	-0.01	-0.01	0.08	-0.02
	EU	–	–	–	–	0.20	0.30	-0.05	-0.01	0.11	-0.03
Slovakia	Total	–	–	–	–	0.05	0.33	0.26	-0.08	0.27	-0.13
	EU	–	–	–	–	0.25	0.38	0.32	0.16	0.28	-0.11
CEECs	*Total*	*-0.10*	*0.07*	*0.11*	*0.42*	*0.13*	*0.33*	*0.16*	*0.07*	*0.11*	*-0.30*
	EU	*0.19*	*0.28*	*0.34*	*0.44*	*0.23*	*0.31*	*0.18*	*0.10*	*0.20*	*-0.02*

* Czechoslovakia before 1993.

Source: Authors' calculations from IMF, *Direction of Trade Statistics*, 2002.

Wang and Winters, 1996). For most of CEECs, the EU has always been the largest potential trading partner and the most important source of capital and technology. However, under central planning, they usually under-traded with the EU and over-traded with the socialist countries (CMEA members). Once the state monopoly over international trade was abolished and economic regimes had become more market- oriented, the EU quickly emerged as their largest trading partner. Thus, a sizeable portion of the readjust ment in trade patterns might be due to a correction of the earlier patterns, which were distorted by political considerations. Figure 1.2 confirms the validity of this hypothesis: there is a clear inverse relationship, though not entirely systematic, between the annual average growth rates over the period and the initial level of export and import volumes with the EU.

While the volume of trade between CEECs and the EU influenced the pace of the trade reorientation at the beginning, the evidence indicates that subsequent trade expansion was driven by industrial restructuring. Hungary, Poland, Slovenia and the Czech Republic are the countries that have been the most integrated with the EU since 1995, and they are also the most advanced in the transition.

The effects of policy-induced integration are difficult to evaluate. The EAs were signed after most CEECs had started their transition processes. Thus, they have been exerting their effects explicitly only since the end of the 1990s, when CEECs started to harmonize their domestic economies and

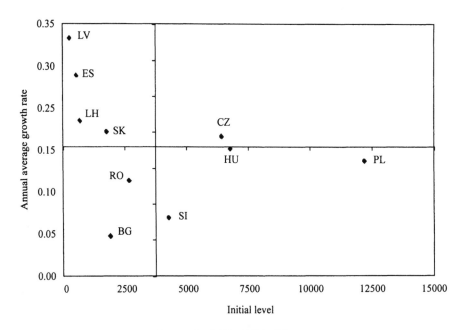

Figure 1.2a   Imports from the EU, 1993–99

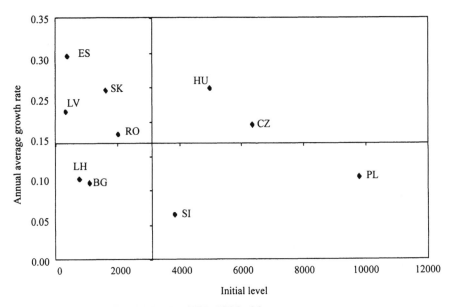

Figure 1.2b   Exports to the EU, 1993–99

Source: Authors' calculation from IMF, *Direction of Trade Statistics*, 2002.

institutions to those of the EU (Kaminski, 2001). However, because the trade provi sions of the EAs entered into force earlier in the 1990s, the EAs have played an indirect role in shaping CEECs' new trade regimes, promoting trade integration with the EU and its main commercial partners.

2.2   FDI Inflows

The above considerations of trade realignment of CEECs also apply to the case of FDI, with minor changes. No investment had been undertaken in CEECs by European investors before 1990.[2] However, shortly after the fall of the Berlin wall, foreign investment was encouraged by virtually all CEECs. Institutional reasons, such as the privatization process and fiscal and other promotional measures, as well as economic factors, such as expanding domestic markets, low labour costs, the endowment of skilled labour forces and other natural resources, are at the base of the CEECs' success in attracting FDI. Analogous to trade, proximity has played an important role in determining FDI flows. According to a number of studies, about two-thirds of the total FDI inflows in CEECs originate from the EU (Eurostat, 2002; Alessandrini, 2000).

At the end of 1999, the cumulative inflows into CEECs of FDI[3] from Western European enterprises amounted to €46 billion, about 7 per cent of all FDI assets held by the EU abroad (Eurostat, 2001). EU FDI levels have increased considerably over the 1990s, from about two billion Ecus in 1992 to 11.2 billion in 1999, with an annual average growth rate of about 57 per cent (see Figure 1.3). Despite that, the CEECs share of total EU FDI outflows shrank over the period, from 11 per cent in 1992 to 4 per cent in 1999. This surprising result implies that total EU FDI outflows abroad have grown more than those directed to CEECs. FDI outflows directed elsewhere recorded an average annual growth rate of about 69 per cent over the period, with growth rates in the second half of the decade higher than those experienced before 1995. The opposite is true for CEECs (see Table 1.3).

The main countries receiving FDI among the CEECs are Poland (41 per cent of the cumulated total of EU FDI flows into CEECs from 1992 to the end of 1999), the Czech Republic (25 per cent) and Hungary (20 per cent), which together obtained more than 80 per cent of the total FDI assets in the CEECs. The concentration of FDI in the aforementioned countries was reduced in the second half of the decade (94 per cent in 1995 and 87 per cent in 1999). The reduction of the share of investment in Hungary and Czech Republic exceeded the increase for Poland during the same period, thus leaving more room for other countries.

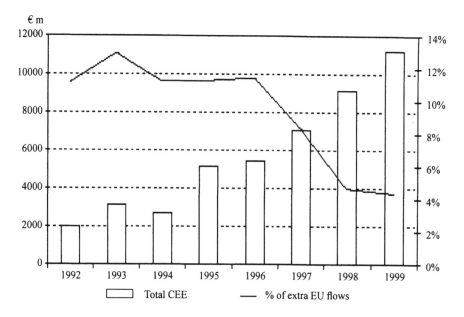

Figure 1.3   FDI inflows from the EU and their share in total extra-EU FDI flows

Source: Authors' calculations from Eurostat, *European Union direct investment yearbook*, 2001 edition.

In particular, Table 1.3 shows that Bulgaria, Estonia, Lithuania and Romania have increased their shares of the total EU FDI assets over the period, while Latvia, Slovenia, and Slovakia have just about maintained their positions. Hungary presents a reduction in FDI received from the EU both in relative and absolute terms. The ECU 2.1 billion registered in 1995, the highest in the region, was the peak of the time series, which reached its bottom in 1999 when Hungary registered a disinvestment of €2 million, not considering reinvested earnings. This trend was driven by the advancement of the privatization process, which was nearly complete in 1999 (UNCTAD, 2001).

More information on dynamics can be obtained by combining flow and stock data. Relating the sum of previous years' FDI flows to the last year's (1999) end stocks (proxied by cumulated flows over the period), it is possible to get an impression of how recently FDI was established. A high ratio indicates that a significant share of the last year's FDI assets was established during the period considered. A lower value indicates that a higher proportion of assets were cumulated in a previous period. Figure 1.4 shows the results of this simple exercise for two periods, 1997–99 and 1999 alone. It is clear that Hungary

Table 1.3 EU FDI* Outflows to CEECs (mil. ECU/EUR)

	1992	1993	1994	1995	1996	1997	1998	1999	Total	% 99	% 95
Bulgaria	9	31	63	9	50	140	172	136	610	0.01	0.00
Czech R.	768	812	974	1,594	1,299	1,916	1,576	2,534	11,å473	0.25	0.31
Estonia					62	73	362	232	729	0.02	0.00
Hungary	989	1,217	839	2,102	1,073	1,565	1,537	-2	9,320	0.20	0.41
Latvia					21	46	45	-6	106	0.00	0.00
Lithuania					57	52	415	245	769	0.02	0.00
Poland	230	758	616	1,132	2,427	2,492	4,189	7,076	18,920	0.41	0.22
Romania	-12	25	49	75	136	409	437	543	1662	0.04	0.01
Slovakia		242	107	139	213	253	271	210	1435	0.03	0.03
Slovenia		31	51	68	64	99	136	226	675	0.01	0.01
CEECs (1)	1,984	3,116	2,699	5,119	5,402	7,045	9,140	11,194	45,699		
Change		0.57	-0.13	0.90	0.06	0.30	0.30	0.22	0.57		
Extra-EU											
15 (2)**	17.88	24.25	24.19	45.6	47.42	84.7	198.2	259.3	701.3		
Change		0.36	0.00	0.89	0.04	0.79	1.34	0.31	0.69		
(1)/(2)	0.11	0.13	0.11	0.11	0.11	0.08	0.05	0.04	0.07		

*   Excluding reinvested earnings.
** Billions of ECU/EUR.

Changes are computed annually from 1993 to 1999; % 99 and %95 refer to country's share of cumulated flows in the respective year .

Source: Eurostat, *European Union Direct Investment Yearbook,* 2001 edition and authors' calculations.

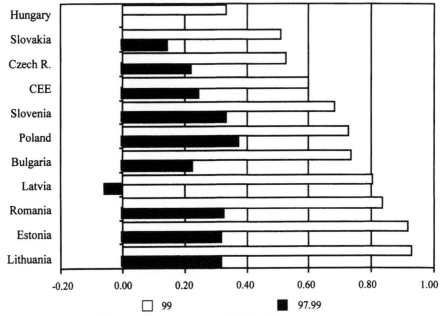

Figure 1.4   FDI flows over stocks, 1997–99

Source: Authors' calculations from Eurostat, *European Union Direct Investment Yearbook,*
2001 edition.

belongs to the second case, since only 33 per cent of its EU FDI was established in 1997–99, as compared with the 80 per cent or more of Latvia, Lithuania, Estonia and Romania. The Czech and Slovak Republics also show relatively early inflows of FDI; only 50 per cent of their FDI was cumulated in the late 1990s. In 1999, Poland and Slovenia also gained momentum, with EU FDI flows representing 37 and 33 per cent of the final assets of that year, respectively.

CEECs also differ according to the importance of FDI in their economies. Figure 1.5 compares the position of CEECs in 1995 and 1999, combining the absolute level of EU FDI flows (indicated by the area of the pies) with two key indicators of FDI penetration in the host economy, i.e., FDI per capita (vertical axis) and FDI flows as a percentage of GDP (horizontal axis) .[4]

The figure shows that in the second half of the 1990s, EU FDI had increased in 1999 in all CEECs not only in absolute terms, (the pies for 1999 are all bigger) but also in terms of its impact on the host economy, as is suggested by the pies positioned along the upwards diagonal. The only exception is Hungary: in 1995 it was ahead of the rest of the CEECs, with the highest inflow of EU FDI – both in absolute and per capita terms – which accounted for about 6

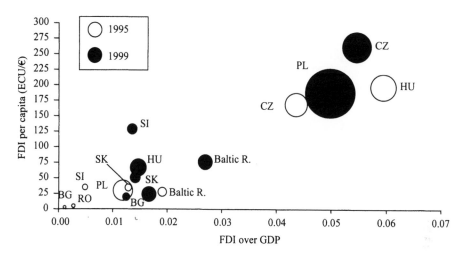

Figure 1.5   EU FDI in CEECs: FDI per capita, FDI over GDP and total values, 1995 and 1999

Source: Authors' calculations from Eurostat, *European Union Direct Investment Yearbook*, 2001 edition.

per cent of its GDP. In 1999, EU FDI in Hungary was negative, leaving the leading positions to the Czech Republic and Poland. With a ratio of FDI to GDP of about 5 per cent each, they are positioned near the advantageous top-right hand portion of the graph, very far from the other countries. Among the 'followers', Slovenia and the Baltic countries have emerged. Slovenia had the highest level of FDI per capita, while the Baltics had the greatest ratio of FDI to GDP. Despite impressive developments, Bulgaria and Romania are in less favourable positions, with low FDI per capita levels and FDI still having a relatively modest impact on GDP.

Overall, these results indicate that the viability of institutional and economic reforms plays a key role in the location of foreign investments. Not surprisingly, countries with stable institutional and economic environments as well as strong trade relationships with the EU benefited more significantly from EU FDI than other countries. The concentration of EU FDI in Poland, the Czech Republic and Hungary may be interpreted as an indication that a country's prospects for accession contribute considerably to its selection as a destination country for FDI. Causality might, however, also work in the opposite direction, with consistent flows of FDI positively contributing to economic stability and trade integration, thus increasing the credibility of new institutions and their prospects for accession.[5]

These considerations do not help us to understand the allocation of FDI within countries or the motivations that drive European firms' decisions to produce in, rather than export to, Central and Eastern Europe. This second topic, though interesting, is not directly linked with the issue being explored in this chapter and in the following chapters. Here, we would like to demonstrate the degree to which Western and Eastern European economies have integrated since the begin ning of the transition process and the possible implications of this process for candidate countries' subnational economic systems. From this standpoint, it is clearly important to know the amount of total FDI inflow, its distribution within countries, and its impact on domestic economies, rather than the motivations for undertaking direct investment.[6]

Turning to the allocation of FDI within countries, we find that the available information is often unsystematic and not homogenous. Generalizing the available, though incomparable, evidence for Hungary, Latvia, Lithuania and Poland in 1997 and 1999, Dohrn (2001) points out that FDI is highly concentrated in capitals and in regions bordering the EU.[7] However, this process is not uniform over time and across sectors. Using the number of foreign firms in each NUTS 3 region in the Czech Republic, Hungary, Poland, and Romania for the period 1990–97 as a unit of measure, Altomonte and Resmini (2002a) demonstrate that the Western European firms that invested in CEECs first located in the capital cities and then spread throughout the rest of the countries, without showing any preference for particular locations.[8] Although this process is common to all economic activities, it is particularly evident for the manufacturing sector. Only in the service sector do capital cities show a higher concentration of FDI than the national average.

3 Spatial Implications of Economic Integration: the Role of Geography

The phenomena briefly described in the previous section pose a major challenge for regional economic analysis. The redirection of international trade flows from East to West and the shift in economic regimes from supply-constrained to demand-constrained raise important questions about the allocation of economic activity over space, not only among countries but also within countries.

Although the economic literature on the spatial implications of the eastern enlargement is rich and full of interesting contributions, which have highlighted the most relevant aspects at both national and, when possible, regional levels

(Baldwin, 1997; Petrakos , 1996; Dohrn 2001), a few studies explicitly consider the role of geography as a determinant of the allocation of economic activity across space (Sachs, 1997).

Why should the enlargement process have a geographical component? First of all, the geography of the ten candidate countries is quite different: only four of them – Poland, Hungary, the Czech Republic and Slovenia – border the core EU member countries, while the others share borders with less developed and more troubled countries that are not presently involved in the enlargement process. Some are landlocked – such as Hungary and the Czech and Slovak Republics – while others have important and well-developed coastal areas.

Second, trade does not occur in a homogeneous space. Factors such as distance, accessibility and centrality thus play an important role in determining the location of economic activities across space, giving countries (and regions) closer to the core EU markets, and/or well endowed with high quality infrastructures, better prospects for growth in the new European economic space created by the pro gressive elimination of trade barriers.

Third, geography also matters for FDI. Several studies have already demonstrated, though in a different context, that accessibility – broadly defined as endowment and quality of infrastructure and closeness to core markets – plays an important role in foreign firms' locational decision process.

Last but not least, the geography of FDI has strong economic implications. It is widely accepted that FDI had a positive impact on transition countries' economic development processes (UN/ECE, 2001). As a matter of fact, FDI brought capital, new technologies, managerial skills and market competencies into CEECs: these factors are all needed in order to overcome the severe restructuring problems of the old industrial system and to speed up the relaunch of the economy and economic growth. Beyond these direct effects, FDI may generate technological spillovers and pecuniary externalities (input-output linkages with domestic firms), if well embedded within local industrial fabrics so that it ignites self-sustained processes of endogenous growth (Blomstrom and Kokko, 1996; Markusen and Venables, 1999; Altomonte and Resmini, 2002b). These considerations indicate that the location of FDI is able to affect both the path and the pace of the restructuring process at national and subnational levels.

All these phenomena might eventually result in a different allocation of economic activity across space, provided that production factors and firms are mobile. This reallocation creates winners and losers and challenges regions' capabilities to face structural change. The process of transformation itself has strong spatial implications that are linked to endogenous factors such as the pace and success of restructuring policies and the existing condition of labour markets and infrastructure (Petrakos, 2000). A number of studies have already discussed the spatial impact of the transition (Petrakos, 1996). Here, we would only like to point out that geography also affects the transition process, since the countries adjacent to the EU, namely Poland, Hungary, the Czech Republic and Slovenia, are leading the transition process and will enter the EU in the first round, while the peripheral ones, such as Bulgaria and Romania, will be integrated more slowly. Eastern European countries, such as Russia, Ukraine, Belarus and Moldova, still face serious economic problems, which seem to further increase their geographical distance from the economic centers of Western Europe.

These facts not only indicate that integration with the EU and transition are strictly interrelated and are often driven by the same forces, but also that the former is likely to be able to compensate for, or, at least, not further exacerbate the spatial implications of the latter. The remainder of this book is devoted to these theses.

4 Aims and Outline of this Book

The present book brings together the results of a research project entitled 'European Integration, Regional Specialization and Location of Industrial Activity in Accession Countries'[9] that attempted to identify, explain and compare the effects of economic integration on patterns of regional specialization and location of industrial activity in five accession countries, namely Bulgaria, Estonia, Hungary, Romania and Slovenia.

The focus is on the changes in their economic geography resulting from integration. To reach this broad and ambitious objective, this research project concentrates on the following issues:

- patterns of regional specialization and geographical concentration of manufacturing activities during the 1990s within the selected countries and also on a comparative basis;
- the determinants of these patterns and the consequences of changes in the allocation of manufacturing activities for regional wages and economic growth;
- potential winners and losers in the enlargement process, both at country and regional levels, and lessons from economic integration with the EU for policy.

The research combines country and comparative analyses. T he project has been carried out in three stages, which are represented in the three parts into which this book is divided.

Part I, 'Analytical Framework', identifies the objectives of the analysis, theoretical hypotheses and predictions, as well as the research methodology. As illustrated in this chapter, CEECs have experienced increasing integration with the EU via trade and FDI since 1990. The anticipated accession of these countries to the EU motivates the need to assess the regional impact of their increasing economic integration with the EU. In order to understand how and to what extent trade and FDI affect a country's economic geography, one needs to identify the underlying forces that shape the spatial distribution of economic activity.

In Chapter 2, Nijkamp, Resmini and Traistaru review economic theories and empirical evidence about the impact of economic integration on specialization patterns and the location of industrial activity. Two important strands of the main economic literature have been analyzed, namely location theory and international trade theory. As suggested by Adam Smith, location and trade are two interrelated phenomena that play a central role in regional economics and economic geography; consequently, it is important to know and understand location's impacts on trade flows as well as trade's impacts on location decisions. The authors provide a critical overview of a wide set of models explaining the nature, motivations and theoretical implications of the location of production, from Von Thünen (1842) to Ohlin (1933) – who considered international trade as a part of general localization theory – as well as the recent attempts by Fujita, Krugman and Venables (1999) to build a 'New Economic Geography' (NEG). Theoretical implications are then discussed in light of recent empirical findings concerning, on the one hand, the widening and deepening of the European integration process (Brü lhart, 2001; Amiti , 1999; Hallet, 2000), and, on the other hand, additional and more recent experiences of economic integration, such as the creation of NAFTA (Hanson, 1996 and 1998).

In the third chapter, Traistaru and Iara describe the data set and measures used for the analysis of specialization and industrial location patterns in the five accession countries included in this study. A unique and original data set (REGSTAT) has been specially generated, containing regional indicators covering the period from 1990–99. This data set includes regional employment by sectors of activity and manufacturing branches, average earnings, unemployment, demographic variables, Gross domestic product (GDP), exports, measures of infrastructure, enter prise intensity, education, research and development (R&D), and public expenditures. The main characteristics of this database are extensively discussed and absolute and relative measures of regional specialization and geographical concentration of industrial activity are defined and explained. The selected measures have two important properties from an economic geography standpoint: they are comparable across industries or locations, and they take into account the overall distribution of activity across sectors (for specialization) and regions (for location), as suggested by Overman, Redding and Venables (2001).

Part II, 'Country Analyses', includes five country studies that address the above-mentioned research questions. In particular, each country study first provides empirical evidence of increased integration with the EU in terms of trade composition and FDI penetration; then, it discusses the main changes that have occurred in regional specialization and industrial concentration patterns due to the previously identified changes. Absolute and relative specialization and concentration measures have been computed using employment, output and export data. Thus, a discussion of similarities and differences as well as the statistical significance of each measure is also provided at a country level. What usually emerges is the reshaping of economic geography, with increasing regional specialization patterns in Bulgaria and Romania, decreasing regional specialization patterns in Estonia, and not significantly changed patterns in Hungary and Slovenia.

Spiridonova finds that in Bulgaria (Chapter 4), during the period 1990–99, absolute and relative specialization has increased in 20 regions out of 28, at annual rates ranging between 1.3 and 1.9 per cent, according to the measure used. Most manufacturing industries have become more concentrated. The most concentrated industries include fuels, basic metals, and paper and printing products, while the least concentrated industries are in traditional sectors, such as food, beverages and tobacco and textiles and apparel. There has been a relative decrease in industrial employment in the big cities (Sofia, Varna, Plovdiv, Gabrovo, Stara Zagora) and traditional industrial centers (the southwest, south-central and north-central regions). GDP per capita is significantly and negatively associated with the regional unemployment rate, suggesting that poor regions are more likely to suffer high unemployment.

In the case of Estonia (Chapter 5), Fainshtein and Lubenets demonstrate that the process of integration with the EU has resulted in substantial structural regional changes. Specialization has increased by 2 to 5 per cent a year, depending on the level of development. Relocation activity occurred from fairly developed regions (Northern Estonia) to the poorly industrialized periphery, as a consequence of improved infrastructure and persisting wage differentials.

These results confirm the predictions of the NEG theory and indicate a tendency towards the reduction of regional differences. The geographical concentration of industry shows, on average, a slight decline. Again, several differences can be found across industries. The most concentrated industries include motor vehicles and transport equipment and the least concentrated include food, beverages and tobacco prod ucts, textiles and clothing, and furniture.

Maffioli analyses Hungary (Chapter 6) and shows that, over the period 1992–99, the manufacturing sector has been influenced by the relocation of activity, mainly from Budapest to the Western regions of the country and, within them, to the regions bordering the EU. Broadly speaking, western regions are richer and more specialized than the rest of the country, because of the relatively higher concentration of economic activity in those sectors that fit best with EU import profiles (machinery and equipment) and because there are more foreign-owned enterprises in these regions than the national average, both in absolute and relative terms. Both these phenomena result from targeted local industrial policies, whose main instruments have been the creation of customs-free trade zones and industrial parks. An econometric analysis confirms that the on going process of economic integration with the EU has had a positive impact on the Western part of the country, which includes not only regions directly bordering the EU, but also regions very close to Budapest (Fejér). This explains why, after the Europe Agreements entered into force, distance to border regions became more relevant as a determinant of wage differentials.

For the case of Romania (Chapter 7), Traistaru and Pauna conclude that, over the period 1991–99, absolute specialization increased in 25 out of 41 regions, while relative specialization increased in 29 regions. The highest degree of regional specialization is found in Bucharest and the northeast, southwest and south regions, and the lowest in the central, west and northwest regions. Most manufacturing industries (ten out of 13) have become more concentrated. The most concentrated industry is electrical machinery and the least concentrated is food, beverages and tobacco. There is evidence of a decrease in the share of manufacturing employment in the capital city and an increase in the share of manufacturing employment in the Central region. Regional GDP is more concentrated than regional GDP per capita. However, while the degree of concentration of regional GDP has remained almost the same throughout the period from 1993–98, the concentration of the regional GDP per capita has increased, indicating a tendency towards income polarization.

Damijan and Kostec investigate the case of Slovenia (Chapter 8), and find that during the period from 1994–98, absolute and relative specialization increased in seven out of 12 regions. The lowest degree of regional specialization was observed in the largest regions (Osrednjeslovenska and Podravska) while specialization was the greatest in the smallest regions (Spodnjeposavska and Zasavska) and in the Pomurska region. There is evidence of the relocation of manufacturing activity between regions. Regions with initially large shares of manufacturing employment have benefited more from economic integration with the European Union, while regions with small shares of manufacturing employment have not gained in the relocation process. Econometric tests suggest that lower growth in regional GDP per capita is associated with higher unemployment rates.

Part III, 'Comparative Analysis and Lessons', focuses on comparative analyses of patterns and determinants of regional specialization and industrial concentration, adjustments in border regions and the impact of economic integration on regional relative wages and their determinants. Given the characteristics of the selected countries in terms of economic development and geography, these analyses are of particular interest.

In Chapter 9, Traistaru, Nijkamp and Longhi compare and explain patterns of regional specialization and geographic concentration of manufacturing in the five countries included in this study. They find that the geographical position of regions and their proximity to EU markets influences the observed patterns of regional specialization. Proximity to EU markets is associated with low specialization in countries that are more advanced in terms of the EU accession process (Estonia, Hungary, Slovenia) and with high specialization in Bulgaria. Proximity to other accession countries is associated with high specialization in Hungary and with low specialization in Bulgaria, Estonia and Romania. Proximity to countries outside the EU enlargement zone is associated with high specialization in Bulgaria, Hungary and Slovenia and low specialization in Romania. Non-border regions are diversified in Bulgaria and Hungary and have a high degree of specialization in Romania and Slovenia. As far as economic performance is concerned (GDP per capita, productivity, unemployment and average wages), highly specialized regions seem to perform worse than the national averages while diversified regions are doing better. Patterns of manufacturing concentration indicate that highly concentrated industries are those with large economies of scale, high levels of technology and high wages, while industries with low technology levels and low wages are dispersed. The estimated econometric models indicate that both factor endowments and geographic proximity to industry centers (capital cities and EU markets) explain the economic geography in EU accession countries.

The extension of the border areas – which account for 66 per cent of land area and 58 per cent of total population in accession countries – and the location of the selected countries along different EU borders, from North (Estonia) to South (Bulgaria), with Hungary and Slovenia close to the core EU countries and Romania far from any EU borders, suggested the need for an in-depth analysis of the adaptation processes in border regions. In Chapter 10, Resmini identifies several types of border regions in the selected five countries: regions bordering the EU, regions bordering other accession countries and regions bordering external countries (non-EU, non-accession countries). In her multi-level comparative analysis, the author shows that regions bordering the EU have been taking advantage of their location since the beginning of the transition process. High wages, skilled labour forces and well-developed service sectors have contributed to increasing the regional share of employment in manufacturing activities, despite the restructuring effect generated by the presence of foreign firms. The location of manufacturing activities in other border regions has been positively affected not only by FDI, but also by the regions' accessibility to industrial centers (capital cities), a skilled labour force and the service sector. Despite trade integration and liberalization, distance from the capital city, which remains at the core of economic activity in several candidate countries, still exerts a negative effect on relative employment. Regional differences seem to be less marked when one considers relative employment growth, which seems to have followed a pattern of convergence within regions and have been led mainly by FDI. Growth prospects indicate that regions gaining the most from economic integration with the EU are border regions. These benefits, however, are distributed unevenly within these regions.

The theoretical advances described in Chapter 2 have important implications for the adjustment of regional relative wages, as investigated by Damijan and Kostevc in Chapter 11. In particular, the authors study whether the response of relative regional wages to a reduction of foreign trade costs is monotonic and leads to strong regional polarization – as suggested by Krugman (1991a, 1991b) under the hypothesis of perfect labour mobility – or U-shaped and associated with less polarization and some convergence, as predicted by more recent NEG models under different hypothesis on labour mobility, regional symmetry and the availability of factors of production (Fujita, Krugman and Venables, 1999; Damjian and Kostevc, 2002). The implications of these three models are tested using panel data techniques for the period 1990–2000. The findings suggest that, in Estonia and Romania, relative regional wages tend to diverge and income per capita to polarize, as predicted by the Krugman models.

On the other hand, regional wage adjustment in Bulgaria and Hungary seems to follow a non-monotonic pattern, more in line with the predictions of the other two NEG models. In Slovenia, there is clear evidence of a catching-up process in the majority of regions, which challenges all theories' predictions.

The variety of phenomena highlighted by empirical evidence from diverse countries, regions, and industrial sectors indicates an increasing need for a regional policy response, which would be evidenced in the implementation of spatially targeted incentives in specifically designated areas. An articulated policy response is necessary, on the one hand, to redirect regions' specialization towards industries with growth potential and, on the other hand, to reduce income polarization. For the first case, policy should concentrate on further restructuring and internationalization of the regional production structure, on attracting foreign direct investment and on enhancing innovative and technological potential. For the second case, support for local entrepreneurship and human resources development, upgrading of infrastructures aimed at capacity-building to deal with structural performance weaknesses, in particular, and expansion of limited marketing horizons, in general, are suggested. Priority in this field should also be given to managing the influence of economic integration with the EU effectively. These issues are pointed out by Traistaru, Nijkamp and Resmini in the concluding chapter (Chapter 12).

Notes

1   This comparison has been made only for the period 1994–99, which is common to all countries' data.

2   Hungary is an exception, given the earlier start of economic reforms (Kaminski, 1999).

3   Cumulative inflows of FDI can be interpreted as a rough proxy for FDI assets.

4   The analysis also includes reinvested earnings. The figures may not, therefore, coincide with those reported in Table 1.3 from 1995 onwards.

5   Despite the growing body of literature on the role and determinants of FDI in CEECs, these issues are still being given much consideration. See, among others, Lankes and Venables (1996) and Resmini (2000) on the relationship between FDI and progress in the transition process; Brenton and Di Mauro (1997) on the relationship between trade and FDI; and Bevan and Estrin (2000) on the relationship between prospects for accession and FDI.

6   The reader interested in this topic has a huge body of literature to explore. See, beyond those quoted in the previous footnote, Campos and Kinoshito (2001), Passarelli and Resmini (1999) and Sheehy (1994).

7   Petrakos (2000), Bachtler and Downes (1999) and Weise, Bachtler, Downes et al. (2001) adopt the same point of view without providing further empirical evidence.

8   This process has been defined as 'hub and spoke' by Contessi (2001).

9   This research project has been undertaken with financial support from the European Community's PHARE ACE Programme, 1998.

References

Aghion, F. and Blanchard, O. (1994), 'On the Speed of Transition in Central Europe', in S. Fischer, and J. Rotemberg (eds), *NBER Macroeconomics Annual*, MIT Press, Cambridge.

Alessandrini, S. (2000), *The EU Foreign Direct Investments in the CEECs*, Milano, Giuffrè.

Altomonte, C. and Resmini, L. (2002a), 'The Geography of Foreign Direct Investment in Transition Countries: A Survey of Evidence', in A. Tavidze (ed), *Progress in International Economics Research*, New York, Nova Science Publishers, Inc.

Altomonte, C. and Resmini, L. (2002b), 'Multinational Corporations as a Catalyst for Local Industrial Development: the Case of Poland', *Scienze Regionali. The Italian Journal of Regional Science*, No. 2, pp. 29–57.

Amiti, M. (1999), 'Specialisation Patterns in Europe', *Weltwirtschaftliches Archiv*, Vol. 135(4), pp. 573–93.

Bachtler J. and Downes, R. (1999), 'Regional Policy in the Transition Countries: a Comparative Assessment', *European Planning Studies*, R. Vol. 7, pp. 793–808.

Baldwin, R., François, J. and Porter, R. (1997), 'EU-enlargement – Small Cost for the West, Big Gains for the East', *Economic Policy*, Vol. 12, pp. 125–76.

Blanchard, O. (1997), *The Economics of Post Communism Transition*, Oxford University Press, Oxford.

Bevan, A. and Estrin, S. (2000), 'The Determinants of Foreign Direct Investment in Transition Economies', CEPR Discussion Paper no. 2638, London, Centre for Economic Policy Research.

Blomstrom, M. and Kokko, A. (1997), *Regional Integration and Foreign Direct Investment*, NBER working paper no. 6019.

Brainard, S. (1993), 'A Simple Theory of Multinational Corporations and Trade with a Trade-off between Proximity and Concentration', NBER working paper no. 4269.

Brainard, S. (1997), 'An Empirical Assessment of the Proximity-Concentration Trade-off between Multinational Sales and Trade', *American Economic Review*, Vol. 87, pp. 520–44.

Brenton, P. and Di Mauro, F. (1997), 'The Potential Magnitude and Impact to FDI Flows to CEECs', CEPS working paper.

Brülhart, M. (2001), 'Evolving Geographical Concentration of European Manufacturing Industries', *Weltwirtschaftliches Archiv*, Vol. 137(2), pp. 215–43.

Campos, N. and Kinoshito, Y. (2001), 'Agglomeration and Determinants of Foreign Direct Investment in Transition Countries', paper presented at the International Conference on Transition Economics, Portoroz, Slovenia, June 24–26.

Contessi, S. (2001), 'Geographical Patterns in the Location of FDI: Evidence from Central European Regions', ISLA working paper no. 03–EE, Milano, 'L. Bocconi' Univeristy.

Damijan, J. and Kostevc, C. (2002), 'The Impact of European Integration on Inter-regional Relocation of Manufacturing and Wage Structure: Does Economic Geography Work in Transition Countries?', mimeo, University of Ljubljana.

Dohrn, R. (1998), 'The New Trade Structures between East and West.: Is Integration already a Matter of Fact?', in W. Potratz and B. Widmaier (eds), *East European Integration and New Division of Labour in Europe*, Graue Reihe des Instituts fur Arbeit und Technik, 1998–99, IAT, Gelsenkirchen, pp. 47–58;

Dohrn, R. (2001), 'The Impact of Trade and FDI on Cohesion', background paper for the 2nd Report on Cohesion, RWI, Essen, April.

Eurostat (2002), *European Union Direct Investment Yearbook*, Brussels, Eurostat.

Fujita, M., Krugman, P. and Venables, A. (1999), *The Spatial Economy. Cities, Regions and International Trade*, MIT Press, Cambridge, Mass.

Hallet, M. (2000), 'Regional Specialisation and Concentration in Europe', Economic Papers No. 141, European Communities, Brussels.

Hanson, G. (1996), 'Regional Adjustment to Trade Integration', *Regional Science and Urban Economics*, Vol. 28, pp. 419–44.

Hanson, G. (1998), 'North American Economic Integration and Industry Location', *Oxford Review of Economic Policy*, Vol. 14, pp. 30–44.

Hoekman, B. and Djankov, S. (1996), ' Intra-Industry Trade, Foreign Direct Investment, and the Reorientation of Eastern European Exports', World Bank, working paper no. 1652.

International Monetary Fund (2002), *Direction of Trade Statistics*, International Monetary Fund, Washington DC.

Kaminski, B. (2001), 'How Accession to the European Union Has Affected External Trade and Foreign Direct Investment in Central European Economies', Washington, DC, World Bank, working paper no. 2578.

Kaminski, B. (1999), 'Hungary's Integration into EU Markets: Production and Trade Restructuring', Washington DC. World Bank, working paper no. 2135.

Kaminski, B., Wang, Z.K. and Winters, A. (1996), *Foreign Trade in the Transition: The International Environment and Domestic Policy*, Series on Economies in Transition, UNDP and World Bank, Washington.

Krugman, P. (1991a), *Geography and Trade*, MIT Press, Cambridge, Mass.

Krugman, P. (1991b), 'Increasing Returns and Economic Geography', *Journal of Political Economy*, Vol. 99, pp. 483–99.

Lankes, H.P. and Venables, A.J. (1996), *Foreign Direct Investment in Economic Transition: the Changing Pattern of Investments*, The Economics of Transition, Vol. 4, No. 2, pp. 331–47.

Landesmann, M. (1995), 'The Patterns of East-West European Integration: Catching-up or Falling Behind?', in R. Dobrinsky and M. Landesmann (eds), *Transforming Economies and European Integration*, Edward Elgar, Aldershot, pp. 116–40.

Markusen, J. and Venables, A.J. (1999), 'Foreign Direct Investment as a Catalyst for Industrial Development', *European Economic Review*, No. 43, pp. 335–56.

Ohlin, B. (1933), *Interregional and International Trade*, Harvard University Press, Cambridge, Mass.

Overman, H.G., Redding, S. and Venables, A.J. (2001), 'The Economic Geography of Trade, Production, and Income: a Survey of Empirics', CEPR Discussion Paper No. 2978, London, Centre for Economic Policy Research.

Passarelli, F. and Resmini, L. (1998), 'Foreign Direct investment in Central and Eastern Europe: the Role of the European Firms', *Economia e Politica Industriale*, No. 99, September, pp. 5–36.

Petrakos, G. (1996), 'The Regional Dimension of Transition in Eastern and Central European Countries: an Assessment', *Eastern European Economics*, Vol. 34, pp. 5–38.

Petrakos, G. (2000), 'The Spatial Impact of East-West Integration in Europe', in G. Petrakos, G. Maier and G. Gorzelak (eds), *Integration and Transition in Europe: The Economic Geography of Interaction*, Routledge, London.

Resmini, L. (2000), 'The Determinants of Foreign Direct Investment in the CEECs: New Evidence from Sectoral Patterns', *The Economics of Transition*, Vol. 8, pp. 665–89.

Ricardo, D. (1817), *On Principles of Political Economy and Taxation,* Ch. VII, London J. Murray, 1st edn [reprinted as Vol. 1 of *The Work and Correspondence of David Ricardo*, ed. P. Sraffa, Cambridge University Press, 1951].

Sachs, J. (1997), 'Geography and Economic Transition', working paper, Center for International Development, Harvard University.

Sheehy, J. (1994), 'Foreign Direct Investment in the CEECs', *European Economy*, No. 6, pp. 129–50.

UNCTAD (2001), *World Investment Report*, United Nations, Geneva.

UN/ECE (2001), 'Economic Growth and Foreign Direct Investment in the Transition Economies', Ch. 5, *Economic Survey of Europe*, No. 1, United Nations, Geneva, pp. 185–225.

Weise, C., Bachtler, J., Downes, R., McMaster, I. and Toepel, K. (2001), 'The Impact of EU Enlargement on Cohesion', Background paper for the 2nd Cohesion Report, DIW and EPRC, Berlin and Glasgow, March.

Von Thünen, J.H. (1842), *Der Isolierte Staat in Beziehung auf Landwirtschaft und Nationalökonomie*, Léopold, Rostock, Germany [trans. Wartenberg, C.M. (1966), *Von Thünen's Isolated State*, Pergamon, Oxford, UK].

Chapter 2

European Integration, Regional Specialization and Location of Industrial Activity: a Survey of Theoretical and Empirical Literature*

Peter Nijkamp, Laura Resmini and Iulia Traistaru

1 Introduction

Over the last two decades, the member states of the European Union (EU) have further deepened their economic integration through the Single European Act and the Economic and Monetary Union. Meanwhile, the EU has continued its widening through several enlargements, including countries that differ with respect to their economic development and geographical location. In response to these changes, Canada, the United States and Mexico have recently set up the North American Free Trade Agreement (NAFTA), while Argentina, Brazil, Paraguay and Uruguay have formed MERCOSUR, an imperfect custom union aiming at gradually creating a common market in the Southern Cone of the American continent.

These recent examples of economic integration have all been accompanied by a substantial increase in intra-bloc trade flows. Among the potential effects associated with increased trade and, more generally, economic integration, what has been most interesting for the current debate is the possibility that economic integration may trigger the location of economic activities, affecting both structures and income levels across and within countries participating in the integration process.[1]

The spatial implications of the integration process are particularly important for EU accession candidate countries. The spatial distribution of their production and income has already been severely affected by the transition to democracy and a market economy. Despite that, formal analyses linking locational dynamics and the integration with the EU are still scarce.[2]

This chapter aims at contributing to filling this gap. We review relevant theoretical literature on the location of economic activity and international trade with the purpose of setting up a theoretical framework for the empirical studies discussed in the following chapters.

The two previously mentioned strands of economic theory reviewed here focus on the spatial distribution of economic activities within and across countries and uncover the role of geography (space) in economics.[3] We examine their hypotheses, main results and predictions in terms of the locational consequences of economic integration.

A few cautions should be noted before reading this chapter. It is not possible to summarize such a vast body of theoretical thinking in a few pages. Consequently, we have made no attempt at performing a comprehensive survey of the literature and have instead selected particularly important examples of how economists have dealt with the location of economic activities, its determinants and its main consequences over time. Although location theory has a long history in economics, especially in regional science and economic geography, it regained momentum in the 1990s, when prominent scholars, such as P. Krugman, A. Venables, P. Romer and M. Fujita, focused on this topic. In what follows, attention is paid to traditional and recent developments in both location theory and international trade theory. More space is devoted to recent contributions since the early contributions have been extensively treated in textbooks and hence are more familiar to most of readers than the recent advances.[4]

2 Location Theories

Location theory addresses the regional and urban dimensions of economic decisions and analyses the behaviour of both firms and of households. The locus, geographic place, is thus of prime interest in location analysis. More specifically, the main question in location theory is: what are the motives for choosing a particular location and what are the spatial implications? Location theory not only investigates the spatial point decision of an individual actor (household, business firm), but also looks into the regularities and consequences (patterns, geographic structures) that emerge from individual actors' decisions. Locational decisions are usually influenced by so-called externalities, which means that decisions made previously by a group of actors impact on the current decisions of new market entrants. This explains why cities emerge as a spatial conglomerate for many individual actors seeking urbanization economies, or why industrial complexes arise as a consequence of agglomeration advantages (such as joint use of infrastructure). Thus, location theory not only has a micro aspect, but also has a macro (collective or social) aspect; locational decisions are, ultimately, usually interactive decisions.

Location theory forms a central element in regional economics and economic geography. Location and trade were already regarded by the grandfather of economics, Adam Smith, as two interwoven phenomena: location impacts on trade flows and trade impacts on location decisions.

In the history of economics a great variety of contributions on the nature, motives, and implications of the location of production can be found. We will offer only a few examples here from this rich history (see also Gorter and Nijkamp, 2001). Location and land use in agricultural production was studied extensively by von Thünen (1842), by using a simple profit maximization scheme for spatial product choices, as a function of revenues, product types, and transport costs. He was able to demonstrate the existence of a set of concentric rings of agricultural products, based on Ricardian comparative cost theory leading to product specialization. Von Thünen's rent theory has formed the basis for the urban land rent theory, which has become a central focus of urban economics.

The location of industrial firms was investigated by Weber (1909) by looking for a cost minimizing location solution for a new firm on the basis of resource inputs (and other material inputs) and product outputs (to be shipped to the market). This intriguing and complex question was solved by Weber (1909) by using a triangular force field of input and output points, within which the optimal location has to be determined.

The spatial concentration of economic activity in a city has also been a source of intense research, starting with the work of Marshall (1925), who argued that agglomeration benefits are central to city formation. He referred to externalities that caused such benefits, in particular better information and skills, trade growth, specialized equipment, and the availability of skilled labour.

The shift towards multiproduct economies and the spatial inter-relationships between multiple products received due attention some 50 years ago in the context of the so-called central place theory, developed by Christaller (1933) and Lösch (1954). The basic idea is that the location of economic activity is subject to agglomeration advantages. As a consequence, economic activities are not spread randomly, but manifest themselves in clusters. Moreover, in the light of product diversity, price elasticity, and the existence of daily and non-daily goods, a spatial hierarchy is the logical outcome of a series of profit-maximizing location decisions made by firms. This hierarchy means that a place with a certain ranking order also includes firms producing goods with a lower order. This theory has become a landmark in location theory and has had far-reaching implications for spatial planning. In the postwar period, a wide range of new and refreshing contributions have also been offered, such as the growth pole theory (Perroux, 1955), the cumulative causation model (Myrdal, 1957), and the forward-backward linkage view of regional development (Hirschman, 1958).

The complexity of location decisions in a continuous space or in a discrete network configuration has also become a rich field within the discipline of operations research (Paelinck and Nijkamp, 1985), where many attempts have been made to derive mathematically the optimal site of public facilities or private goods, especially in the context of a multiproduct location decision. It should be noted that, in a discrete space with a finite number of location options, the number of combinatorial possibilities may be formidable, so that rather advanced mathematical software had to be developed.

Locational decisions are usually not stand-alone choices of a firm, but are part of a broader set of firm decisions, for example, about market areas, marketing channels, technology to be used, image, and so on Moreover, in an open multi-region or multi-country system, various types of spatial agglomeration patterns may emerge, depending inter alia on transport costs, forward and backward linkages, and the immobility of resources. In this context, it is clear that there is a need for a more integrated and dynamic approach to spatial equilibrium phenomena. This broader perspective should involve international trade and growth theory, as suggested inter alia by Krugman (1991b), as well as transportation science and industrial organization. This also means that factor mobility, transportation costs, and transaction costs become an integral part of modern theories on location and urban or regional growth. This wider perspective seems to be a necessary condition to a rigorous way forward.

In particular, against the background of the current globalization process, location theory is positioned in a global force field. This force field leads to many rapid responses and behavioural adjustments for business firms, so that stable and robust locations are increasingly replaced by nomadic types of business behaviour. Rapidly emerging technological changes (including ICT) encourage location theory to become a battlefield for industrial forces. Clearly, this dynamic evolution process has both advantages and disadvantages, and much empirical work would be needed to find out what the pros and cons of these dynamic changes are (see also Cuadrado Roura and Parellada, 2002 and Atalik and Fischer, 2002).

3 International Trade Theory

3.1 Neo-classical Trade Models

Conventional trade theory has focused on differences in productivity (technology) or endowments between countries and regions and explained specialization patterns through differences in relative production costs termed 'comparative advantages'.

The neo-classical models introduced by Ricardo (1817), Heckscher (1919) and Ohlin (1933) assume perfect competition, homogeneous products and non-increasing returns to scale. In such a world, location is determined exogenously by the inherited spatial distribution of natural resource endowments, technologies and/or factors of production. Thus, the emerging location pattern is one of inter-industry specialization and, in the absence of any differences between regions and/or countries, economic activities are predicted to locate uniformly across space.

Empirical tests of trade models (reviewed by Leamer and Levinsohn, 1995) indicate that while relevant, comparative advantage is not sufficient to explain recent developments in the world economy. As pointed out by Venables (1998), the Heckscher-Ohlin theory does not provide an appropriate explanation for the location of industry across countries and regions with similar factor endowments (for example, Western Europe) or within areas with highly mobile factors of production (for example, the United States of America) or for the geographical concentration of industrial activity in developing countries and their trade performance. To understand the uneven distribution of economic activity across space, one needs to look beyond comparative advantage.

3.2 New Trade Theory

During the 1980s, new trade models have been developed to provide an analytical framework for the explanation of intra-industry trade, that is, trade with differentiated goods within similar product categories (Krugman, 1979, 1980; Helpman and Krugman, 1985; Krugman and Venables, 1990). Intra-industry trade makes up a large proportion of the trade between industrialized countries and regions with similar factor endowments that differ only in size. The main features of new trade theory models are increasing returns to scale, consumers' preference variety and imperfect competition.[5] These activity-specific features have been labeled 'second nature' characteristics, while geography, factor endowments and technology pertain to the 'first nature' of regions and/or countries (Krugman, 1993).

The new trade theory models underscore the geographical advantage of large regions and regions with good market access as an explanation for specialization and the location of activity patterns. Both intra- and inter-industry trade are predicted. As noted by Venables (1998), the logic of these models is that firms with increasing returns to scale tend to concentrate their production in a few locations, whereas firms with constant or decreasing returns can be dispersed in many locations. The concentration of production makes large regions and, more generally, regions with good market access particularly attractive as production locations and export bases. The presence of many firms in these regions will increase the demand for labour and drive up wages. Thus, in equilibrium, regions with good market access are expected to experience a combination of higher wages and net exports of manufacturing products.[6]

The model by Krugman and Venables (1990) suggests the implications of trade liberalization and imperfect competition for industrial location. The model includes two production sectors: one sector (agriculture) operates under perfect competition and constant returns to scale and produces a homogeneous commodity and the other (manufacturing) operates under monopolistic competition and increasing returns to scale and produces differentiated goods. While agricultural goods are produced for local consumption, the manufactured goods can be exchanged against a trade cost. The firms are assumed to be mobile and they can choose their location across two regions/countries with uneven initial factor endowments: the 'core region', with larger endowments of factors or production compared to the other, the 'periphery'. The model assumes immobility of production factors and no comparative advantage effects.[7]

Manufacturing firms will prefer to locate in the large market, the 'core region', because of the presence of increasing returns to scale and trade costs. The size of the trade costs, that is, the cost of transporting goods from one region to another, is decisive in determining the distribution of manufacturing between the core and periphery. At very high trade costs (in autarky), manufactured goods are essentially non-tradable and firms have to locate their production in the region they supply. As trade costs fall, due to increasing returns to scale, the larger core region will attract manufacturing firms, which will enjoy larger profits. High profits will attract more manufacturing firms to the core region, which will then increase its share in world manufacturing and will become a net exporter of manufactured goods to the periphery. Further reductions of trade barriers will cause an increase in the demand for factors of production in the core region, relative to the periphery, and factor prices, including wages, will be driven up. The relationship between trade costs and the share of manufacturing has an inverse U-shape. The geographical advantage of the core region will be greatest at intermediate trade costs. When trade barriers and transport costs are small enough, the geographical advantage of the core region will be offset by factor costs considerations. At this stage, some firms will move to the periphery.

The prediction of new trade theory regarding the distribution of economic activity between the core and periphery is relevant in the case of the accession of Central and East European countries to the European Union. The current economic integration situation could be seen as one with 'intermediate trade costs'. Further integration could result in the relocation of manufacturing towards these countries due to factor costs considerations (Hallet, 1998).

Although these are powerful ideas, especially when predicting the effects of economic integration, they do not offer an exhaustive explanation of the linkages between trade and location. In particular, the new trade theory fails to explain why countries with similar structures end up with different production structures, why firms in particular sectors locate close to each other, leading to regional specialization patterns, and, finally, why industries spread successively from country to country (Ottaviano and Puga, 1998). The 'New Economic Geography' (NEG) also addresses these questions, making the geographical advantage of large markets endogenous.

3.3   The New Economic Geography

The new economic geography models[8] assume that the geographical advantage of large markets is endogenous and suggest that specialization patterns are the result of the spatial agglomeration of economic activities (Krugman, 1991a, 1991b). Krugman's analysis is based on a model similar to the two sectors-two regions model of Krugman and Venables (1990). In this case, the two regions have the same size. Each sector (agriculture and manufacturing) uses a specific factor of production and it is not only firms that move between regions/ countries. Industrial workers (the factor specific to manufacturing) can also move across regions. The central feature of this model is the agglomeration of economic activity triggered by the relocation of firms and workers from one region to another. The relocation of firms from one region to the other will bring about an increased variety of goods (due to monopolistic competition), and increased wages (due to increased labour demand) and local expenditure (due to demand linkages) in the receiving region. The assumption of labour mobility in manufacturing implies that industrial workers will follow firms into the receiving region. The initial relocation of firms and workers generates cumulative effects leading to the concentration of manufacturing in the 'receiving' region at th expense of the 'donor' region. This scenario could develop, however gradually, following the lowering of trade costs. Initially when trade costs are very high, each region produces manufacturing goods for its own local market and thus the manufacturing sector is evenly split between the two regions. The fall of trade costs will allow firms to serve non-local markets and will thus encourage the agglomeration of manufacturing due to demand linkages. The place where agglomeration occurs could be determined merely by history. This model has dramatic spatial implications for the European Union. The fall of trade barriers will reinforce the benefits for regions with an initial scale advantage in particular sectors, while the rest of the regions will de-specialize in those sectors.

The mechanism of agglomeration at the basis of the aforementioned models depends on the relationship between a mobile labour force and the demand for goods. However, empirical evidence demonstrates that workers are not as mobile as assumed in the theoretical models.[9] Despite that, agglomeration may still increase, provided that workers may move between sectors and that firms have 'supply-side linkages': manufacturing firms benefit from locating in a region where they have access to suppliers providing a range of specialized inputs (Krugman and Venables, 1995; Venables, 1996). In this case, the relation between integration and industry location does not change. What changes is the mechanism of agglomeration: while in Krugman (1991a, 1991b) type models, an increase in the number of firms in a location increases demand for locally produced goods through the expenditures of workers who have migrated there from other locations, in Krugman and Venables (1995) type models, the higher demand arises from the intermediate expenditures of newly established firms.[10]

The results discussed so far have important spatial implications with respect to increased economic integration. It is therefore essential to understand how much they depend on the particular analytical framework employed (Fujita and Thisse, 2002). For example, Krugman and Venables (1995) assume that the elasticity of labour supply in manufacturing with respect to agricultural wages is infinite. However, as Puga (1999) points out, when a finite elasticity of labour supply is allowed, inter-regional wage differentials will prevent manufacturing in the donor region from collapsing at a certain level of trade costs. In this case, the fall of trade costs will produce both agglomeration and degglomeration of manufacturing in the core as well as the periphery. Limited mobility between sectors will lead to a rise in wages in the agricultural sector at a level sufficient to determine the location of firms in the industry with lower wages. Ludema and Wooton (1997) obtain a similar result in the case of imperfect labour mobility across regions in the presence of demand-side linkages. Although reducing trade costs generates agglomeration effects in one region (manufacturing moves to the 'core'), the process is incomplete and is reversed with a further decrease in trade costs.[11]

In summary, new economic geography explains location and trade within a theoretical framework that is simultaneously tractable and solidly micro-founded.[12] Indeed, it has an 'intellectual debt' to location theory and regional science, though it has been developed in the tradition of trade theory (Neary, 2001).[13] This is why it seems to be the most appropriate tool for analysing the impact of economic integration with the EU – a phenomenon which is international in its own character – on industry location and regional specialization – a topic which has a national dimension – in EU accession countries. Cross-fertilization is, thus, not only expected but also necessary for further interesting advances in economics. This could be an opportunity to follow the path that Adam Smith and Bertil Ohlin traced a long time ago.

4 Empirical Evidence

Compared to the theoretical literature, empirical analyses of the impact of economic integration on regional specialization and geographic concentration of industries is still scarce. Most studies have focused on economic integration experiences in the United States and the European Union. Existing empirical studies can be classified into two groups. A first (larger) group underlines the usefulness of economic geography for understanding patterns of regional specialization and industry location. The second and more recent group attempts to discriminate between alternative theories for understanding the above-mentioned phenomena. In what follows, we focus mainly on the first group of empirical studies, since they better fit the objective of the present work.[14]

Krugman (1991a) compares four regions in the United States with four large countries in the EU and shows that geographic concentration of manufacturing is higher in the US than in Europe. The most concentrated industries are textile industries, while high-technology sectors are less concentrated.

Ellison and Glaeser (1997) analyse the geographic concentration of US manufacturing industries. Using a model that controls for industry characteristics, they find that almost all industries seem to be localized.

Many industries are, however, only slightly concentrated, and some of most concentrated industries are related to natural advantages.

In a series of papers, Hanson (1996, 1997a, 1997b, 1998a and 1998b) looks at the US-Mexico integration process and assesses the locational forces identified by the new economic geography models. He finds that integration has shifted industry towards border regions both in the US and in Mexico and that demand and cost linkages have become more important as determinants of industrial location. This result is also supported by evidence about the decreasing importance of the distance from the capital city and the increasing importance of distance from the border between the US and Mexico in explaining inter-regional wage differentials.

With respect to Europe, Brülhart and Torstensson (1996) study the evolution of industrial specialization patterns in 11 EU countries (all except Luxembourg and the more recent member states, Austria, Finland, and Sweden) between 1980 and 1990. They find support for the U-shaped relationship between the degree of regional integration and spatial agglomeration predicted by NEG models when labour mobility is low: activities with larger scale economies were more concentrated in regions close to the geographical core of the EU during the early stages of European integration, while concentration in the core has fallen in the 1980s.

Using production data in current prices for 27 manufacturing industries, Amiti (1999) finds that there was a significant increase in specialization between 1968 and 1990 in Belgium, Denmark, Germany, Greece, Italy, and the Netherlands; no significant change in Portugal; and a significant decrease in specialization in France, Spain and the UK. There was a significant increase in specialization between 1980 and 1990 in all countries. With more disaggregated data (65 industries), the increase in specialization is more pronounced: the average increase is 2 per cent for all countries except Italy, compared to one per cent in the case with 27 manufacturing industries.[15] However, analyses based on trade data indicate that EU Member States have a diversified rather than specialized pattern of manufacturing exports (Sapir, 1996; Brülhart, 2001).

With respect to the geographical concentration of industries, Brülhart (1996) finds that between 1980 and 1990, 14 of the 18 industries considered became more geographically concentrated in Europe (as measured by Gini coefficients). Sectors characterized by large economies of scale have shown larger increases in concentration. Amiti (1997) shows that 17 out of 27 industries experienced an increase in geographical concentration, with an average increase of 3 per cent per year in leather products, transport equipment and textiles. Only six industries experienced a fall in concentration, with paper and paper products and 'other chemicals' showing particularly marked increases in dispersion.

Brülhart and Torstensson (1996) find a positive correlation between the geographical concentration of industries in the European Union (measured by industry Gini coefficients) and centrality indices (calculated as in Keeble et al., 1986) both for 1980 and 1990. This result indicates an industry bias towards European core regions. Similar results are provided by Brülhart (1998).

Discriminating between different trade theories is not easy, since they have several overlapping aspects, as we have discussed in the previous section. Despite that, some interesting contributions have been published, although it is too early to understand whether these theories are supported by empirical evidence. Kim (1995) found that empirical evidence is inconsistent with external economies, thus rejecting Krugman (1991a and 1991b) type models. Meanwhile, scale economies (proxied by average plant size) explain industry localization patterns over time and raw material intensity (used as a proxy for neoclassical models) explains localization patterns across industries. Davis and Weinstein (1996, 1998 and 1999) focus on home market effects to discriminate between neo-classical and new trade models. They found mixed evidence: only in the last two papers did they find that the home market effect was more important than comparative advantage in determining manufacturing output and trade patterns, thus supporting Krugman (1980) type models. Halland et al. (1999) simultaneously test alternative trade theories to explain industry localization patterns in the European Union. The main result is that localization of expenditure is by far the most important factor affecting industry localization, while Midelfart-Knarvik, Overman and Venables (2001) confirm the importance of supply linkages.[16]

A number of recent papers look at the effects of trade policy on agglomeration (Martin and Ottaviano, 1996a; Ottaviano 1996; Puga and Venables 1997; Walz, 1997), while Trionfetti (1997) analyses the implications of different procurement policies on industry location. A common finding in these papers is that the design of trade agreements and infrastructure networks shapes locational advantage in terms of access to world markets. This is applied by Puga (2001) to discuss the implications of the new economic geography for European regional policy.

With respect to EU accession countries, most of the research on regional issues has focused on patterns of disparities with the aim of identifying policy needs at a regional level.[17] It has been claimed that the processes of internationalization and structural change in accession countries tend to favour metropolitan and western regions, as well as regions with a strong industrial base (Petrakos, 1996). In addition, at the macro-geographical level the process of transition is expected to increase disparities at the European level, favouring countries near the East-West frontier (Petrakos, 2000). Increasing core-periphery differences in Estonia are documented in Raagmaa (1996). Altomonte and Resmini (2002) point out the role of foreign direct investment in shaping regional specialization in accession countries.

Yet, to date, there is no comprehensive study on the impact of economic integration with the European Union on regional specialization and location of industrial activity in accession countries.[18]

Notes

*   This research was undertaken with support from the European Community's PHARE ACE Programme 1998. The content of the publication is the sole responsibility of the authors and it in no way represents the views of the Commission or its services.
1   To recall the scientific debate on the spatial implications of economic integration alone, see Krugman and Venables (1990), Krugman (1991a, 1993), Greenaway and Hine (1991), Hanson (1996, 1997a, 1998), Baer, Haddad and Hewings (1998), Brülhart (2001), Amiti (1998, 1999), Venables (1998), and Overman et al. (2001).
2   Petrakos (1996, 2000) has made two excellent contributions to this area of research.
3   A few illustrations of economists' interest in the role of space in economics are given here: 'The question how the economy fits into space not only opens a new field but leads in the final analysis to a new formulation of the entire theory of economics' (Lösch, 1954). 'The theory of international trade is only a part of a general localisation theory' (Ohlin, 1933). 'Space: the final frontier' (Krugman, 1998). 'Relying on new theoretical tools, this 'new economic geography' has quickly emerged as one of the most exciting areas of contemporary economics' (Fujita, Krugman and Venables, 2000).
4   Location theory is extensively treated in most intermediate and advanced textbooks of regional science, such as Nijkamp (1986) and Armstrong and Taylor (2000). Neo-classical trade theories, represent the bulk of textbooks in international economics (Krugman-Obsfeld, 2002; Jones and Kenen, 1984; Grossman and Rogoff, 1995). In contrast, new economic geography, when considered, is discussed in no more than one chapter.
5   The first trade models including increasing returns to scale were based on the concept of Marshallian external economies. They therefore maintained the assumption of perfect competition (Panagariya, 1981 and 1986; Markusen and Melvin, 1981; Ethier, 1982). The most influential approach, however, considers increasing returns to scale that are internal to the firms and that allow for imperfect competition modeled *à la* Spence (1976) and Dixit and Stiglitz (1977). See Wolfmayr-Schnitzer (1999) for a detailed discussion of the early new trade models and Helpman and Krugman (1985) for alternative approaches.
6   This idea, labeled 'home market' effect, was originally developed by Krugman (1980, 1995). Davis (1998) demonstrates that its functioning is strictly related to the asymmetry

in trade costs between homogeneous and differentiated goods. 'When transport costs are identical for both types of goods, the home market effect vanishes. This holds regardless of the magnitude of the market size differences' (Davis, 1998, p. 1265).

7   As pointed out by Ottaviano and Puga (1998), the difference in factor endowments reflects better access to markets from the core than from the periphery rather than differences in actual size. This distinction, however, becomes relevant only in empirical investigations (Davis and Weistein, 1996, 1998 and 1999).

8   NEG models are discussed in more details in Wolfmayr-Schnitzer (1999), European Commission (2000), Neary (2001), Puga (2002).

9   This is especially true for European countries. See Eichengreen (1993) and Obstfeld and Peri (1998).

10   Labour mobility and the mobility of firms demanding intermediates are the two main agglomeration mechanisms developed by NEG scholars (Fujita, Krugman and Venables, 2000). However, they are not the only ones. Baldwin (1999) shows that factor accumulation can play the same role as migration in generating agglomeration through demand linkages. Martin and Ottaviano (1996b) develop an inter-temporal mechanism of agglomeration by considering input-output linkages in the presence of a third innovative sector.

11   The limits deriving from the choice of the mathematical formalization have been highlighted and extensively discussed from different perspectives by Neary (2001) and Martin (1999). In a very recent paper, Fujita and Thisse (2002) point out that the qualitative results of NEG models, which at first sight look like 'a collection of examples' are 'fairly robust'.

12   This does not mean that NEG theory has no limits and shortcomings. Space is one-dimensional and spatial units of analysis are not defined. Consequently, regions, countries and/or cities used throughout models by new economic geographers can not be univocally identified. Firms cannot engage in industrial strategies, such as vertical integration, out-sourcing and cross-border horizontal mergers, because of the assumption of Chamberlinian imperfect competition. Finally, technological externalities and local public goods are not considered, though they are responsible for agglomeration processes at the local level (Audretsch and Feldman, 1996). See Neary (2001) and Martin (1999) for in-depth discussions of NEG's weaknesses.

13   Neary (2001) identifies two features of NEG models peculiar to international trade traditions: the simultaneous consideration of both trade and location, as suggested by Ohlin (1933), and the emphasis on a single cause for both phenomena. As Ricardo focused on technology, Hecksher and Ohlin on factor endowments, and, among recent contributions, Helpman and Krugman (1985) on imperfect competition, NEG scholars concentrate on pecuniary externalities.

14   This survey does not consider theoretical and empirical works on regional growth and convergence. Again, our motivation lies, on the one hand, in the fact that this topic is only marginally related to our general objective; on the other hand, that there is no agreement among scholars on whether or not these studies actually belong to economic geography, since they usually do not consider issues related to location. Interested readers may refer to Barry (1996) and, for a more critical approach, to Martin (1999).

15   Other evidence of increasing specialization in EU countries in the 1980s is provided by Hine (1990) and Greenway and Hine (1991). Using data on gross value added for 17 branches of economic activities between 1980 and 1995, Hallet (2000) concludes that the spatial effects of European integration are, however, less dramatic than it seems initially.

16   See Overman, Redding and Venables (2001) for a comprehensive survey of empirics on trade and location.

17 See, for instance, Spiridonova (1995 and 1999) for Bulgaria, Nemes-Nagy (1994 and 1998) for Hungary, Constantin (1997) for Romania and Buchtler and Downes (1999) for a comparative assessment.
18 Several studies have been carried out as background for the Second Report on Economic and Social Cohesion of the European Commission. See, for instance, Weise et al. (2001) and Dorhn et al. (2001).

References

Armstrong, H. and Taylor, J. (2000), *Regional Economics and Policy*, 3rd edn, Blackwell, Oxford.

Audretsch, D. and Feldman, M.P. (1996), 'R&D Spillovers and the Geography of Innovation and Production', *American Economic Review*, Vol. 86, No. 3, pp. 630–40.

Altomonte, C. and Resmini, L. (2002), 'The Geography of Foreign Direct Investment in Transition countries: a Survey of evidence', in A. Tavidze (ed.), *Progress in International Economics Research*, Vol. I, Nova Science Publisher, Inc., New York.

Amiti, M. (1998), 'New Trade Theories and Industrial Location in the EU: a Survey of Evidence', *Oxford Review of Economic Policy*, Vol. 14 (2), pp. 45–53.

Amiti, M. (1999), 'Specialisation Patterns in Europe', *Weltwirtschaftliches Archiv*, Vol. 134(4), pp. 573–93.

Atalik, G. and Fischer, M.M. (eds) (2002), *Regional Development Reconsidered*, Springer-Verlag, Berlin.

Baer, W., Haddad, E. and Hewings, G. (1998), 'The Regional Impact of Neo-liberal Policies in Brazil', *Economia Aplicada*, Vol. 2, No. 2.

Bachtler, J. and Downes, R. (1999), 'Regional Policy in the Transition Countries: a Comparative Assessment', *European Planning Studies*, Vol. 7, No. 6, pp. 793–807.

Baldwin, R. (1999), 'Agglomeration and Endogenous Capital', *European Economic Review*, Vol. 43, No. 2, pp. 253–80.

Barry, F. (1996), 'Peripherality in Economic Geography and Modern Growth Theory', *World Economy*, Vol. 19, No. 3, pp. 345–65.

Brülhart, M. (1996), 'Commerce et spécialisation géographique dans l'Union Européenne', *Economie Internationale*, Vol. 65, pp. 169–202.

Brülhart, M. and Torstensson, J. (1996), 'Regional Integration, Scale Economies and Industry Location', CEPR Discussion Paper No. 1435, London.

Brülhart, M. (1998), 'Economic Geography, Industry Location and Trade: the Evidence', *The World Economy*, Vol. 21, No. 6, pp. 775–801.

Brülhart, M. (2001), 'Evolving Geographical Specialisation of European Manufacturing Industries', *Weltwirtschaftliches Archiv*, Vol. 137(2), pp. 215–43.

Christaller, W. (1933), *Die Zentralen Orte in Süddeutschland, Fischer*, Jena, Germany [trans. Baskin, C.W. (1966), *Central Places in Southern Germany*, Prentice-Hall, Englewood Cliffs, NJ].

Constantin, D. (1997), 'Institutions and Regional Development Strategies and Policies in the Transition Period: the Case of Romania', paper presented at the 37th European Congress of the European Regional Science Association, Rome.

Cuadrado Roura, J.R. and Parellada, M. (eds) (2002), *Regional Convergence in the European*

Union, Springer-Verlag, Berlin.

Davis, D. (1998), 'The Home Market, Trade, and Industrial Structure', *The American Economic Review*, Vol. 88, No. 5, pp. 1264–76.

Davis, D. and Weinstein, D. (1996), 'Does Economic Geography Matter for International Specialisation?', NBER working paper 5706.

Davis, D. and Weinstein, D. (1998), 'Market Access, Economic Geography and Comparative Advantage: an Empirical Assessment', NBER working paper 6787.

Davis, D. and Weinstein, D. (1999), 'Economic Geography and Regional Production Structure: an Empirical Investigation', *European Economic Review*, Vol. 43, No. 2, pp. 379–407.

Dixit, A.K. and Stiglitz, J.E. (1977), 'Monopolistic Competition and Optimum Product Diversity', *The American Economic Review*, Vol. 67, No. 3, pp. 297–308.

Döhrn, R., Milton, A.R. and Radmacher-Nottelmann, N. (2001), 'The Impact of Trade and FDI on Cohesion', background paper for the Second Cohesion Report, RWI, Essen.

Eichengreen, B. (1993), 'Labour Markets and the European Monetary Unification', in P.R. Masson and M.P. Taylor (eds), *Policy Issues in the Operation of Currency Unions*, Cambridge University Press, Cambridge.

Ellison, G. and Glaeser, E. (1997), 'Geographic Concentration in U.S. Manufacturing Industries: a Dartboard Approach', *Journal of Political Economy*, Vol. 105, No. 5, pp. 889–927.

Ethier, W.J. (1982), 'National and international Returns to Scale in the Modern Theory of International Trade', *The American Economic Review*, Vol. 72, No. 3, pp. 950–59.

European Commission (2000) *The Impact of Economic and Monetary Union on Cohesion*, Luxembourg: Office for Official Publications of the European Communities.

Fujita, M. and Thisse, J.F. (2002), 'Agglomeration and Market Interaction', CEPR discussion paper no. 3362, London.

Fujita, M., Krugman, P. and Venables, A. (2000), *The Spatial Economy: Cities, Regions, and International Trade*, MIT Press, Cambridge, Mass.

Gorter, C. and Nijkamp, P. (2001), 'Location Theory', in *Encyclopedia of the Behavioral Sciences*, Elsevier, Amsterdam, pp. 9013–19.

Greenway, D. and Hine, R.C. (1991), 'Intra-Industry Specialisation, Trade Expansion and Adjustment in the European Economic Space', *Journal of Common Market Studies*, Vol. 29, No. 6, pp. 603–22.

Grossman G. and Rogoff, K. (1995), *Handbook of International Economics*, Vol. III, North-Holland, Amsterdam.

Haaland, J.I., Kind, H.J., Knarvik, K.H. and Torstensson, J. (1999), 'What Determines the Economic Geography of Europe', CEPR discussion paper no. 2072, London.

Hallet, M. (1998), 'The Regional Impact of the Single Currency', paper presented at the 38th Congress of the European Regional Science Association, Vienna, September.

Hallet, M. (2000), 'Regional Specialisation and Concentration in the EU', Economic Papers No. 141, European Communities, Brussels.

Hanson, G.H. (1996), 'Economic Integration, Intra-industry Trade, and Frontier Regions', *European Economic Review*, Vol. 40, No. 3–5, pp. 941–9.

Hanson, G.H. (1997a), 'Localization Economies, Vertical Organization, and Trade', *The American Economic Review*, Vol. 86, No. 5, pp. 1266–78.

Hanson, G.H. (1997b), 'Increasing Returns, Trade, and the Regional Structure of Wages', *The Economic Journal*, Vol. 107, No. 1, pp. 113–33.

Hanson, G.H. (1998a), 'Market Potential, Increasing Returns, and Geographic Concentration', Working Paper No. 6429, National Bureau of Economic Research.

Hanson, G.H. (1998b), 'Regional Adjustment to Trade Liberalization', *Regional Science and Urban Economics*, Vol. 28, No. 4, pp. 419–44.

Heckscher, E. (1919), 'The Effect of Foreign Trade on Distribution of Income', *Economisk Tidskrift*, pp. 497–512, reprinted in H.S. Ellis and L.A. Metzler (eds) (1949), *A.E.A. Readings in the Theory of International Trade*, Blakiston, Philadelphia, pp. 272–300.

Helpman, E. and Krugman, P. (1985), *Market Structure and Foreign trade: Increasing Returns, Imperfect Competition and the International Economy*, Harvester Wheatsheaf, Brighton.

Hine, R.C. (1990), 'Economic Integration and Inter-Industry Specialisation', CREDIT Research Paper 89/6, University of Nottingham.

Hirschman, A. (1958), *The Strategy of Economic Development*, Yale Univeristy Press, New Haven, Conn.

Jones, R. and Kenen (1984), *Handbook of International Economics*, Vol. I, North-Holland, Amsterdam.

Keeble, D., Offord, J. and Thompson, C. (1986), *Peripheral Regions in a Community of Twlve Member States*, Commission of the EC, Luxembourg.

Kim, S. (1995), 'Expansion of Markets and the Geographic Distribution of Economic Activities: the Trend in US Regional Manufacturing Structure, 1860–1987', *Quarterly Journal of Economics*, Vol. 110, No. 4, pp. 881–908.

Krugman, P. (1979), 'Increasing Returns, Monopolistic Competition and International Trade', *Journal of International Economics*, Vol. 9, No. 4, pp. 469–79.

Krugman, P. (1980), 'Scale Economies, Product Differentiation, and the Pattern of Trade', *American Economic Review*, Vol. 70, No. 5, pp. 950–59.

Krugman, P. (1991a), *Geography and Trade*, MIT Press, Cambridge, Mass.

Krugman, P. (1991b), 'Increasing Returns and Economic Geography', *Journal of Political Economy*, Vol. 99, No. 3, pp. 484–99.

Krugman, P. (1993), 'First Nature, Second Nature, and Metropolitan Location', *Journal of Regional Science*, Vol. 33, No. 1, pp. 129–44.

Krugman, P. (1995), 'Increasing Returns, Imperfect Competition and the Positive Theory of International Trade', in G. Grossman and K. Rogoff (eds), *Handbook of International Economics*, Vol. III, North Holland, Amsterdam, pp. 1243–77.

Krugman, P. (1998), 'Space: the Final Frontier', *Journal of Economic Perspectives*, Vol. 12, No. 2, pp. 161–74.

Krugman, P. and Obstfeld, M. (2002), *International Economics. Theory and Policy*, 5th edn, Addison-Wesley, Reading, Mass.

Krugman, P. and Venables, A. (1990), 'Integration and the Competitiveness of Peripheral Industry', in C. Bliss and J. Braga de Macedo (eds), *Unity with Diversity in the European Community*, Cambridge University Press, Cambridge, pp. 56–77.

Krugman, P. and Venables, A. (1995), 'Globalisation and the Inequality of Nations', *Quarterly Journal of Economics*, Vol. 110, No. 4, pp. 857–80.

Leamer, E.E. and Levinsohn, J. (1995), 'International Trade Theory: the Evidence', in G.M. Grossman and K. Rogoff (eds), *Handbook of International Economics*, Vol. III, Amsterdam, Elsevier-North Holland, pp. 1339–94.

Lösch, A. (1954), *The Economics of Location*, Yale University Press, New Haven.

Ludema, R.D. and Wooton, I. (1997), 'Regional Integration, Trade and Migration: Are Demand Linkages Relevant in Europe?' CEPR discussion paper no. 1656.

Markusen, J.R. and Melvin, J.R. (1981), 'Trade, Factor Prices and Gains from Trade with Increasing Returns to Scale', *Canadian Journal of Economics*, Vol. 14, No. 3, pp. 450–69.

Marshall, A. (1925), *Principles of Economics*, Macmillan, London.

Martin, R. (1999), 'The New "Geographical Turn" in Economics: Some Critical Reflections', *Cambridge Journal of Economics*, Vol. 23, No. 1, pp. 65–91.

Martin, P. and Ottaviano, G.I.P. (1996a), 'La géographie de l'Europe multi-vitesses', *Economie Internationale*, No. 67, pp. 45–65.

Martin, P. and Ottaviano, G.I.P. (1996b), 'Growth and Agglomeration', CEPR discussion paper no. 1529, London.

Midelfart, K.H., Overman, G. and Venables, A. (2001), 'Comparative Advantage and Economic Geography: Estimating the Location of Production in the EU', mimeo, LSE, London.

Myrdal, G. (1957), *Economic Theory and Underdeveloped Regions*, Duckworth, London.

Neary, P. (2001), 'Of Hype and Hyperbolas: Introducing the New Economic Geography', *Journal of Economic Literature*, Vol. XXXIX, June, pp. 536–61.

Nemes-Nagy, J. (1994), 'Regional Disparities in Hungary during the Period of Transition to a Market Economy', *GeoJournal*, Vol. 32, No. 4, pp. 363–8.

Nemes-Nagy, J. (1998), 'The Hungarian Spatial Structure and Spatial Processes', *Regional Development in Hungary, 15–26*, Ministry of Agriculture and Regional Development, Budapest.

Nijkamp, P. (ed.) (1986), *Handbook of Regional and Urban Economics: Regional Economics*, Vol. I, North-Holland, Amsterdam.

Obstfeld, M. and Peri, G. (1998), 'Asymmetric Shocks: Regional Non-adjustment and Fiscal Policy', *Economic Policy*, Vol. 13, No. 26, pp. 206–59.

Ohlin, B. (1933), *Interregional and International Trade*, Harvard University Press, Cambridge, Mass.

Ottaviano, G.I.P. (1996), 'The Location Effects of Isolation', *Swiss Journal of Economics and Statistics*, Vol. 132, pp. 427–40.

Ottaviano, G.I.P. and Puga, D. (1998), 'Agglomeration in the Global Economy: a Survey of the "New Economic Geography"', *World Economy*, Vol. 21, No. 6, pp. 707–31.

Overman, G., Redding, S. and Venables, A. (2001), 'The Economic Geography of Trade, Production and Location: a Survey of Empirics', CEPR discussion paper no. 2978, Center for Economic Policy Research, London.

Paelinck, J.H.P. and Nijkamp, P. (1985), *Operational Theory and Method in Regional Economics*, Avebury, Aldershot.

Panagariya, A. (1981), 'Variable Returns to Scale in Production and Patterns of Specialisation', *American Economic Review*, Vol. 71, No. 1, pp. 221–30.

Panagariya, A. (1986), 'Increasing Returns and the Specific Factors Model', *The Southern Economic Journal*, Vol. 53, No. 1, pp. 1–17.

Perroux, F. (1955), 'Note sure la notion de pôle de croissance', *Economique Appliqué*, No. 2, pp. 307–20.

Petrakos, G. (1996), 'The Regional Dimension of Transition in Eastern and Central European Countries: an Assessment', *Eastern European Economics*, Vol. 34, No. 5, pp. 5–38.

Petrakos, G. (2000), 'The Spatial Impact of East-West integration', in G. Petrakos, G. Maier and G. Gorzelak (eds), *Integration and Transition in Europe: the Economic Geography of Interaction*, Routledge, London.

Puga, D. (2002), 'European Regional Policy in the Light of Recent Location Theories', *Journal of Economic Geography*, Vol. 2, No. 4, pp. 373–406.

Puga, D. (1999), 'The Rise and Fall of Regional Inequalities', *European Economic Review*, Vol. 43, No. 2, pp. 303–34.

Puga, D. and Venables, A. (1997), 'Preferential Trading Arrangements and Industrial Location', *Journal of International Economics*, Vol. 43, No. 3–4, pp. 347–68.

Raagmaa, G. (1996), 'Shifts in Regional Development in Estonia during the Transition', *European Planning Studies*, Vol. 4, No. 6, pp. 683–703.

Ricardo, D. (1817), *On the Principles of Political Economy and Taxation*, London.

Sapir, A. (1996), 'The Effects of Europe's Internal Market Programme on Production and Trade: a First Assessment', *Weltwirtschaftliches Archiv*, Vol. 132, No. 3, pp. 457–75.

Spence, M. (1976), 'Product Selection, Fixed Costs, and Monopolistic Competition', *Review of Economic Studies*, Vol. 43, pp. 217–35.

Spiridonova, J. (1995), *Regional Restructuring and Regional Policy in Bulgaria*, National Center for Regional Development and Housing Policy, Sofia.

Spiridonova, J. (1999), 'Depressed Regions in Bulgaria', in *Proceedings of the 5th Conference of the Network of Spatial Planning Institutes of Central and East European Countries*, Krakow.

Thünen, J.H. von, (1842), *Der Isolierte Staat in Beziehung auf Landwirtschaft und Nationalökonomie*, Léopold, Rostock, Germany [trans. Wartenberg, C.M. (1966), *Von Thünen's Isolated State*, Pergamon, Oxford, UK].

Traistaru, I. (1999), 'Regional Patterns of Private Enterprise Development in Romania', Working paper, Center for European Integration Studies, University of Bonn.

Trionfetti, F. (1997), 'Public Expenditure and Economic Geography', *Annales d'Economie et de Statistique*, No. 47, pp. 101–25.

Venables, A. (1996), 'Equilibrium Locations of Vertically Linked Industries', *International Economic Review*, Vol. 37, No. 2, pp. 341–59.

Venables, A. (1998), 'The Assessment: Trade and Location', *The Oxford Review of Economic Policy*, Vol. 14, No. 2, pp. 1–6.

Walz, U. (1997), 'Growth and Deeper Regional Integration in a Three Country Model', *Review of International Economics*, Vol. 5, No. 3, pp. 492–507.

Weber, A. (1909), *Über den Standort der Industrien*, Tübingen, Germany, [trans. Friedrich, C.F. (1929), *Alfred Weber's Theory of Location of Industries*, University of Chicago Press, Chicago].

Weise, C., Bachtler, J., Downes, R. et al. (2001), 'The Impact of EU Enlargement on Cohesion'. Background paper for the Second Report on Economic and Social Cohesion of the EC, DIW and EPRC, Berlin and Glasgow.

Wolfmayr-Schnitzer, Y. (1999), 'Economic Integration, Specialization and the Location of Industries: a Survey of the Theoretical Literature', WIFO Working Paper No. 120, WIFO-Austrian Institute of Economic Research, Vienna.

Chapter 3

Data and Measurement*

Iulia Traistaru and Anna Iara

1 Data Set REGSTAT

1.1 Overview

Patterns of regional specialization and geographical concentration of industrial activity have been identified and analysed in five accession countries[1] on the basis of employment data at a regional level, disaggregated by manufacturing branches. To this purpose, we have generated a special data set, REGSTAT, which includes, apart from employment, other variables at the regional level used in our analysis such as: demographic variables, average earnings, Gross Domestic Product (GDP), measures of infrastructure, enterprise intensity, education, research and development (R&D) and public expenditure. In general, the period covered is 1990–99. However, in the case of Slovenia, official statistics are available starting with 1994 for average earnings, number of domestic firms, number of firms with foreign participation and R&D, and starting with 1997 for the remaining variables. In addition to data from official statistics, the database for Slovenia includes data compiled from companies' balance sheet reports.[2] Regional variables are available at NUTS 3 and NUTS 2 levels. The database's content is presented in Tables 3.A1–3.A15.

Table 3.1 gives an overview of the sizes of regions in the five accession countries included in this study. Estonia and Slovenia are classified as single regions at the NUTS 2 level. Except for Estonia and Slovenia, NUTS 2 regions only serve for the purposes of statistics and planning, whereas in most countries, the NUTS 3 level consists of meso-level units of state administration with rights to self-government.

Most data is taken from the regular publications of national statistical offices. Data that is not officially reported has been collected from other sources. In particular, among others, some countries' labour market data are from Labour Offices or similar institutions whereas firm level data has been partly collected from commercial registers. As mentioned above, in the case of Slovenia, due to the lack of availability of data from official sources, the data set is extended by the inclusion of data gathered from companies' balance sheets.

Table 3.1   NUTS 2 and NUTS 3 units in Bulgaria, Estonia, Hungary, Romania and Slovenia

Country	Number of units	Administrative status	Population 1999 (thousands) Average	st. dev.
		NUTS 2		
Bulgaria	6	Planning region	1,365.1	635.5
Estonia	1	State	1,439.2	0
Hungary	7	Planning region	1,441.7	655.1
Slovenia	1	State	1,987.8	0
Romania	8	Planning region	2,807.3	599.2
		NUTS 3		
Bulgaria	28	–	292.5	220.2
Estonia	5	–	287.8	143.3
Hungary	20	Intermediate administrative unit	504.6	364.2
Slovenia	12		165.6	139.4
Romania	41	Intermediate administrative unit	547.8	325.1

Source: REGSTAT data set.

1.2   Employment Data

Employment data is used in this study to proxy regional industrial structures. Measures of regional specialization and spatial concentration are calculated on the basis of regional employment data on manufacturing branches.

For all countries, employment data[3] is available disaggregated by sectors[4] on the NUTS 3 level; employment in manufacturing is further disaggregated by branches. In the case of Slovenia, there exist two sets of data on employment; the first was supplied by the official statistical authorities and the second was compiled from company balance sheet reports.

The industrial classification used in this study is the NACE rev. 1 two-digit classification.[5] However, for Bulgaria and Hungary employment data have been collected according to their national classifications. For these two countries, aggregations have been made to bring these classifications as close as possible to the NACE rev. 1 two-digit classification.

Data on regional average wages is used to calculate regional relative wages, which is the dependent variable in regressions estimating the impact of trade liberalization and the role of transport costs on the regional structure of wages. Data on GDP is used in the analysis of the relationship between regional specialization and growth.

1.4   Structural Regional Characteristics

While the variables introduced so far are used for descriptive purposes or as dependent variables in econometric analysis, the following are used with the purpose of controlling for various demographic and economic characteristics of the regions in the econometric analyses:

- the distance between pairs of county capitals;[6]
- numbers of domestic firms;
- number of firms with foreign participation;
- number of self-employed;
- density of national public roads;
- number of personal cars;
- number of students enrolled in higher education;
- number of telephone lines;
- population by age groups (distinguishing those aged 0–14, 15–24, 15–65 and 66 and above);
- public expenditure.

2   Measurement of Trade Performance

2.1   Description of Trade Patterns

The measurements used to analyse the performance of the five accession countries in their trade with the EU include the trade coverage index, revealed comparative advantage and the intra-industry index.

Trade coverage index   The trade coverage index (TCI) is the ratio of exports (X) to imports (M) as shown below:

$$TCI_i = \frac{X_i}{M_i}$$

i = the manufacturing branch or product group

Revealed comparative advantage index The revealed comparative advantage (RCA)[7] is the ratio of the share of a product group i for country j and the share of the exports of the product group i in the total export for a group of countries: The RCA index indicates whether or not country j is a preferential supplier for a certain product group i. Values higher than one indicate a comparative advantage.

$$ RCA_i = \frac{x_{if}}{X_j} \bigg/ \frac{\sum_j \sum_i x_i^f}{\sum_j X_j} $$

x_{ij} = the export of product group i in country j; Xj = the total export of country j.

The Grubel-Lloyd index of intra-industry trade The size of intra-industry trade indicates the extent of the economic integration of a country. To this effect, we use in our analysis the Grubel-Lloyd[8] (GL) index calculated as follows:

$$ GL_i = 1 - \frac{|X_i - M_i|}{(X_i + M_i)} $$

The Grubel-Lloyd index takes values between zero and one. The higher the value, the greater the extent of intra-industry trade and the degree of economic integration.

3 Measures for Regional Specialization and Geographic Concentration of Industries

Regional specialization and geographic concentration of industries are defined in relation to production structures.[9] Regional specialization is defined as the distribution of the shares of an industry i in total manufacturing in a specific region j compared to a benchmark. A region j is found to be specialized in a specific industry i if this industry has a high share in the manufacturing employment of region j. The manufacturing structure of a region j is 'highly specialized' if a small number of industries have a large combined share in total manufacturing.

Geographic concentration measures the distribution of the shares of regions in a specific industry i. A specific industry i is said to be 'concentrated' if a large part of production is carried out in a small number of regions.

Specialization and concentration could be assessed using absolute and relative measures. Several indicators are proposed in the existing literature, each having certain advantages as well as shortcomings. The notations for and definitions of the indicators used in our analysis are presented in Box 3.1.

Box 3.1   Indicators of regional specialization and geographic concentration of industries[10]

$E =$ employment

$s =$ shares

$i =$ industry (sector, branch)

$j =$ region

$s_{ij}^{S} =$ the share of employment in industry i in region j in total employment of region j

$s_{ij}^{C} =$ the share of employment in industry i in region j in total employment of industry i

$s_i =$ the share of total employment in industry i in total employment

$s_j =$ the share of total employment in region j in total employment

$$s_{ij}^{S} = \frac{Eij}{Ej} = \frac{Eij}{\sum_i Eij} \qquad\qquad s_{ij}^{C} = \frac{Eij}{Ei} = \frac{Eij}{\sum_j Eij}$$

$$s_i = \frac{Ei}{E} = \frac{\sum_j Eij}{\sum_i \sum_j Eij} \qquad\qquad s_j = \frac{Ej}{E} = \frac{\sum_i Eij}{\sum_i \sum_j Eij}$$

The Herfindahl index

Regional specialization measure   Geographical concentration measure

$$H_j^S = \sum_i (s_y^S)^2 \qquad\qquad H_i^C = \sum_j (s_y^C)^2$$

The dissimilarity index

Specialization measure   Concentration measure

$$DSR_j = \sum_i \left| s_y^S - s_i \right| \qquad\qquad DCR_i = \sum_j \left| s_y^C - s_j \right|$$

Box 3.1 cont'd

Krugman specialization index **Spatial separation measure**
Regional specialization measure

$$K_{kl} = \sum_i \left| S_{ik}^s - S_{il}^s \right|$$

$$SP^j = C\sum_{k=1}^{n}\sum_{l=1}^{n}(S_{kj}^c\, 110_{lj}^c\, \delta_{kj}^c)\, ,\, \delta_{kl}$$

is the distance between two regions k and l, and C is a constant

Gini coefficients
Gini coefficient for regional specialization

$$GINI\,{}_i^c = \frac{2}{n^2 R}\left[\sum_{i=1}^{n}\lambda_i \left| R_i - \bar{R}\right|\right] \text{ where } R_i = \frac{S_{ij}^s}{S_i},\ \bar{R}=\frac{1}{n}\sum_{i=1}^{n}R_i\,;$$

λ_i indicates the position of the industry i in the ranking of R, in descending order Gini coefficient for concentration of industries

$$GINI\,{}_i^c = \frac{2}{m^2 C}\left[\sum_{j=1}^{m}\lambda_j \left| C_j - \bar{d}\right|\right] \text{ where } C_j = \frac{S_{ij}^c}{S_j},\ \bar{C}=\frac{1}{m}\sum_{j=1}^{m}R_i\,;;$$

λ_i indicates the position of the region in the ranking of C_j in descending order

The Herfindahl index of regional specialization This is an absolute measure of specialization often used in industrial economics. It sums up the squares of industry shares in the total activity in the region. It takes values between zero and one and is positively related to regional specialization. Given the absolute nature of the Herfindahl index, the sum of the squares of shares is biased towards the larger regions.

Krugman specialization index This index is a relative measure of regional specialization and was introduced in international economics by Krugman (1991), comparing the industrial structures of the US and Europe. It sums up the absolute difference of the industrial structures of two regions, k and l:[11].

$$K_{kl} = \sum_i \left| S_{ik}^s - S_{il}^s \right|$$

The index is zero if the two regions have the same industrial structures. Its maximum value is two, reached if the two regions do not have common industries.

Dissimilarity index of regional specialization   The dissimilarity index is a relative measure of regional specialization. It compares the industrial structure of one region with an average distribution across regions. It is constructed by summing up absolute differences between shares of industries in economic activity (that is, employment) in a particular region and their average value across counties by branch. It takes values between zero and two and is positively related to regional specialization. The minimum value indicates maximum diversity, and the maximum value, maximum specialization.

Gini coefficient of regional specialization   Particularly in empirical work on income distribution, Gini indices are widely used to describe patterns of inequality. The Gini index is known as the area between the Lorenz concentration curve for a given distribution and a hypothetical curve for equal distribution of the factor under scrutiny.[12] It constitutes a measure of relative concentration, assuming the value of zero for equal distribution.

In the context of regional specialization, the degree of equality of the distribution of industries in a given region is to be assessed. Reasonably, the industrial structure in that particular region is set in relation to an average value for the units to be compared (that is, the nationwide distribution) rather than to an equal distribution of industries.[13] Providing a measure of relative specialization, this index takes values between zero and one.[14] The disadvantage of Gini indices is that the same value may stem from different distributions: comparable index values do not allow inferences about those.

The following equivalents of the indices of specialization of regions presented above have been applied in the study of spatial concentration of industries in the countries included in this research project.

Herfindahl index of geographical concentration   As a measure of absolute concentration of activity in an industry i across regions $j_1...j_m$, the Herfindahl index of geographical concentration is calculated as the sum of the regions' shares in national employment in the particular industry: It is positively related with the geographical concentration of industries.

Dissimilarity index of geographical concentration The dissimilarity index of geographical concentration is a relative measure of relative spatial concentration of industries.

It sums up the absolute differences between the regional and the national-level value of shares in total employment in a particular industry, which applies the national-level share of an industry in total employment as the norm to which regional-level values are related.

The index is zero in case of equal distribution of the particular industry across regions and increases the more the respective industry is concentrated in a few regions.

Gini coefficient of geographical concentration of industries (locational Gini coefficient) The 'sectoral' equivalent of the above-described Gini coefficient of regional specialization is the locational Gini index.[15]

This index takes values between zero and one[16] and is positively related to the concentration of industries.

Spatial separation The indices of geographical concentration presented above do not directly take into account the spatial dimension, that is, they do not process information on distance between locations. Therefore, from them, it cannot be distinguished if, say, an industry is located in several neighboring regional units – implying concentration on a higher spatial level – or whether it is dispersed across the same number of regional units across the country. To assess this aspect of location patterns of industries, Midelfart-Knarvik et al. (2000) propose a measure of spatial separation of an industry *j*. The index is zero if industrial production is concentrated to a single location. Higher values of the measure indicate greater spatial separation.

Notes

* This research was undertaken with support from the European Community's PHARE ACE Programme 1998. The content of the publication is the sole responsibility of the authors and it in no way represents the views of the Commission or its services.
1 The countries included in this study are Bulgaria, Estonia, Hungary, Romania and Slovenia.
2 The balance sheets were obtained from the Slovenian Payment Transactions Agency and refer to companies with more than ten employees.
3 While for Bulgaria, Estonia and Romania, data on employment refer to persons employed, data for Hungary and Slovenia include employees only.

4   Sectoral decomposition is done along the lines of the NACE rev. 1 one-digit sections in the following manner: Primary sector: NACE sections A+B (that is, agriculture, hunting, forestry, fishing); Secondary sector: NACE sections C-F (that is, mining, manufacturing, electricity/gas/water supply and construction); Tertiary sector: Other activities.

5   That is, data are grouped into the two-digit subsections DA-DN of manufacturing.

6   This variable is only available at the NUTS 3 level. Since in Hungary, Romania and Bulgaria, NUTS 2 regions are only 'artificial' units for statistical and planning purposes (whereas in the case of the other two countries, the whole state territory forms a NUTS 2 region) without proper regional capitals, this variable could not be determined for the NUTS 2 level.

7   This index is known as the Balassa index, named after the economist who first used it.

8   On theoretical considerations concerning intra-industry trade and the Grubel-Lloyd index, respectively, see Wolfmayr-Schnitzer (1998) and Brülhart (2001).

9   See Aiginger (1999), for a survey of theoretical and empirical literature on regional specialization and geographic concentration of industries.

10   The indicators are defined following Krugman (1991), Aiginger (1999), Devereux et al. (1999) and Midlefart-Knarvik et al. (2000).

11   This measure is similarly employed by Brülhart (2001) to compare industrial structures across EU countries, while Midelfart-Knarvik et al. (2000) present an extended version of the Krugman dissimilarity index to several regions that makes use of absolute differences between activity shares of one particular region and an average value calculated for all *other* regions.

12   For an explanation of the Lorenz concentration curve, see Brülhart (2001).

13   See Devereux et al. (1999).

14   Analogously to the property of the 'Locational' Gini index highlighted by Devereux et al. (1999), the lower bound for the above index is $(1-m/n)$ if the number of regions m is less than the number of industries n.

15   See Devereux et al. (2001).

16   If fewer industries than regions are investigated, the lower bound for the index is $(1 - n/m)$.

References

Aiginger, K. et al. (1999), 'Specialisation and (Geographic) Concentration of European manufacturing', *Enterprise DG Working Paper No. 1*, Background Paper for the 'The competitiveness of European Industry: 1999 Report', Brussels.

Amiti, M. (1999), 'Specialisation Patterns in Europe', *Weltwirtschaftliches Archiv*, 135(4), pp. 573–93.

Brülhart, M. (2001), 'Marginal Intra-Industry Trade: Towards a Measure of Non-Disruptive Trade Expansion', University of Lausanne, mimeo.

Devereux, M., Griffith, R. and Simpson, H. (1999), 'The Geographic Distribution of productive Activity in the UK', *Institute for Fiscal Studies Working Paper*, W99/26.

Devereux, M., Griffith, R. and Simpson, H. (2001), 'The Geography of Firm Formation', *Institute for Fiscal Studies*, London, mimeo.

Krugman, P. (1991a), *Geography and Trade*, MIT Press, Cambridge, Mass.

Midelfart-Knarvik, K., Overman, H.G. and Venables, T. (2000), 'The Location of European Industry', *Economic Papers No. 142*, European Commission, DG for Economic and Financial Affairs.

Wolfmayr-Schnitzer, Y. (1998), 'Intra-Industry Trade of CEECs', in OECD (ed.), *The Competitiveness of Transition Economies*, Paris.

Appendix: REGSTAT Database Content

BG = Bulgaria
EST = Estonia
HU= Hungary
RO= Romania
SLO= Slovenia.

Table 3.A1 REGSTAT: Employment

Code*	Time span	Definition	Spatial disaggregation	Data source
BG_EMP BG_EACT BG_EIND	1990–99	Number of persons employed	NUTS 3	National Statistical Institute
EST_EMP EST_EACT EST_EIND	1989–99	Number of persons employed	NUTS 3	Statistical Office of Estonia
HU_EMP HU_EACT HU_EIND	1990–99	Number of employees	NUTS 3	Hungarian Central Statistical Office
RO_EMP RO_EACT RO_EIND	1990–99	Number of persons employed	NUTS 3	National Institute of Statistics
SLO_EMP SLO_EACT SLO_EIND	1997–99	Number of full-time employees	NUTS 3	Statistical Office of the Republic of Slovenia
SLO_EMP SLO_EACT SLO_EIND	1994–98	Number of employees	NUTS 3	PaymentTransaction Agency

* XX_EMP = total regional employment; XX_EACT = regional employment by sectors (primary, secondary, tertiary); XX_EIND = regional manufacturing employment by branches.

Table 3.A2   REGSTAT: Average earnings

Code	Time span	Definition	Spatial disaggregation	Data source
BG_WAGE	1990–99	Average of net yearly earnings wages and salaries of the employed, additional benefits/bonuses included, in domestic currency	NUTS 3	National Statistical Institute
EST_WAGE	1992–99	Average of gross wages of employees, in domestic currency	NUTS 3	Statistical Office of Estonia
HU_WAGE	1992–99	Average of gross monthly wages and salaries of employees, additional benefits/bonuses included, in domestic currency	NUTS 3	National Labour Methodology Centre
RO_WAGE	1992–99	Average of gross monthly wages and salaries of employees, additional benefits/bonuses included, in domestic currency	NUTS 3	National Institute of Statistics
SLO_WAGE_A	1997–99	Average of net yearly wages and salaries of employees, additional benefits/bonuses included, in domestic currency	NUTS 3	Statistical Office of the Republic of Slovenia
SLO_WAGE_B	1994–98	Average of net yearly wages and salaries of employees in domestic currency	NUTS 3	Payment Transaction Agency

Table 3.A3 REGSTAT: GDP

Code	Time span	Definition	Spatial disaggregation	Data source
BG_GDP	1995–98	GDP in EUR, current prices	NUTS 3	National Statistical Institute
EST_GDP	1996–98	GDP in USD, constant PPI-deflated market prices of 1996		Statistical Office of Estonia
HU_GDP	1994–98	GDP in USD, current prices	NUTS 3	Hungarian Central Statistical Office
RO_GDP	1993–98	GDP per capita in PPS	NUTS 2	Romanian Statistical Institute
SLO_GDP	1997	GDP in USD PPS, current prices	NUTS 3	National Institute of Statistics

Table 3.A4 REGSTAT: Distance between pairs of county capitals

Code	Definition	Regional level	Data source
BG_DIST	Distance between pairs of county capitals as by road, in km	NUTS 3	Ministry of Transportation
EST_DIST	Distance between pairs of county capitals as by road, in km	NUTS 3	Road atlas of Estonia, Tallinn 1996
HU_DIST	Distance between pairs of county capitals as by road, in km	NUTS 3	n.a.
RO_DIST	Distance between pairs of county capitals as by road, in km	NUTS 3	Publirom Publishing House (ed.): Map of Romania
SLO_DIST	Distance between pairs of county capitals as by road, in km	NUTS 3	Interactive Atlas of Slovenia 2000

Table 3.A5 REGSTAT: Number of domestic firms

Code	Time span	Definition	Spatial disaggregation	Data source
BG_ENTD	1990–2001	Number of domestic firms	NUTS 3	Information Service Ltd.
EST_ENTD	1993–99	Number of firms	NUTS 3	Central Commercial Register
HU_ENTD	1991–99	Number of registered corporations and unincorporated enterprises	NUTS 3	Hungarian Central Statistical Office, Tax and Auditing Office of the Ministry of Finance
RO_ENTD	1990–99	Number of domestic registered firms	NUTS 3	Romanian Chamber for Industry and Trade
SLO_ENTD	1994–98	Number of registered firms in the manufacturing industry	NUTS 3	Payment Transaction Agency

Table 3.A6 REGSTAT: Number of firms with foreign participation

Code	Time span	Definition	Spatial disaggregation	Data source
BG_ENTF	1990–2000	Number of firms with foreign participation	NUTS 3	Information Service Ltd.
EST_ENTF	1993–99	Number of firms owned by persons who are foreigners by private law	NUTS 3	Central Commercial Register
HU_ENTF	1991–99	Number of firms with foreign participation	NUTS 3	Hungarian Central Statistical Office, Tax and Auditing Office of the Ministry of Finance
RO_ENTF	1990–99	Number of firms with foreign participation	NUTS 3	Romanian Chamber for Industry and Trade
SLO_ENTF	1994–98	Number of registered firms in manufacturing sector with at least 10% of capital owned by foreign subjects	NUTS 3	Payment Transaction Agency

Table 3.A7 REGSTAT: Number of self-employed

Code	Time span	Definition	Spatial disaggregation	Data source
BG_SELF	n.a.	n.a.	n.a.	n.a.
EST_SELF	1996–99	Number of self-employed, annual average	NUTS 3	Statistical Office of Estonia
HU_SELF	1991–99	Number of self-employed (i.e., free-lance professions)	NUTS 3	Hungarian Central Statistical Office, Tax and Auditing Office of the Ministry of Finance
RO_SELF	1991–99	Number of self-employed	NUTS 3	Romanian Chamber for Industry and Trade
SLO_SELF	1997–99	Number of self-employed	NUTS 3	Statistical Office of the Republic of Slovenia

Table 3.A8 REGSTAT: Density of national public roads

Code	Time span	Definition	Spatial disaggregation	Data source
BG_ROAD	1990–99	Density of public roads, km/km^2 of territory	NUTS 3	National Statistical Institute
EST_ROAD	1992–99	Density of public roads, km/km^2 of territory	NUTS 3	Counties in Figures, Statistical Office of Estonia
HU_ROAD	1991–99	Density of public roads, km/km^2 of territory	NUTS 3	Ministry of Transport, Telecommunication and Water Management
RO_ROAD	1990—99	Density of public roads, km/km^2 of territory	NUTS 3	National Institute of Statistics
SLO_ROAD	n.a.	n.a.	n.a.	n.a.

Table 3.A9   REGSTAT: Number of personal cars

Code	Time span	Definition	Spatial disaggregation	Data source
BG_CAR	1990–99	Numbers	NUTS 3	National Statistical Institute
EST_CAR	1992–99	End-year number of private passenger cars	NUTS 3	Counties in Figures, Statistical Office of Estonia
HU_CAR	1990–99	Number of passenger cars registered in the country	NUTS 3	Data Processing Office of the Ministry of Interior
RO_CAR	1990–99	Number of personal cars	NUTS 3	National Institute of Statistics
SLO_CAR	n.a.	n.a.	n.a.	n.a.

Table 3.A10   REGSTAT: Number of students

Code	Time span	Definition	Spatial disaggregation	Data source
BG_STD	1990–99	Number of students enrolled in higher education	NUTS 3	National Statistical Institute
EST_STD	1994–99	Number of students enrolled in higher education	NUTS 3	Statistical Office of Estonia
HU_STD	1990–99	Number of full-time students enrolled in	NUTS 3	Ministry of Education
RO_STD	1990–99	Number of students enrolled in higher education	NUTS 3	National Institute of Statistics
SLO_STD	1995–98	Number of students enrolled in higher education	NUTS 3	Statistical Office of the Republic of Slovenia

Table 3.A11 REGSTAT: Number of telephone lines

Code	Time span	Definition	Spatial disaggregation	Data source
BG_TEL	1990–99	Total number of telephone lines	NUTS 3	National Statistical Institute.
EST_TEL	1992–99	Number of telephone sets/main telephone lines, as of 1 January	NUTS 3	Counties in Figures, Statistical Office of Estonia
HU_TEL	1990–99	Number of main telephone lines including public stations	NUTS 3	Ministry of Transport, Telecommunication and Water Management
RO_TEL	1990–99	Number of telephone lines	NUTS 3	National Institute of Statistics
SLO_TEL	n.a.	n.a.	n.a.	n.a.

Table 3.A12 REGSTAT: Population by age groups

Code	Time span	Definition	Spatial disaggregation	Data source
BG_POP	1990	Population by age groups	NUTS 3	National Statistical Institute
EST_POP	1992–99	Resident population by age groups, as of 1 January	NUTS 3	Statistical Office of Estonia
HU_POP	1990–99	Resident population by age groups	NUTS 3	Central Registration and Electoral Office
RO_POP	1990–99	Population by age groups, as of 1 July	NUTS 3	National Institute of Statistics
SLO_POP	1996–99	Population by age groups	NUTS 3	Statistical Office of the Republic of Slovenia

Table 3.A13 REGSTAT: Public expenditure

Code	Time span	Definition	Spatial disaggregation	Data source
BG_EXP	1990–99	Public investment in local currency	NUTS 2	National Statistical Institute
EST_EXP	1992–99	Public expenditure in local currency	NUTS 3	Counties in Figures, Statistical Office of Estonia
HU_EXP	1990–99	Current operation expenditures in local currency	NUTS 3	Regional Administrative and Information Service of the Government
RO_EXP	1990–99	Public expenditure in local currency	NUTS 3	National Institute of Statistics
SLO_EXP	n.a.	n.a.	n.a.	n.a.

Table 3.A14 REGSTAT: Research and development activity

Code	Time span	Definition	Spatial disaggregation	Data source
SLO_RD	1994–98	R&D expenditure of enterprises in local currency, constant prices of 1994	NUTS 3	Payment Transaction Agency
BG_RD	1990–99	size of personnel of R&D units	NUTS 3	n.a.
EST_RD	n.a.	n.a.	n.a.	n.a.
HU_RD	1995–99 except 1996	size of personnel of R&D units	NUTS 3	Office of the Doctoral Council Secretary of the Hungarian Academy of Sciences
RO_RD	n.a.	n.a.	n.a.	n.a.

Table 3.A15 REGSTAT: Unemployment

Code	Time span	Definition	Spatial disaggregation	Data source
BG_UNE	1991–99	Unemployment rate	NUTS 3	National Employment Office
EST_UNE	1995–2000	Number of registered unemployed, annual average	NUTS 3	Labour Market Board
HU_UNE	1990–99	Number of registered unemployed, end-year	NUTS 3	National Labour Methodology Centre
RO_UNE	1991–99	Number of registered unemployed	NUTS 3	National Institute of Statistics
SLO_UNE	1997–99	% of registered unemployed in the active population, monthly average of end-month figures	NUTS 3	Statistical Office of the Republic of Slovenia

PART II
COUNTRY STUDIES

Chapter 4

The Emerging Economic Geography in Bulgaria*

Julia Spiridonova

1 Introduction

The integration drive and its orientation within the different periods of Bulgarian development have had a definitive impact on regional dynamics and spatial patterns.

The implications of the former integration of the country, namely the extreme dependence of the national economy on CMEA, are still tangible. These are related, above all, to the development of heavy and strongly material-intensive industries, despite a background of limited national energy and ore resources, and to technological backwardness because of the system's relative isolation from global competition. The result has been an accelerated process of territorial concentration in the eastern direction and the emergence of a national periphery along the western and southern borders of the country.

In the nineties the country underwent fundamental socio-economic changes, related to the transition to a market economy and adaptation to the requirements of the European and world markets. The new economic orientation towards the European Union (EU) imposed different standards for the competitive capacities and survival of the national industries.

The past ten years of economic transition and EU-orientation were a period that needs to be evaluated from the point of view of mobility of economic activities and possible re-location of industries, the behaviour of the individual regions, the dynamics of regional discrepancies and the stability of the territorial structures. During that period, Bulgaria passed through different economic crises and phases in its relations with the EU and the individual regions manifested different paces of transformation. The objective of this chapter is to study the new trends of European integration, regional specialization and location of industrial activities and their spatial implications for Bulgaria.

The first section gives an overview of the data set, definitions, sources and time period.

The second section presents the main driving forces for increasing economic integration with the EU and provides evidence about trade liberalization and increasing openness through growth in trade with the EU and inward FDI flows.

The patterns of regional specialization and industrial location before and after the liberalization of trade with the EU are discussed in the next section. A descriptive analysis, using different specialization and concentration indices (Herfindahl index, Gini coefficients, Balassa-Hoover index), depicts the changes in regional and manufacturing structures and disparities. The trend model estimations elucidate the changes in regional specialization and geographical concentration.

The following sections are devoted to econometric estimations of the impact of economic integration on the regional wage structure and on regional specialization and growth.

The winning and losing regions that have emerged from the unevenly distributed economic integration gains, as well as the policies needed to mitigate the great regional imbalances, are described in the last section.[1]

2   The Data

The data set includes total employment and employment by branches of economic activity, unemployment rate, population by age groups, wage rates, FDI, GDP, number of domestic companies, number of companies with foreign capital, number of students, R&D personnel, number of telephone posts, vehicles, road network density and public expenditure. The time period comprises 1990–99 (for several types of data, 2000 was included as well). Most data is presented at NUTS 3 and NUTS 2 levels. The data have been obtained from the publications of the National Statistical Institute, both traditional and electronic. Other sources have been used as well, for instance, the Foreign Investments Agency, the National Employment Office, and so on.

Data collection involved overcoming significant problems emerging from the transformations carried out during the transition:

- One of these changes was the introduction of the Statistical Classification of Economic Sectors (NACE Rev. 1), the National Product Nomenclature (CPA) and the Nomenclature of Industrial Production in replacement of the Classification of Branches of National Economy (CBNE'86) from 1986.

- Another significant change occurred in the number and spatial scope of the regions since 1999. The administrative regions at the NUTS 2 level were dismantled and 28 NUTS 3 level districts replaced them. In 2000, six planning regions (NUTS 2 level) were created for planning purposes.

The collection of data on employment required the greatest effort, since these data bear the direct impact of the above-described transformations. To this end, data based on primary information from municipalities (NUTS 4 level) was used. The resolution of the second problem, namely the transition from CBNE'86 to NACE, was initially carried out through expert estimates at the NUTS 4 level. In order to ensure a higher level of reliability, primary information from companies was used in the process of transforming the data.

Regional data for the indicators of the GDP have existed since 1995. For the period 1995–99, regional data about GDP in PPS (according to EUROSTAT) are presented as well.

3   Economic Integration with the European Union since 1990

3.1   Trade Liberalization and Trade Performance

Trade agreements   Including the Europe Agreement, Bulgaria became a member of the World Trade Organization in December 1996 and applies a liberal foreign trade regime that meets the WTO requirements.

In March 1993 the country signed the Europe Agreement of Association, which entered into force on 1 February 1995. The Interim Agreement on Trade and Trade Related Matters covering trade components entered into force on 31 December 1993. In accordance with the Agreement of Association, customs duties between Bulgaria and EU countries on industrial goods are being dismantled and should be completely eliminated by 2002 at the latest.

Since 1993, according to the Agreement between Bulgaria and the European Free Trade Association, preference in trade with EFTA countries (Switzerland, Norway, Iceland and Liechtenstein) is granted on almost the same terms and conditions as those pursuant to the Europe Agreement.

Bulgaria has been a member of the Central European Free Trade Agreement since July 1998. In accordance with the above agreement, Bulgaria embarked on a process of liberalization of trade of industrial and agricultural goods with CEFTA countries (Poland, the Czech Republic, Slovakia, Hungary, Romania and Slovenia), which was completed by 1 January 2002.

Free trade agreements with Turkey and Macedonia came into force on 1 January 1999 and 1 January 2000, respectively. Customs duties with Turkey will be reduced gradually until 2002 and with Macedonia until 2005.

Redirection of trade towards the EU Since 1990 there has been a definitive reorientation of Bulgaria's foreign trade patterns towards the EU. Its share in Bulgaria's total trade turnover has increased each year during the period from 1990–99: it grew from 22 per cent in 1989 to 32 per cent in 1990 and exceeded 50 per cent in 1999. The change in Bulgaria's geographic trading patterns was more significant on the export side than on the import side. While in 1990 the EU's share of the country's total exports was 8.42 per cent, in 1999 this share was 51.86 per cent (see Table 4.1).

Table 4.1   Main indicators of trade with the EU (per cent)

Year	Total turnover	Exports US$m	Imports US$m	EU share in turnover %	EU share in exports %	EU share in imports %
1990	26,567.9	13,439.6	13,128.3	13.12	8.42	17.94
1991	6,133.0	3,432.6	2,700.4	21.22	17.40	26.08
1992	8,521.3	3,991.7	4,529.6	32.07	28.86	34.90
1993	8,889.0	3,768.7	5,120.3	31.15	29.84	32.12
1994	8,206.6	3,935.1	4,271.5	37.18	37.54	36.84
1995	10,983.1	5,334.9	5,638.2	37.01	37.60	36.51
1996	9,964.1	4,890.2	5,073.9	35.99	38.82	33.26
1997	9,871.7	4,939.7	4932	39.94	43.11	36.76
1998	9,318.0	4,303.5	5,014.5	47.59	49.36	46.07
1999	9,444.1	3,973.8	5,470.3	50.17	51.86	47.21

Source: National Statistical Institute (NSI).

Bulgaria's major trading partners within the EU are Germany (9.1 per cent of total exports and 24.3 per cent of total imports), Italy (14.3 per cent of exports and 8.5 per cent of imports), Greece (7.8 per cent of exports and 4.9 per cent of imports), France (4.8 per cent of exports and 4.9 per cent of imports), Belgium (6.1 per cent of exports and 1.3 per cent of imports) and the UK (2.4 per cent of exports and 2.1 per cent of imports) (see Table 4.A1).

Trade specialization The change in the composition of EU-oriented exports over the period of 1990–99 was mainly the result of a much faster increase in exports of manufactured products than of agricultural ones. Textiles, clothing and footwear, ferrous and non-ferrous metallurgy, machinery and equipment, and fertilizers and chemicals comprise the biggest share of industrial goods exports (see Table 4.A2). The time profile of shares in EU imports points to shifts in Bulgaria's specialization: from food products to agricultural materials and textile fibers, and from capital equipment to light industries such as clothing, textiles, furniture and footwear as well as metallurgy.

Skilled-labour-intensive products account for a significantly lower share of EU-oriented exports than unskilled-labour-intensive products: their share in EU external imports is also significantly lower. The share of natural-resource-intensive products declined throughout the 1990s: it dropped from 47 per cent in 1989 to 30 per cent in 1998. So did the share of capital-intensive-products, albeit at a slower pace. Bulgarian suppliers of capital- and human capital- (skilled labour) intensive products remain, however, at a comparative disadvantage in EU markets. Although the share of natural-resource-intensive products has been on the decline, the aggregate share of unskilled labour- and resource-intensive products in EU-oriented exports has stayed at roughly the same level of around 62 per cent since 1989 (see Table 4.A3).

In terms of processing, there was a very significant shift to intermediate-stage products and, to a lesser extent, to finished products. The Bulgarian presence in EU markets for commodity chains has increased because of expansion in exports of processed commodities. Primary stage products accounted for less than 10 per cent of commodity chains EU-oriented exports in 1998, down from 32 per cent in 1989. The share of intermediate-stage products rose from 26 per cent in 1989 to 51 per cent in 1998, mainly as a result of a dramatic increase in exports of semi-processed copper. The shift towards intermediate-stage products has resulted in an increase in the share of EU-external imports of these products. The share of finished products dropped slightly from 42 to 40 per cent over this period, but their share in EU external imports stayed at the same level over the period from 1994–98 (see Table 4.A3).[2]

The calculation of *trade coverage indices (TCI)*[3] as a ratio of countries' exports of a given commodity chain to the countries' imports of the same commodity chain is often used to assess countries' comparative advantage in trade.

The assessment of commodity chains according to values of TCI shows ten groups, according to the Harmonized System (HS) nomenclature, with an indices greater than one in 1999 (see Table 4.2).

Increasing values in the period 1995–99 have been noted for: Vegetable products, food, beverages and tobacco, textiles and textile articles, footwear and accessories and precise and optical equipment.

Relative value stabilization is characteristic of: plastics and rubber, leather and leather products, wood and articles of wood, cellulose, paper and articles thereof, and non-metallic minerals.

Unstable value dynamics are shown by: mineral products, base metals and articles thereof, and vehicles and transport equipment.

Decreasing values of TCI over the period are displayed by: live animals, animal products, animal or vegetable fats and oils, chemical or allied products and machinery and equipment.

The *export specialization index (RCA – revealed comparative advantage)*[4] in trade relations with the EU gives information about the comparative position of certain products.[5]

The top 12 commodity chains with the *highest values of RCA*, according to HS commodity groups with values greater than one, are: live animals, animal products, leather and leather products, animal or vegetable fats and oils, non-metallic minerals, textiles and textile articles, cellulose, paper and articles thereof and food, beverages and tobacco (see Table 4.3).

From 1995–99 the comparative advantage on the EU market increased for the flowing commodity groups: live animals, animal products, animal or vegetable fats and oils, leather and leather products, cellulose, paper and articles thereof, textiles and textile articles, vehicles, transport equipment and precise and optical equipment (see Table 4.3).

Comparing the TCI and RCA values in different sectors we can conclude that the groups with values of both indices greater than one possess the greatest export potential (option one). When the TCI value of a group is smaller than one and its RCA value is greater than one, it means that an economy has a strong position in the European market for these goods even if it does not have an internal comparative advantage (imports exceed exports) – option two. If the TCI value is higher than the unit and the RCA is smaller than the unit, the economy does not have a strong position in the European market, even if it has gained an internal comparative advantage in the given group, since the share of exports in this group is smaller than the respective share of imports to the EU from the rest of the world (option three).

The commodity groups with both internal and external comparative advantage in the EU market (TCI and RCA over one) are: live animals, animal products, food, beverages and tobacco, textiles and textile articles, footwear and accessories and non-metallic minerals.

Table 4.2 Trade coverage indices (TCI) in trade relations with the EU

No.	BRANCH (according to HS)	1995	1996	1997	1998	1999
	TOTAL:	0.95	1.07	1.14	0.94	0.78
I	Live animals, animal products	2.22	3.11	1.34	0.98	1.33
II	Vegetable products	3.79	3.85	1.63	5.17	4.62
III	Animal or vegetable fats and oils	0.51	0.04	0.13	0.03	0.02
IV	Food, beverages and tobacco	1.42	1.94	1.92	1.62	1.83
V	Mineral products	1.82	3.53	4.00	1.00	1.12
VI	Chemical or allied products	0.87	0.93	1.08	0.51	0.36
VII	Plastics and rubber	0.66	0.79	0.82	0.66	0.47
VIII	Leather and leather products	1.01	0.70	0.51	0.53	0.59
IX	Wood and articles of wood	4.60	3.80	3.81	3.84	4.58
X	Cellulose, paper and articles thereof	0.26	0.25	0.36	0.22	0.22
XI	Textiles and textile articles	0.85	0.98	1.00	1.06	1.14
XII	Footwear and accessories	1.89	2.07	2.42	2.40	2.69
XIII	Non-metallic minerals	1.30	1.22	1.42	1.38	1.35
XIV	Base metals and articles thereof	4.31	3.47	4.49	4.59	3.13
XV	Machinery and equipment	0.38	0.45	0.36	0.35	0.29
XVI	Vehicles, transport equipment	0.09	0.43	0.36	0.10	0.04
XVII	Precise and optical equipment	0.12	0.15	0.18	0.19	0.25
XVIII	Miscellaneous manufactured articles	0.99	1.12	1.13	1.00	1.06

Source: Author's calculations based on data provided by the National Statistical Institute.

Table 4.3   Specialization indices (RCA) in trade relations with the EU

No.	BRANCH (according to HS)	1995	1996	1997	1998	1999
	TOTAL:	2.34	2.91	1.17	1.04	1.71
I	Live animals, animal products	4.00	3.60	1.42	5.50	5.94
II	Vegetable products	0.54	0.03	0.12	0.03	0.03
III	Animal or vegetable fats and oils	1.50	1.82	1.67	1.73	2.35
IV	Food, beverages and tobacco	1.91	3.30	3.50	1.06	1.44
V	Mineral products	0.92	0.87	0.95	0.54	0.46
VI	Chemical or allied products	0.69	0.74	0.71	0.70	0.60
VII	Plastics and rubber	1.07	0.65	0.45	0.56	0.76
VIII	Leather and leather products	4.84	3.56	3.33	4.08	5.88
IX	Wood and articles of wood	0.27	0.24	0.31	0.24	0.29
X	Cellulose, paper and articles thereof	0.89	0.91	0.87	1.13	1.46
XI	Textiles and textile articles	1.99	1.94	2.12	2.55	3.45
XII	Footwear and accessories	1.37	1.14	1.24	1.47	1.73
XIII	Non-metallic minerals	4.54	3.25	3.92	4.88	4.02
XIV	Base metals and articles thereof	0.40	0.42	0.32	0.37	0.38
XV	Machinery and equipment	0.10	0.40	0.31	0.11	0.05
XVI	Vehicles, transport equipment	0.13	0.14	0.15	0.20	0.33
XVII	Precise and optical equipment	1.05	1.05	0.98	1.07	1.36

Source: Author's calculations based on data provided by the National Statistical Institute.

Table 4.4 Intra-industry trade indices in trade relations with the EU

No.	BRANCH (according to HS)	1995	1996	1997	1998	1999
I	Live animals, animal products	0.62	0.49	0.86	0.99	0.86
II	Animal or vegetable fats and oils	0.42	0.41	0.76	0.32	0.36
III	Food, beverages and tobacco	0.67	0.07	0.24	0.06	0.05
IV	Mineral products	0.83	0.68	0.69	0.76	0.71
V	Chemical or allied products	0.71	0.44	0.40	1.00	0.94
VI	Plastics and rubber	0.93	0.96	0.96	0.68	0.53
VII	Leather and leather products	0.79	0.88	0.90	0.79	0.64
VIII	Wood and articles of wood	0.99	0.82	0.68	0.69	0.74
IX	Cellulose, paper and articles thereof	0.36	0.42	0.42	0.41	0.36
X	Textiles and textile articles	0.41	0.41	0.53	0.36	0.36
XI	Footwear and accessories	0.92	0.99	1.00	0.97	0.94
XII	Non-metallic minerals	0.69	0.65	0.58	0.59	0.54
XIII	Base metals and articles thereof	0.87	0.90	0.83	0.84	0.85
XV	Machinery and equipment	0.38	0.45	0.36	0.36	0.48
XVI	Vehicles, transport equipment	0.55	0.62	0.53	0.52	0.45
XVII	Precise and optical equipment	0.17	0.60	0.53	0.18	0.07
XVIII	Miscellaneous manufactured articles	0.22	0.26	0.30	0.31	0.41

Source: Author's calculations based on data provided by the National Statistical Institute.

Good export potential (option two) can be identified for the following groups: animal or vegetable fats and oils, mineral products, leather and leather products, cellulose, paper and articles thereof and precise and optical equipment.

Vegetable products, wood and articles of wood and base metals and articles thereof represent the third option.

Intra-industry trade Trade in industrial spare parts and components has been growing rapidly over the last decade – much faster than trade in finished goods. As a consequence, the inter-industry division of labour has become increasingly marginalized by the more complex specialization implicit in *intra-industry trade*, presently enriched by *intra-product specialization*, which extends the division of labour to parts and components of products. This trade has several advantages: it is frequently accompanied by the transfer of technology and managerial know-how; it offers direct access to larger markets, allowing the exploitation of economies of scale; and it boosts exports without firms incurring marketing costs.

The main factors determining the level of intra-industry trade (IIT[6]) are vertical specialization, that is, the location of different stages of production in different countries, including subcontracts, and horizontal intra-industry trade, involving finished products.

Bulgarian industry has taken little advantage of these trends in a limited number of sectors. The commodity groups with highest indices of intra-industry trade with the EU, from 1995–99, ranked by index values in 1999 are: chemical or allied products, footwear and accessories, live animals, animal products, base metals and articles thereof, wood and articles of wood, mineral products and leather and leather products (see Table 4.4).

3.2   Foreign Direct Investment

According to the Bulgarian Foreign Investment Agency (BFIA), throughout the decade investment flows in Bulgaria have consistently increased, from US$34 million in 1992 to more than US$1,040 million in 2000, representing over 6 per cent of the GDP

Considering the difficulties associated with the unfavourable external environment (for example, a period of armed conflict in neighbouring countries and the effect of the Russian crisis on investment), Bulgaria has achieved great success in consistently attracting new capital flows. Over 52 per cent of this investment has taken the form of joint ventures, greenfield investment, reinvested earnings and credits, while 32 per cent is attributed to privatization and 8 per cent to capital market investment in 2000 (see Figure 4.1).

Two phases may be identified in foreign capital penetration. The first phase extends until 1997, when FDI inflows were very low. The per capita cumulative inflows of US$54 over the 1990–96 period were some of the lowest among CE candidates for membership in the EU.

In the second phase (1997–2001), FDI inflows increased quite impressively, with the bulk of them going to the industrial sector. One should take account of the fact that this was a time period marked by dynamic policy reforms, currency board enforcement and financial stabilization. The share of investments contributing to the development of an environment facilitating trade has been also on the increase, with considerable foreign investment in the banking sector, and in communications and transport.

A significant amount of FDI was associated with the privatization of state-owned enterprises (SOEs). Privatization-related FDI accounted for around 32 per cent of total foreign investment inflows over the 1992–2001 period. The share reached its highest level, 66.2 per cent, in 1997, dropped to 25.1 per cent in 1998, and increased to 27.7 per cent and 36.6 per cent in 1999 and 2000, respectively.

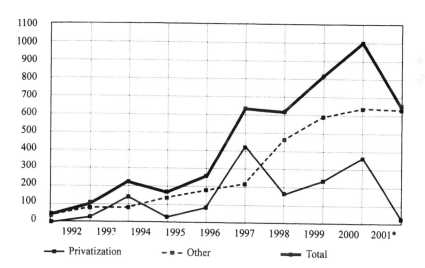

Figure 4.1   Foreign direct investment inflows by years in US$m

'*Non-privatization*' – Greenfield investment + additional investment in companies with foreign participation + reinvestment + joint ventures.

Source: Bulgarian Foreign Investment Agency (BFIA).

Most companies recently investing in Bulgaria have opted for majority ownership rather than joint ventures. This indicates a significant improvement in the business climate, as foreign investors do not seem to need local partners to navigate the administrative environment.

Priority among foreign investment types shall be given to greenfield FDI reinvestment into Bulgarian-registered companies with foreign equity and infrastructure projects in transport, power generation and communications.

The EU member states are the biggest investors in Bulgaria, with a share of almost two-thirds. The five largest investors in the country are from the EU: Germany (US$563.2 million), Greece (US$541.7 million), Italy (US$451.4 million), Belgium (US$415.9 million) and Austria (US$351.2 million).[7] Expectations and confidence in the business sector have gradually and consistently improved since early 1999. By 2000 the Business Confidence Indicators for Bulgaria were growing steadily

In 2000,[8] foreign capital penetration was greatest in the financial sector, with a total amount of US$688.9 million (22.0 per cent), followed by trade – US$462.5 million (14.78 per cent), petroleum, chemical, rubber and plastic products – US$289.8 million (9.26 per cent), mineral products (cement, glass) – US$249.2 million (7.96 per cent), telecommunications – US$229.6 million (7.34 per cent), metallurgy – US$139.7 million (4.46 per cent), mechanical products – US$120.6 million (3.85 per cent), food products – US$118.8 million (3.80 per cent), wood products and paper – US$105.6 million (3.38 per cent), tourism – US$104.8 million (3.35 per cent), textiles and clothing – US$0.5 million (2.89 per cent) and electrical engineering, electronics, computers and communication equipment – US$73 million (2.33 per cent).

The territorial localization of foreign capital in Bulgaria confirms our conclusions about the disproportionate concentration of foreign investment in the metropolitan area and the polarizing pattern of economic growth and its dependence on the economic, geographical, functional and demographic characteristics of regions at the national and international-European level.[9] More than 75 per cent of the foreign investments in the country are located in the district areas of the largest cities in the country. For the city of Sofia and the Sofia region their share is 57 per cent, for Varna – 11 per cent, for Bourgas – 6.3 per cent, and for Plovdiv – 5 per cent. In this respect, small cities and peripheral regions do not seem to have an equal share of the benefits of openness. In fact, if domestic resources closely match the location pattern of foreign capital, many of them may be further marginalized.

4 Regional Specialization Patterns

4.1 Regional Structure and Disparities

Bulgaria is divided into 262 municipalities[10] (with local self-government) and 28 districts[11] (whose administration is appointed by the government). In addition, six planning regions[12] have been defined as a basis for the planning, application and monitoring of the regional interventions in a decentralized manner, corresponding to the regional policy practices in the EU.[13]

Bulgarian regions at the NUTS 2 level are relatively balanced in terms of territory and population (see Table 4.5). The average size of a region's area is 20080 km^2, and of the population is 1.521 thousand persons. The largest region, the south central, is 1.9 times the size of the smallest, the north central. In terms of population, the most populous region, the southeast, is 2.6 times the size of the smallest, the northwest.

The economic differences between the regions are slightly deeper, although still tolerable. Measured by the indicator of GDP per capita (see Table 4.6), the differences are not significant, except in the case of the northwest region, which lags significantly behind the remaining planning regions. In 1999 the ratio of regions between the maximum and minimum GDP per capita was 1.65. The differences in the unemployment rate are much more pronounced – about three times greater.

Considerably larger disparities exist among the 28 districts. The average territory size for a district is 3,964 km^2 and the ratio between the size of the largest district (Bourgas) and the smallest one (Sofia City) is 5.7. In terms of population, the average size is 293,000 people and the ratio between the figures for the biggest district (Sofia City) and the smallest one (Vidin) is 8.7.

At the district level, differences in GDP per capita are characterized by changing dynamics. For instance, this ratio was 1.76 in 1995. By 1999 the ratio had already increased to 3.11. The ratio between the unemployment levels is 6.5 (see Table 4.7). Sofia City, Bourgas and Stara Zagora rank steadily in the first three places in the group of districts with the best development indicators (measured by GDP per capita and the unemployment rate).

Comparisons on a general European level show that the regional differences in Bulgaria are among the smallest when they are compared with EU member states and the remaining candidate states.[14] None of the other countries manifests such convergence at the NUTS 2 level (with respect to GDP per capita), which would be very acceptable at a higher national development level.

Table 4.5 **Planning regions – NUTS 2, 1999**

Planning region	Territory ('000 km²)	%	Population Number	%	No districts/ municipalities	GDP/cap Euro	Index GDP/cap %
Northwest	10.6	9.6	585,512	7.1	3/33	925.2	64.9
North central	18.0	16.2	1,226,052	15.0	5/40	1,244.3	87.3
Norteast	20.0	18.0	1,343,382	16.4	6/49	1,414.8	99.2
Southwest	14.7	13.2	824,491	10.1	3/22	1,517.5	106.4
South central	27.5	24.8	2,068,739	25.3	6/66	1,155.0	81.0
Southeast	20.3	18.3	2,142,700	26.2	5/52	1,900.2	133.3
Bulgaria	111.0	100.0	81,908,76	100.0	28/262	1,426.0	100.0

Source: NSI.

Table 4.6 **Regional differences at the NUTS 2 level, measured by coefficient of variation and ratio between the maximum and minimum value**

NUTS_2 regions	1990	1991	1995	1999
Coefficient of variation -%				
GDP	–	–	42.46	62.28
GDP per capita	–	–	11.71	29.72
Unemployment	–	–	28.47	30.52
Expenditure per capita	48.18	77.86	104.02	127.21
Wage	6.22	6.08	8.15	6.25
Telephones/1,000 persons	14.54	14.47	13.51	11.99
Cars/1,000 persons	18.49	18.17	17.52	16.39
Students	119.05	117.21	105.59	95.53
Population	45.20	45.16	45.50	46.56
R&D personnel	153.02	157.01	161.96	157.19
Maximum/minimum				
GDP	–	–	3.87	7.52
GDP per capita	–	–	1.37	2.05
Unemployment	–	–	2.34	2.81
Expenditure per capita	4.36	5.94	11.49	23.78
Wage	1.19	1.19	1.22	1.19
Telephones/1,000 persons	1.37	1.37	1.35	1.35
Cars/1,000 persons	1.68	1.66	1.63	1.57
Students	822.27	448.47	48.61	57.37
Population	3.44	3.44	3.51	3.66
R&D personnel	43.93	46.72	33.72	41.09

Source: author's calculations based on REGSTAT.

The national differences and the differences in regions (between Bulgaria and the 15 EU and candidate states) mark Bulgaria as the country with the lowest GDP per capita, while Bulgarian regions appear to be some of the least developed regions in the EU and CEE.[15] The GDP per capita was 28 per cent of the EU average in 1999.

4.2 *Specialized and Diversified Regions*

In this section we examine how specialization patterns have evolved in the regions over the transition period and during the country's orientation into

Table 4.7 Regional differences at the NUTS 3 level, measured by coefficient of variation and ratio between the maximum and minimum value

NUTS_3 regions	1990	1991	1995	1999
Coefficient of variation				
Territory				38.45
Population				75.28
GDP	–	–	80.22	133.19
GDP per capita	–	–	11.71	29.72
Unemployment	24.78	30.23	32.40	
Wage	5.15	7.91	12.85	12.67
Telephones/1,000 persons	21.74	21.25	19.77	18.52
Cars/1,000 persons	20.94	20.31	19.49	18.38
Students	204.21	191.92	173.50	179.92
Population	69.92	70.36	72.22	75.28
R&D personnel	341.32	349.18	348.99	340.64
Maximum/minimum				
Territory				5.74
Population				8.73
GDP	–	–	9.89	24.56
GDP per capita	–	–	1.76	3.11
Unemployment	2.35	2.87	5.49	6.52
Wage	1.25	1.39	1.62	1.53
Telephones/1,000 persons	3.44	3.28	2.73	2.70
Cars/1,000 persons	2.58	2.44	2.29	2.10
Students	783.15	420.85	165.07	151.90
Population	7.74	7.83	8.10	8.73
R&D personnel	484.57	475.66	376.18	545.65

Source: Author's calculations based on REGSTAT.

the EU. Regional specialization is defined as the distribution of the shares in a sector industry *i* in total manufacturing in a specific region *j*.

For the purposes of the present study, the level of regional specialization is measured by the Herfindahl index, the Krugman (dissimilarity) index and by Gini coefficients. These indexes have been calculated using employment data for NUTS 2 and NUTS 3 regions for the period 1990–99.

Herfindahl index[16] The absolute degree of specialization, measured by the

Herfindahl index, outlines all NUTS 2 regions as low specialized ones and, it is interesting to note, as having the same level of specialization as in 1999 (0.10, except for the southwest region, whose level is 0.09) (see Table below).

According to the values of this index for the period 1990–99, a very slight increase in the specialization rate is observed in the following NUTS 2 regions: northwest, north central and south central. The most developed region, the southwest region, which includes the Sofia City region, shows some decrease in its specialization rate and a trend towards increased diversification of manufacturing. The other two regions – northeast and southeast – demonstrate some minor changes during the period, but had the same values in 1999 as they had in 1990.

The conclusion that may be made on the basis of this index is that the NUTS 2 regions, which are less developed and were harder hit by the crisis, manifest an increased degree of regional specialization. The reason for this is that a large portion of the less competitive sectors has deteriorated or has been almost liquidated as a result of its strong vulnerability. The process of European integration acts in a similar fashion, since it encourages the development of relatively competitive manufacturing and blocks the development of inefficient and less competitive manufacturing.

The picture at the district (NUTS 3) level is more mixed: 20 districts have increased their degree of specialization, four districts (Rousse, Bourgas, Shumen and Sofia District) have maintained the same degree of specialization and another four districts (Sofia City, Smolyan, Varna and Pazardjik) have reduced their degree of specialization from 1990–99. This process is most pronounced for the Sofia City District (see Table 4.9).

Another conclusion that can be drawn is that there is a certain regional variation in the degree of specialization (the values of the index range between 0.07 and 0.16), but a low level of specialization in all fields can be clearly identified.

4.3   Dissimilarity Index[17]

The values of the Dissimilarity index for the NUTS 2 regions for 1990–99 are relatively low (below 0.45), which means that there are no considerable differences between the regional industrial structures and the national average.

Significant increases in the degree of relative specialization during the period under review are characteristic of the southeast, northwest and southwest regions. There is a certain decrease in specialization in the northeast region and

Table 4.8   Herfindahl index for regional specialization, NUTS 2

NUTS 2 regions	1990	1991	1992	1993	1994	1995	1996	1997	1998	1999
Northwest	0.08	0.08	0.08	0.09	0.09	0.09	0.09	0.09	0.09	0.10
North central	0.09	0.09	0.10	0.10	0.09	0.10	0.11	0.11	0.10	0.10
Northeast	0.10	0.09	0.09	0.09	0.09	0.10	0.10	0.10	0.10	0.10
Southeast	0.10	0.09	0.08	0.09	0.09	0.09	0.09	0.09	0.09	0.10
South central	0.08	0.08	0.08	0.09	0.09	0.09	0.09	0.09	0.10	0.10
Southwest	0.10	0.10	0.09	0.10	0.10	0.10	0.10	0.09	0.09	0.09

Source: Author's calculations based on REGSTAT.

Table 4.9   Herfindahl index for regional specialization, NUTS 3 level

NUTS 3 regions	1990	1991	1992	1993	1994	1995	1996	1997	1998	1999
Blagoevgrad	0.11	0.10	0.09	0.09	0.08	0.09	0.10	0.11	0.13	0.15
Bourgas	0.13	0.12	0.11	0.13	0.12	0.13	0.12	0.13	0.13	0.13
Varna	0.12	0.11	0.10	0.11	0.10	0.11	0.12	0.12	0.12	0.12
Veliko Tarnovo	0.10	0.11	0.11	0.12	0.12	0.13	0.14	0.15	0.15	0.15
Vidin	0.14	0.15	0.16	0.17	0.16	0.20	0.20	0.20	0.20	0.16
Vtratza	0.09	0.09	0.10	0.11	0.11	0.11	0.12	0.12	0.13	0.14
Gabrovo	0.13	0.13	0.14	0.14	0.13	0.15	0.15	0.14	0.14	0.14
Dobrich	0.10	0.09	0.10	0.11	0.10	0.12	0.12	0.12	0.12	0.13
Kardjali	0.09	0.10	0.10	0.11	0.12	0.12	0.13	0.14	0.16	0.16

Table 4.9 cont'd

NUTS 3 regions	1990	1991	1992	1993	1994	1995	1996	1997	1998	1999
Kustendil	0.09	0.09	0.09	0.10	0.10	0.10	0.11	0.11	0.11	0.12
Lovech	0.10	0.10	0.10	0.11	0.10	0.11	0.11	0.11	0.11	0.10
Montana	0.09	0.09	0.09	0.10	0.10	0.10	0.11	0.10	0.10	0.10
Pazardjik	0.08	0.08	0.08	0.08	0.07	0.08	0.07	0.07	0.07	0.07
Pernik	0.12	0.12	0.12	0.13	0.13	0.14	0.14	0.15	0.15	0.14
Pleven	0.11	0.11	0.11	0.12	0.11	0.12	0.13	0.12	0.12	0.12
Plovdiv	0.10	0.10	0.11	0.12	0.12	0.13	0.13	0.13	0.13	0.12
Razgrad	0.10	0.11	0.11	0.12	0.11	0.13	0.13	0.13	0.13	0.13
Russe	0.08	0.08	0.08	0.08	0.08	0.08	0.08	0.08	0.08	0.08
Silistra	0.10	0.10	0.11	0.11	0.10	0.11	0.12	0.11	0.12	0.12
Sliven	0.11	0.11	0.12	0.12	0.11	0.13	0.14	0.15	0.14	0.12
Smolyan	0.17	0.18	0.20	0.20	0.19	0.18	0.17	0.17	0.15	0.12
Sofia	0.15	0.16	0.15	0.17	0.17	0.16	0.15	0.14	0.13	0.13
Sofia region	0.08	0.08	0.07	0.07	0.07	0.07	0.07	0.07	0.07	0.08
Stara Zagora	0.12	0.13	0.13	0.14	0.14	0.14	0.15	0.15	0.16	0.16
Targoviste	0.10	0.11	0.11	0.12	0.12	0.13	0.14	0.15	0.16	0.16
Haskovo	0.09	0.09	0.09	0.09	0.09	0.10	0.10	0.10	0.11	0.11
Shumen	0.11	0.10	0.10	0.11	0.12	0.11	0.12	0.11	0.11	0.10
Yambol	0.11	0.11	0.11	0.12	0.13	0.14	0.13	0.13	0.13	0.12

Source: author's calculations based on REGSTAT.

Table 4.10 Dissimilarity index for regional specialization, NUTS 2

NUTS 2 regions	1990	1991	1992	1993	1994	1995	1996	1997	1998	1999
Northwest	0.35	0.35	0.34	0.36	0.36	0.38	0.39	0.39	0.39	0.38
North central	0.29	0.29	0.28	0.29	0.26	0.31	0.32	0.30	0.27	0.27
Northeast	0.35	0.35	0.36	0.35	0.33	0.37	0.41	0.39	0.35	0.33
Southeast	0.35	0.35	0.35	0.39	0.36	0.41	0.42	0.45	0.44	0.45
South central	0.22	0.22	0.23	0.24	0.24	0.25	0.24	0.25	0.23	0.22
Southwest	0.33	0.33	0.33	0.33	0.29	0.35	0.35	0.37	0.35	0.33

Source: Author's calculations based on REGSTAT.

its level is constant in the south central and north central regions (see Table 4.10). We can also note that there is greater variation in the index values in 1999 (from 0.27 to 0.45) than there was in 1990 (from 0.29 to 0.35).

At the district level (NUTS 3), the industrial employment structure is similar to the national average and this process evolves dynamically. In 1990 the value of this index was higher than 0.5 in 21 districts (out of a total of 28), while in 1999 it was only 50 per cent, that is, it is self-evident that the process of relative specialization is gaining speed (see Table 4.11).

During the 1990–99 period, a clearly manifested trend of increase in the number of regions (NUTS 3) in which the value of this index was rising has been observed. The value of the index decreases, however insignificantly, in only six districts (Smolyan, Bourgas, Shumen, Yambol, Pleven and Montana).

Gini coefficients[18] Gini coefficients measure summed up differences in specialization rates by cumulating the differences in the shares of a region and the shares of the norm (national average), after ranking the industries according to their specialization ratios. The index lies between zero and one. An index close to zero indicates that the distribution of production *i* matches the overall distribution in the country, whereas an index close to one indicates that a region is completely specialized in one industry with a small overall share in manufacturing in the country.

The values of the Gini coefficients define NUTS 2 regions as relatively diversified (that is, within the limits of 0.22 and 0.49 in 1999). The trend goes in the direction of increase of the degree of regional specialization. The only exception is the north central region, where this process has been observed to follow a different dynamic pattern and in 1999 the region was noted to be much more strongly specialized than at its 1990 level (see Table 4.12).

According to the constructed coefficients for NUTS 3 regions, one may note an increase in the regional specialization level (in 18 of a total of 28 districts) during the period 1990–99. It is also interesting to note that until 1998 all districts (with the exception of Bourgas, which retained its 1990 level, and Gabrovo, where a slight drop was registered) have shown increases in their level of specialization, quite significant ones in certain cases (see Table 4.13).

Changes in regional specialization By regressing the log of the specialization index on a time trend we test the significance of the changes in regional specialization, using a trend model of the form:[19]

Table 4.11 Dissimilarity index for regional specialization, NUTS 3

NUTS 3 regions	1990	1991	1992	1993	1994	1995	1996	1997	1998	1999
Blagoevgrad	0.55	0.56	0.54	0.58	0.54	0.55	0.54	0.58	0.62	0.64
Bourgas	0.65	0.64	0.63	0.67	0.66	0.68	0.69	0.70	0.65	0.63
Varna	0.52	0.51	0.51	0.52	0.49	0.54	0.61	0.61	0.57	0.55
Veliko Tarnovo	0.47	0.50	0.49	0.50	0.46	0.51	0.52	0.53	0.53	0.53
Vidin	0.70	0.74	0.80	0.83	0.79	0.91	0.90	0.94	0.92	0.75
Vtratza	0.45	0.51	0.55	0.54	0.55	0.56	0.57	0.57	0.62	0.64
Gabrovo	0.62	0.65	0.62	0.62	0.58	0.64	0.67	0.64	0.61	0.60
Dobrich	0.41	0.45	0.44	0.46	0.41	0.47	0.49	0.55	0.55	0.53
Kardjali	0.63	0.64	0.59	0.63	0.60	0.65	0.70	0.75	0.73	0.70
Kustendil	0.44	0.48	0.53	0.54	0.55	0.59	0.66	0.67	0.65	0.69
Lovech	0.55	0.57	0.57	0.60	0.64	0.67	0.66	0.62	0.61	0.60
Montana	0.38	0.39	0.36	0.36	0.36	0.39	0.38	0.37	0.36	0.35
Pazardjik	0.52	0.50	0.50	0.51	0.56	0.53	0.55	0.55	0.54	0.53
Pernik	0.68	0.70	0.68	0.70	0.58	0.74	0.79	0.80	0.81	0.76
Pleven	0.44	0.46	0.40	0.39	0.38	0.43	0.45	0.42	0.41	0.37
Plovdiv	0.28	0.33	0.41	0.40	0.40	0.42	0.42	0.44	0.43	0.39
Razgrad	0.63	0.63	0.63	0.64	0.65	0.69	0.73	0.74	0.74	0.75
Russe	0.32	0.31	0.31	0.32	0.32	0.34	0.34	0.34	0.40	0.39
Silistra	0.43	0.43	0.45	0.42	0.39	0.47	0.49	0.48	0.49	0.46
Sliven	0.60	0.61	0.66	0.68	0.67	0.78	0.77	0.78	0.74	0.70
Smolyan	0.81	0.86	0.94	0.94	0.94	0.94	0.92	0.93	0.88	0.72
Sofia	0.49	0.53	0.53	0.55	0.49	0.56	0.56	0.60	0.58	0.55
Sofia region	0.44	0.48	0.46	0.48	0.49	0.50	0.51	0.48	0.53	0.47
Stara Zagora	0.58	0.61	0.60	0.62	0.64	0.64	0.61	0.64	0.65	0.66

Table 4.11 cont'd

NUTS 3 regions	1990	1991	1992	1993	1994	1995	1996	1997	1998	1999
Targoviste	0.36	0.44	0.46	0.50	0.52	0.56	0.62	0.64	0.62	0.63
Haskovo	0.48	0.47	0.54	0.54	0.53	0.55	0.54	0.54	0.52	0.50
Shumen	0.58	0.62	0.61	0.68	0.71	0.67	0.68	0.67	0.62	0.57
Yambol	0.45	0.49	0.49	0.51	0.55	0.57	0.53	0.51	0.49	0.40

Source: Author's calculations based on REGSTAT.

Table 4.12 Gini coefficients for regional specialization, NUTS 2

NUTS 2 regions	1990	1991	1992	1993	1994	1995	1996	1997	1998	1999
Northwest	0.38	0.38	0.39	0.40	0.42	0.42	0.42	0.42	0.41	0.38
north central	0.24	0.24	0.23	0.23	0.20	0.24	0.25	0.23	0.23	0.22
Northeast	0.38	0.38	0.39	0.39	0.38	0.41	0.43	0.43	0.42	0.41
Southeast	0.44	0.44	0.45	0.47	0.46	0.48	0.48	0.47	0.48	0.49
south central	0.24	0.24	0.25	0.26	0.25	0.26	0.26	0.26	0.25	0.25
Southwest	0.25	0.25	0.26	0.26	0.22	0.26	0.27	0.28	0.28	0.27

Source: Author's calculations based on REGSTAT.

Table 4.13 Gini coefficients for regional specialization, NUTS 3

NUTS 3 regions	1990	1991	1992	1993	1994	1995	1996	1997	1998	1999
Blagoevgrad	0.40	0.41	0.44	0.45	0.42	0.46	0.46	0.47	0.49	0.47
Bourgas	0.64	0.64	0.65	0.66	0.65	0.66	0.66	0.65	0.64	0.64
Varna	0.49	0.52	0.54	0.55	0.52	0.57	0.60	0.60	0.60	0.58
Veliko Tarnovo	0.41	0.42	0.41	0.40	0.37	0.41	0.42	0.43	0.44	0.44
Vidin	0.78	0.78	0.80	0.81	0.77	0.84	0.84	0.84	0.84	0.73
Vtratza	0.45	0.48	0.50	0.50	0.52	0.51	0.53	0.54	0.56	0.57
Gabrovo	0.53	0.54	0.52	0.51	0.45	0.52	0.53	0.51	0.51	0.48
Dobrich	0.49	0.48	0.48	0.50	0.45	0.50	0.52	0.52	0.52	0.53
Kardjali	0.59	0.61	0.60	0.61	0.62	0.61	0.65	0.63	0.62	0.63
Kustendil	0.41	0.44	0.47	0.48	0.48	0.51	0.54	0.55	0.55	0.57
Lovech	0.52	0.52	0.53	0.56	0.53	0.60	0.63	0.60	0.61	0.59
Montana	0.31	0.30	0.29	0.29	0.30	0.32	0.32	0.31	0.32	0.29
Pazardjik	0.41	0.41	0.42	0.44	0.44	0.46	0.48	0.49	0.49	0.51
Pernik	0.61	0.61	0.61	0.62	0.56	0.63	0.67	0.67	0.67	0.66
Pleven	0.43	0.42	0.38	0.38	0.34	0.39	0.38	0.44	0.44	0.41
Plovdiv	0.29	0.32	0.38	0.37	0.37	0.37	0.37	0.38	0.36	0.32
Razgrad	0.55	0.56	0.57	0.59	0.58	0.61	0.67	0.67	0.66	0.68
Russe	0.39	0.37	0.39	0.40	0.40	0.42	0.43	0.42	0.43	0.43
Silistra	0.46	0.47	0.46	0.45	0.41	0.47	0.48	0.48	0.52	0.52
Sliven	0.54	0.54	0.56	0.59	0.55	0.63	0.65	0.66	0.65	0.63
Smolyan	0.70	0.71	0.74	0.76	0.74	0.76	0.76	0.77	0.77	0.69
Sofia	0.42	0.44	0.44	0.43	0.38	0.43	0.43	0.45	0.45	0.41
Sofia region	0.34	0.33	0.35	0.37	0.37	0.39	0.40	0.40	0.43	0.47
Stara Zagora	0.55	0.57	0.56	0.56	0.55	0.56	0.56	0.57	0.58	0.59

Table 4.13 cont'd

NUTS 3 regions	1990	1991	1992	1993	1994	1995	1996	1997	1998	1999
Targoviste	0.42	0.43	0.42	0.44	0.43	0.46	0.54	0.55	0.54	0.53
Haskovo	0.45	0.44	0.46	0.46	0.43	0.47	0.48	0.49	0.46	0.44
Shumen	0.53	0.55	0.56	0.58	0.62	0.58	0.60	0.59	0.57	0.53
Yambol	0.49	0.50	0.51	0.52	0.55	0.61	0.56	0.55	0.53	0.47

Source: author's calculations based on REGSTAT.

Log SPEC$_{ij}$ = μ + α_{ij} + β time + ε_{ij} (1)
where:
SPEC is the specialization index (Herfindahl, Dissimilarity and Gini index, alternatively).
t (time) = 1, 2, ...n;
n = the number of years

The variable 'time' is a re-scaling of the year to which the indicator refers, re-scaled to have values starting from one (by subtracting the value 1989 from each observed year). The variable 'time' is not expressed in logs.

j = region
i = industry
100*β = the annual percentage change of the specialization measure.

Since we have a panel of ten periods and 28 regions (NUTS 3), we used the fixed effect estimator. To be exact, the number of observations is 280 and the number of groups is 28.

The results of the three regressions are shown in Table 4.14. The first column (Ln Herfindahl) contains the results of the model in which the dependent variable is the natural log of the Herfindahl index for specialization. The second column (Ln Dissimilarity) contains the results of the model in which the dependent variable is the natural log of the dissimilarity index. The third column (Ln Gini) contains the results of the model in which the dependent variable is the natural log of the Gini index.

Since all coefficients seem to be statistically significant (the numbers in parenthesis are the standard errors, and '***' means that they are all significantly different from zero at the 1 per cent level), from the previous table we may conclude that specialization has increased in Bulgaria over the period analysed.

According to the model estimations, the annual percentage change over the period 1990–99 is 1.9 per cent for the degree of specialization measured by the Herfindahl index, 1.6 per cent as measured by the Dissimilarity index and 1.3 per cent as measured by the Gini index.

Table 4.14   Regression results on changes in regional specialization

	Ln Herfindahl	Ln Dissimilarity	Ln Gini
Time (beta coefficient)	0.01944	0.01576	0.01312
	(0.00180)***	(0.00156)***	(0.00115)***
Constant (alpha)	-2.26501	-0.67776	-0.75770
	(0.01119)***	(0.00967)***	(0.00712)***
Observations	280	280	280
Number of observations	28	28	28
R-squared	0.31624	0.28970	0.34257

Standard errors in parentheses.
* Significant at 10 per cent; ** significant at 5 per cent; *** significant at 1 per cent. The numbers in brackets are the standard errors of the estimates.

5   Location and Concentration of Industrial Activity

5.1   The Manufacturing Structure

During the more than a decade of transition, the patterns of growth have been uneven across sectors. The highest growth sectors during this period were communications, transport, agriculture and forestry. Industry took the lead in 1998. Changes in the dynamics of the different industries have resulted in significant changes in the very structure of manufacturing. The sectors with the highest shares in manufacturing output in 1999 were food, beverages and tobacco (29.7 per cent), followed by metal products, machinery and equipment casting (14.3 per cent) and basic metals except casting of metals (12.3 per cent). Coke, refined petroleum products and nuclear fuel (0.8 per cent) and leather, leather and fur clothes, footwear and products (1.5 per cent) have the lowest shares in manufacturing (see Table 4.A4).

Studying the changes in the manufacturing structure between 1990–99, one may observe that the electrical and electronics industry is the sector that has suffered the greatest loss of positions. This industry has diminished its employment share in absolute and relative terms from 14.7 per cent in 1990 to barely 6 per cent in 1999. The food and beverages industry had the highest relative share (29.7 per cent) in 1999, while in 1999 this industry accounted for 13.5 per cent.

At the NUTS 2 level, employment in manufacturing is unevenly distributed. The south central (26.4 per cent) and southwest (26.8 per cent) regions have the largest share, retaining their leading positions during the entire period.

The main changes in the sectoral structure of the regions (NUTS 2) over the period 1990–99 demonstrate similar trends: a severe slump in the share of employment in the electrical and electronics industry (most acutely manifested in the southwest region, whose share has dropped from 22 per cent in 1990 to 8 per cent in 1999), and increases in the share of employment in the food and beverages industry (3 to 6 per cent) and the textile and apparel industry (2 to 12 per cent). A reduction in its relative share of employment in all regions has also been registered for wood and products of wood and cork, and plaiting materials (see Table 4.A4).

As a background to the severe decline in the number of employed, the individual districts (NUTS 3) manifest minor differences with respect to their participation in the territorial distribution of the structure of employment in manufacturing. The individual industries show, to a large extent, a consistent distribution by districts, as well as minor changes in their structural participation from 1999 to 1990.

With respect to the regions (NUTS 2), the distribution of manufacturing shares, as compared to population shares, was more balanced, particularly in 1990, when the ratio of the regional coefficients was 1.3. In 1999 the ratio was already 1.5 (see Table 4.15). The highest value for this coefficient was recorded in the north central region (1.15), which features the highest industrial activity rate and the most highly skilled labour force. It is followed by the value for the southwest region, which includes the industrial agglomeration of Sofia.

The distribution of manufacturing shares, as compared to population shares, is unbalanced at the NUTS 3 level, where the ratio in 1999 was 3.0 (as compared to 2.3 in 1990). The values are quite high in the industrially-developed regions (Gabrovo, Lovech, Bourgas, Plovdiv, Blagoevgrad, Pernik, Stara Zagora, and Veliko Tarnovo). At the same time, the ratio's values have diminished between 1990 and 1999 in the districts with drops in industrial development (Vidin, Silistra, Pazardjik, Montana, Dobrich, Razgrad, Smolyan and Yambol). The majority of districts made only a small move up or down the scale or remained the same (see Table 4.16).

5.2   Concentrated and Dispersed Industries

For the purposes of this study, the level of geographic concentration is measured by the Herfindahl index,[20] the Dissimilarity index[21] and Gini coefficients.[22] These indexes have been calculated on the basis of employment

Table 4.15 Ratio of industrial employment shares to population shares, NUTS 2

NUTS 2 regions	1990	1991	1992	1993	1994	1995	1996	1997	1998	1999
Northwest	1.00	0.98	0.96	0.95	0.90	0.88	0.89	0.91	0.91	0.84
North central	1.09	1.07	1.07	1.07	1.10	1.09	1.12	1.14	1.14	1.15
Northeast	0.83	0.87	0.84	0.81	0.80	0.82	0.79	0.84	0.85	0.82
Southeast	0.89	0.91	0.90	0.87	0.86	0.86	0.86	0.84	0.81	0.78
South central	1.02	1.04	1.02	1.03	1.04	1.04	1.07	1.07	1.08	1.07
Southwest	1.07	1.04	1.09	1.12	1.11	1.11	1.08	1.03	1.03	1.09

Source: author's calculations based on REGSTAT.

Table 4.16 Ratio of industrial employment shares to population shares, NUTS 3

NUTS 3 regions	1990	1991	1992	1993	1994	1995	1996	1997	1998	1999
Total	1.00	1.00	1.00	1.00	1.00	1.00	1.00	1.00	1.00	1.00
Blagoevgrad	1.08	1.10	1.04	0.99	0.93	0.99	0.98	1.02	1.11	1.17
Bourgas	0.87	0.89	0.89	0.87	0.88	0.88	0.89	0.90	0.87	0.90
Varna	0.92	0.97	0.92	0.92	0.93	0.97	1.00	1.03	1.05	1.00
Veliko Tarnovo	1.08	1.06	1.02	0.97	1.00	0.99	1.01	1.06	1.08	1.10
Vidin	0.98	0.93	0.93	0.92	0.88	0.85	0.84	0.82	0.81	0.61
Vtratza	1.08	1.09	1.05	1.04	0.96	0.98	1.00	1.03	1.02	1.01
Gabrovo	1.43	1.41	1.48	1.57	1.64	1.63	1.70	1.70	1.75	1.79
Dobrich	0.80	0.83	0.81	0.75	0.67	0.70	0.70	0.69	0.71	0.70
Kardjali	0.62	0.63	0.65	0.64	0.62	0.62	0.58	0.61	0.64	0.66

Table 4.16 cont'd

NUTS 3 regions	1990	1991	1992	1993	1994	1995	1996	1997	1998	1999
Kustendil	1.18	1.15	1.21	1.25	1.23	1.19	1.25	1.28	1.33	1.38
Lovech	1.20	1.14	1.13	1.11	1.16	1.19	1.19	1.21	1.23	1.22
Montana	0.90	0.88	0.86	0.84	0.83	0.78	0.79	0.83	0.83	0.79
Pazardjik	1.05	1.02	1.00	0.93	0.91	0.93	0.92	0.95	0.93	0.88
Pernik	1.32	1.34	1.36	1.39	1.33	1.34	1.37	1.39	1.38	1.43
Pleven	0.96	0.94	0.97	0.97	0.98	0.98	0.99	1.00	0.94	0.92
Plovdiv	1.03	1.05	0.98	1.02	1.04	1.03	1.14	1.11	1.09	1.11
Razgrad	0.79	0.85	0.78	0.77	0.76	0.77	0.77	0.78	0.78	0.76
Russe	0.98	0.99	0.98	1.01	1.00	0.98	1.00	1.05	1.03	1.05
Silistra	0.80	0.78	0.74	0.71	0.67	0.67	0.64	0.60	0.60	0.59
Sliven	0.96	0.95	0.95	0.92	0.90	0.92	0.91	0.85	0.80	0.69
Smolyan	1.18	1.14	1.11	1.08	1.09	1.06	0.96	1.00	1.04	0.93
Sofia City	1.00	0.97	1.07	1.13	1.15	1.12	1.06	0.96	0.92	0.98
Sofia region	1.14	1.05	1.03	0.97	0.99	1.01	0.99	1.02	1.05	1.07
Stara Zagora	1.25	1.33	1.34	1.37	1.46	1.49	1.48	1.52	1.55	1.53
Targoviste	0.77	0.78	0.77	0.68	0.66	0.67	0.66	0.72	0.74	0.76
Haskovo	0.93	0.93	0.93	0.93	0.91	0.90	0.93	0.90	0.95	0.90
Shumen	0.79	0.89	0.88	0.85	0.85	0.85	0.66	0.92	0.90	0.84
Yambol	0.87	0.88	0.84	0.79	0.78	0.74	0.73	0.68	0.70	0.62

Source: Author's calculations based on REGSTAT.

indicators for all NUTS 2 level and NUTS 3 level regions for the period 1990–99

An analysis of the obtained values of various concentration measures provides grounds for the following conclusions (see Tables 4.A5 and 4.A6):

- Relatively low values of the Herfindahl index of absolute concentration for the entire period are characteristic of all manufacturing industries (excepting coke, refined petroleum products and nuclear fuel), though this is more evidence of spatial dispersal than of strong spatial concentration. This conclusion is emphasized particularly with respect to the district level (NUTS 3). In the case of the planning regions (NUTS 2), the degree of concentration is higher.
- The highest level of absolute concentration has been observed for the manufacture of basic metals except casting of metals (0.51 in 1999 for NUTS 2), pulp, paper and paper products and publishing and printing (0.34) and transport equipment (0.31). None of these can be called newly developed sectors.
- The least concentrated industries are the manufacturing of food products, beverages and tobacco products (0.20), textiles (0.20), apparel (0.22), wood and wood products (0.21) and non-metallic products (0.20). The values for the remaining industries are close to the above figures and range between 0.22 and 0.28.
- A process of increase in the degree of concentration of industries is taking place. The industries for which the process of geographical concentration is most pronounced are: basic metals except casting of metals, transport equipment, leather, leather and fur clothes, footwear and products, and rubber and plastic products.

If we consider relative geographic concentration, the highest values for manufacturing are found for coke, refined petroleum products and nuclear fuel, transport equipment, and mining and quarrying – the mining of ore and coal, and the extraction of petroleum and natural gas. It is interesting to note that the industries with the highest geographic concentration in 1990 do not have the highest geographic concentration in 1999. In 1990, the lowest relative concentrations were noted for food, beverages and tobacco, apparel, machinery and equipment, electrical and electronic industry, and wood and products of wood. By 1999, only the first of these maintained a relatively homogeneous spatial distribution (see Tables 4A.5 and 4A.6). The tables shows that 15 out of 19 industries in NUTS 2 and NUTS 3 regions demonstrate an increase in geographical concentration, with an average increase of about 2 per cent.[23] Only four industries experienced a fall in concentration.

In order to check whether industrial concentration has changed significantly over the period 1990-1990, we have tested the following trend model:

$$\log CONC_{it} = \alpha + \beta t + \varepsilon_{it} \tag{2}$$

CONC = concentration measure (Herfindahl, Dissimilarity, Gini)
i = industry ; $t = 1, 2, n$ n = the number of years
$100*\beta$ = the annual percentage change of the specialization/ concentration measure.

Table 4.17 shows the results of the estimated trend model[24] (2) corresponding to three different measures of concentration. The first column (Ln Herfindahl) shows the results of the regression in which the dependent variable is the Herfindahl index for geographical concentration, the second column (Ln Dissimilarity) – the dependent variable is the Dissimilarity index, the third column (Ln Gini) – the dependent variable is the Gini location index.

Since all estimated coefficients are statistically significant, we can conclude that the geographical concentration of industries has increased in Bulgaria over the analysed period.

According to the model estimations the annual percentage change over the period 1990–99 is 1.9 per cent for the concentration measured by Herfindahl index, 1.4 per cent for that measured by the Dissimilarity index and 1.6 per cent as measured by the Gini index.

Table 4.17   Geographical concentration of manufacturing in Bulgaria, 1990–99

	Ln Herfindahl	Ln Dissimilarity	Ln Gini
Time (beta coefficient)	0.01923	0.01438	0.01623
	(0.00295)***	(0.00256)***	(0.00256)***
Constant (alpha)	−2.43757	−0.52917	−0.87832
	(0.01831)***	(0.01588)***	(0.01587)***
Observations	200	200	200
Number of id. no.	20	20	20
R-squared	0.19165	0.15006	0.18374

Standard errors in parentheses.
* significant at 10 per cent; ** significant at 5 per cent; *** significant at 1 per cent.

6 The Impact of Economic Integration on the Regional Wage Structure

The models outlined in this chapter test the impact of economic integration on regional relative wages. The following three hypotheses have been tested:

1) regional relative wage levels decrease with distance from the capital city;
2) trade liberalization eliminates distance effects;
3) after the entering into force of the EU agreements, distance effects for western or south(west)ern border regions and the other regions will converge to similar levels.

The first hypothesis is based on the assumption that in a closed economy regional wages are decreasing according to the value of transport costs to the industrial center, while in an open economy regional wages are determined by access to foreign markets.[25] For Bulgaria we test this hypothesis for the western[26] regions, but also for the south(west)ern[27] and eastern border regions (Black Sea Coast) (in this case, with Greece, since Greece is the only EU member state bordering Bulgaria). We expect to find that regional relative wages decrease with distance from the capital city and that the wages in western and/or south(west)ern border regions are high relative to other regions.

The second hypothesis is based on the assumption that trade liberalization between accession countries and European Union has determined the relocation of production activities from traditional industrial centers to border regions.

The third hypothesis suggests that access to western markets has become important in determining regional wage differentials in accession countries.

6.1 Model Specifications

Hypothesis 1: regional relative wage levels decrease with distance from the capital city.

$(\beta t < 0)$

$$\log (WAGE_{jt}/WAGE_{ct}) = \alpha + \beta_t \log(DIST_j) + \gamma_t BORD_j + \lambda t (\log DIST_j \times BORD_j) ++ \mu_t (\log DIST_j \times YEAR) + \nu_t (\log DIST_j \times BORD_j \times YEAR) + \varepsilon_{jt} \tag{3}$$

$WAGE_j$ = the wage in region j

$WAGE_c$ = the wage in the capital city

DISTj = the distance between region i (county capital) and capital city BORD = dummy variable for border regions $BORD_j$ = one if region j is a border region and zero otherwise (it captures time and distance-invariant factors specific to western border regions that influence relative wages such as possibilities for cross-border cooperation and work).

For Bulgaria it is important to test this hypothesis not only for the western border region, but for the south(west)ern and eastern border regions with Greece, since Greece is the only EU member state bordering Bulgaria.

1.1 – $BORD_j$: EU border regions (Blagoevgrad, Smolyan and Kardjali);
1.2 – SOUTH: EU border regions and southern border regions (Blagoevgrad, Pazardjik, Smolyan, Kardjali and Haskovo);
1.3 – EAST: eastern border regions (Bourgas, Varna and Dobrich);
1.4 – WEST: western border regions (Vidin, Montana, Vratsa and Sofia regions, and Pernik, Kustendil and Blagoevgrad).
2 – YEAR: since the EU association agreements were signed in 1993 and entered into force on February, 1, 1995, the dummy variable has a value of zero from 1990 to 1994 and a value of one from 1995 to 1999.
3.1 – $DIST_j$: this is the distance between each region and the country's capital.

Hypothesis 2: Trade liberalization eliminates the distance effects.

$(\mu_t = 0)$
$$\log (WAGE_{jt}/WAGE_{ct}) = \alpha + \beta_t \log(DIST_j) + \gamma_t \, BORD_j + \lambda t \, (\log DIST_j \times BORD_j) + \mu_t \, (\log DISTj \times YEAR) + \varepsilon_{jt} \qquad (4)$$

Hypothesis 3: After the entering into force of the EU agreements, distance effects for western border regions and other regions converge to similar levels.

$(\beta_t + \mu_t = \beta_t + \lambda_t + v_t)$

$$\log (WAGE_{jt}/WAGE_{ct}) = \alpha + \beta_t \log(DIST_j) + \gamma_t \, BORD_j + \lambda t \, (\log DISTj \times BORDj) + \mu_t \, (\log DIST \times YEAR) + v_t \, (\log DIST \times BORD \times YEAR) + \varepsilon_{jt} \qquad (5)$$

6.2 Estimation Issues

We estimated Model 3 by pooling all data (across regions and across years), using White-corrected robust standard errors (see Table 4.18).

Table 4.18 Results of the estimates for Model 3

	(1)	(2)
LnDistCapital	−0.04625	−0.04792
	(0.00168)***	(0.00211)***
Border EU	0.21785	0.22744
	(0.31402)	(0.31462)
Dist * Border	−0.06292	−0.06466
	(0.06182)	(0.06194)
South	−0.53894	−0.54821
	(0.22227)**	(0.22278)**
Dist * South	0.11012	0.11178
	(0.04471)**	(0.04481)**
East	6.05951	6.05025
	(0.87619)***	(0.87783)***
Dist * East	−0.95852	−0.95686
	(0.14550)***	(0.14577)***
West	−0.12794	−0.13721
	(0.02668)***	(0.02751)***
Dist * West	0.03078	0.03244
	(0.00723)***	(0.00736)***
Constant		0.00927
		(0.00651)
Observations	280	280
R-squared	0.83933	0.43707

White robust standard errors in parentheses.
* Significant at 10 per cent; ** significant at 5 per cent; *** significant at 1 per cent.

We can easily see that the distance from the country's capital always has a negative significant coefficient, meaning that the more distant the region is from the country's capital, the less its wages are, relative to those in the country capital.

The coefficient for the EU border regions dummy is never significant, while the dummies for the south, east and western borders are always significant. The interaction variables are always significant, with the only exception of the variable 'DISTx BORDr' which is significant only in the third column.

We may note that the four dummies corresponding to Border EU, South, East and West regions are not mutually exclusive: some Bulgarian regions do not belong to any of these categories; at the same time, some regions belong to more than one group. Thus, in principle, we have no reason to omit the intercept term from our estimations (this is equivalent of imposing a zero intercept for those regions not belonging to any of the four groups). In the second column we have repeated the model estimations using the constant term as well. It can easily be seen that the estimates of the coefficients and their level of significance do not change much when we allow the intercept term to be different from zero; furthermore, the intercept itself does not seem to be significantly different from zero in the first two estimates. However, the R-squared drops from 0.84 to only 0.44 when we allow the constant term to be different from zero.

Table 4.19 presents the estimations of Model 4. The only difference between the two models (and therefore between Table 4.18 and Table 4.19) is the interaction variable 'Ln DIST*YEAR', which combines the distance of each region from the country capital and the year of the entering into force of the EU association agreements.

We found that the coefficient of the variable 'Ln DIST*YEAR' is negative. We should comment that, before the entering into force of the EU Association Agreements, the coefficient beta (the coefficient of the variable 'distance from country capital') had a certain value (negative). After the entering into force of the EU Association Agreements, we also estimate the variable 'Ln DIST*YEAR'. Since its coefficient is negative, this means that the negative coefficient for beta (the coefficient of the variable 'distance from country capital') becomes even more negative. From an econometric perspective (if the coefficient is not significant) we might have said that nothing changed before and after the entering into force of the EU Association Agreement.

To check this, we have attempted to make a regression using only observations from before the year of the entering into force of the EU Association Agreements and a regression using only observations from after the year of the entering into force of the EU Association Agreements. In the first regression the beta (the coefficient of the variable 'distance from country capital') was −0.040; in the second regression it was −0.052. Obviously, in these two last regressions we did not use the variable 'Ln DISTt*YEAR'.

In conclusion, we would say that our results confirm that wages decrease with distance from the country capital. However, these changes are very slow,

Table 4.19 Results of the estimates for Model 4

	(1)	(2)
LnDistCapital	−0.04115	−0.04282
	(0.00211)***	(0.00242)***
Border EU	0.21785	0.22744
	(0.30349)	(0.30413)
Dist * Border	−0.06292	−0.06466
	(0.05930)	(0.05942)
South	−0.53894	−0.54821
	(0.22402)**	(0.22452)**
Dist * South	0.11012	0.11178
	(0.04464)**	(0.04474)**
East	6.05951	6.05025
	(0.87033)***	(0.87197)***
Dist * East	−0.95852	v0.95686
	(0.14397)***	(0.14425)***
West	−0.12794	−0.13721
	(0.02679)***	(0.02755)***
Dist * West	0.03078	0.03244
	(0.00715)***	(0.00725)***
Ln Dist*Year	−0.01020	−0.01020
	(0.00231)***	(0.00231)***
Constant		0.00927
		(0.00623)
Observations	280	280
R-squared	0.85029	0.47548

White robust standard errors in parentheses.
* Significant at 10 per cent; ** significant at 5 per cent; *** significant at 1 per cent.

and our time series might still be too short to capture this effect.

The results of Model 5 (Table 4.20) tend to confirm the results of the previous two models, although some coefficients seem to be closer than zero when more explanatory variables are introduced into the model. The majority of the newly introduced explanatory variables seem not to be significantly different from zero.

The interesting hypothesis we may test in this case is whether the distance effects for western border regions and the other regions converge to similar levels after the entering into force of the EU agreements.

$$\beta_t + \mu_t = \beta_t + \lambda_t + v_t \qquad (6)$$

Table 4.20   Results of the estimates for Model 5

	(1)	(2)
LnDistCapital	−0.04040	−0.04207
	(0.00247)***	(0.00277)***
Border EU	0.21785	0.22744
	(0.29880)	(0.29944)
Dist Border	−0.05325	−0.05498
	(0.05922)	(0.05935)
South	−0.53894	−0.54821
	(0.22117)**	(0.22168)**
Dist South	0.10694	0.10861
	(0.04549)**	(0.04559)**
East	6.05951	6.05025
	(0.86781)***	(0.86947)***
Dist East	−0.96313	−0.96146
	(0.14406)***	(0.14433)***
West	−0.12794	−0.13721
	(0.02694)***	(0.02771)***
Dist West	0.02744	0.02911
	(0.00766)***	(0.00777)***
Ln Dist * Year	−0.01169	v0.01169
	(0.00315)***	(0.00315)***
LnDist*Bord*Year	−0.01936	−0.01936
	(0.00858)**	(0.00860)**
LnDist*South*Year	0.00635	0.00635
	(0.00704)	(0.00705)
LnDist*East*Year	0.00921	0.00921
	(0.00582)	(0.00583)
LnDist*West*Year	0.00668	0.00668
	(0.00750)	(0.00751)
Constant		0.00927
		(0.00627)
Observations	280	280
R-squared	0.85389	0.48811

White robust standard errors in parentheses.
* Significant at 10 per cent; ** significant at 5 per cent; *** significant at 1 per cent.

Table 4.21 Results of the test, based on estimates of Model 6

	(1)	(2)
Border F test	1.11	1.17
Prob > F	0.2924	0.2799
South F test	8.25	8.44
Prob > F	0.0044 *	0.0040 *
East F test	42.76	42.45
Prob > F	0.0000 *	0.0000 *
West F test	23.70	25.30
Prob > F	0.0000 *	0.0000 *

(*) the null hypothesis that $\beta_t + \mu_t = \beta_t + \lambda_t + v_t$ is rejected at 5 per cent or lower.

Although all models give the same estimated coefficients, the different standard errors may yield different results of the test. The results are summarized in Table 4.21, in which each column refers to the column with the same number in the table showing the estimates.

The hypothesis is not rejected only for border regions.

7 Regional Specialization and Growth

The model specified in this chapter analyses the impact of regional specialization and various regional characteristics on regional economic growth. The tested question is whether regional specialization is able to explain the GDP changes and if so, in what proportion. We also attempt to determine the relationship between regional specialization and qualitative factors and their importance for regional growth.

7.1 Model Specification

$$\text{Ln} \, (y_{j,t+1} \, / \, y_{jt}) = \alpha + \beta_t \, \ln(\text{SPEC}_{jt}) + \gamma_{it} \, \Sigma \, X_{ijt} + \varepsilon_{jt} \qquad (7)$$

Control variables:
- X1: share of employment in the secondary sector in region j
- X2: share of employment in services in region j
- X3: number of firms with foreign capital per 100,000 inhabitants
- X4: share of population aged 15–65 in region j
- X5: number of telephone lines per 100,000 inhabitants in region j

7.2   Estimation Issues

We use panel data to estimate this model. Since the effect of the control variables X_{ijt} is probably picked up by the fixed effects, in the following table, two kinds of estimations are presented. In columns (1), (3) and (5) of Table 4.9 are model estimates (alternatively using the Herfindahl, Dissimilarity or Gini, specialization indices) using only fixed effects, and not controlling for the regional characteristics in the matrix X. Columns (2), (4) and (6) show the results of the same model in which we also include the control variables contained in X.

The table shows that the specialization coefficient is generally not significant for explaining changes in growth; only when we use the Gini index do we obtain significant results. In this last case it seems that greater specialization in year t may cause a larger positive change in growth.

The control variables X are generally not significantly different from zero. The Wald test shows that these coefficients are generally jointly significant (note that the R-squared is higher when the control variables are included).

8   Regional Winners and Losers and Policy Implications

8.1   Winners and Losers in Industrial Activity Relocation

Previous sections have shown that economic integration gains are unevenly distributed across regions and have caused the relocation of manufacturing activity. It seems, therefore, that spatial effects are related to the intensification of polarization and a geographically divided pattern of development. In general, regions with a more diversified economic structure have experienced higher economic growth and employment, while declining monostructural regions and underdeveloped regions have faced serious and lasting difficulties. Their prospects for recovery are not identical.

Winners in the process of the economic integration with the European Union are the metropolitan region, central regions and regions with a diversified production base. These regions have gained a favourable position and get more foreign investment, which in addition to traditional assets can offer an attractive innovative economic environment and also the related institutional network for economic development.

Losers in the process of European integration are regions with a declining monostructure, rural areas, and certain peripheral regions.

Table 4.22 Estimation of the growth model

	D (1)	D (2)	H (3)	H (4)	Gini (5)	Gini (6)
D	1.95354	1.56544				
	(1.03467)*	(1.05909)				
H			-1.20358	1.70802		
			(3.60467)	(3.78941)		
Gini					4.01752	3.26662
					(1.44713)	(1.50716)
					***	**
X1		-1.44955		-1.03432		-0.30597
		(2.43511)		(2.44953)		(2.39226)
X2		2.61715		2.18092		2.60617
		(2.40008)		(2.41358)		(2.34822)
X3		148.34779		187.18733		169.59143
		(90.09132)		(92.07717)		(86.73210)
				**		*
X4		2.60151		2.89981		2.61783
		(12.15183)		(12.71716)		(11.93198)
X5		-2.84259		-2.88166		-3.26951
		(3.06679)		(3.15348)		(3.03098)
Const.	-1.17694	-2.47220	0.14927	-1.96547	-2.16651	-3.55237
	(0.62324)*	(8.43329)	(0.45080)	(8.75935)	(0.78034)	(8.30660)
					***	
Obs.	112	112	112	112	112	112
No. reg.	28	28	28	28	28	28
R-square	0.04118	0.20126	0.00134	0.18102	0.08497	0.22553

Standard errors in parentheses. D = Dissimilarity; H = Herfindahl. * Significant at 10 per cent; ** significant at 5 per cent; *** significant at 1 per cent.

Different adaptation patterns are observed in border regions.

As a result of the import of raw materials, mainly from the former USSR, a shift in the location of manufacturing facilities from the west to the east took place during the previous period. This has further increased the potential of the traditionally well-developed regions and cities along the Black Sea coast and has promoted the development of the eastern parts of the country (Bourgas, Varna and Dobrich districts). At present these regions are not losing any of their attractiveness as regions for the location of manufacturing activities. One of the reasons for this is that some of the biggest ports on the Black Sea are located there, a fact that makes these regions direct contact points of integration, since Bulgaria has no common western frontier with the EU. Another point worth noting is the boom in the construction of tourist facilities there.

Unlike in the other Eastern and Central European states, the effect of the 'western frontier' does not manifest itself in Bulgaria.

The new processes of integration attach new qualities to regions located along the southern and southwestern border. This is Bulgaria's sole border with an EU member state – Greece. These border regions feature a high density of foreign companies and joint ventures (mainly with Greek capital), predominating in clothing, textiles, wood and wood processing industries, fur and leather production and marble extraction. Their development is encouraged through programs supporting cross-border cooperation.

The situation in the western border regions is quite different. A degradation process is underway there. This process is very strongly manifested in the districts of Kyustendil, Montana and Vidin. The municipality of Blagoevgrad is the sole exception. The reasons for this are conflicts in the states which Bulgaria borders to the west – the Federal Republic of Yugoslavia and Macedonia.

One may definitely note the relocation of economic activities along the east-west transport corridors, less so than along the north-south oriented corridors, which make the regions located along international transport corridors more effective and dynamic.

Geographical factors such as distance, accessibility and centrality emerge as important elements in the spatial organization of activities. A considerable part of the EU's regional differences are associated with inter-country rather than intra-country disparities. Regions can have a strategic or central function that is derived not only from their relative position within their country, but also from the position of the country within the emerging hierarchy of the European economic space[28].

Bulgaria is situated at the periphery of the European space. It is not within the proper scope of the west European centers and axes of development of business and technology. This unfavourable factor does not allow an intensive stream of innovations, goods and people to enter from all directions, important for Bulgaria's integration in the European space and for surmounting the country's serious economic problems. It does not create potential opportunities for the improvement of its efficiency by direct use of the advantages of neighbouring economies that emerge from business cooperation, new financial instruments, networks for implementing innovations, mobility of highly qualified personnel, and so on.

In this respect, the relatively great distance from the EU's core and a lack of adjacency make integration slower and geographically more selective. On the other hand, the geographic factor may, to a large extent, be compensated for by development of the information society, information and communication technologies and the quality and value of human potential. One should add to this the advantages related to the geo-strategic location of the country as a bridge between the East and West, traversed by five of the European transport corridors.

8.2 Specialization, Unemployment and Economic Growth

The relocation of industrial activities between regions, observed over the period of 1990–99, has had a strong impact on employment and unemployment in the various regions. These shifts are not related to adaptation of the population and labour force in the regions that have lost manufacturing employment. Population mobility does not follow the movement of manufacturing activities. Loss of manufacturing employment cannot usually be compensated for by alternative employment. Lower growth of GDP per capita is therefore associated with higher unemployment rates.

To test whether declining regions have experienced permanently higher unemployment, we have regressed the unemployment rates on GDP per capita. The time period is five years (1995–99), since regional GDP per capital data are only available for the years since 1995.

The regression shows that there is a significant negative association between GDP per capita and the rate of unemployment at the regional level (the regression results are presented in Table 4.23). Hence, poor regions are more likely to suffer from unemployment. However, the relationship between the two variables is very weak (the adjusted R^2 is 0.049).

Table 4.23   Results of the regression of GDP per capita on the rate of unemployment

	Coeff.	t-stat
Const.	19.728	10.303
GDP pc	−0.005	−2.890
Adj. R2	0.049	
No. of obs.	145	

Dependent variable: rate of unemployment.

8.3   Policy Implications

Bulgaria confronts the necessity of searching for a compromise between the classical objectives of regional policy – regional growth – or economic and social cohesion.

This policy implies priority interventions within the areas identified for growth and development and, if necessary, within other regions and towns with potential to attain substantial and sustainable economic growth based on their competitive advantages. This does not exclude intervention within the peripheral regions, especially in the case of promising potential or severe social problems.

For the winning regions (big industrial centers and regions), this policy should deal mainly with the restructuring and internationalization of the regional productive system through fostering small and medium size enterprise (SME) development, attracting foreign investment and enhancing innovative and ecological potential, improving competitiveness, and supporting environmental improvements to enhance the quality of life.

In the losing regions – predominantly declining monostructural areas, regions with a less qualified labour force, less developed infrastructure and lesser undertaking and innovative capacity – the policy measures should aim at: enhancing the diversification of the local economy, supporting local entrepreneurship and attracting external investments and supporting human resource development, mainly through capacity-building activities and increased employment opportunities. This will enable the regions to cope with structural weaknesses in performance, in particular, and the limited marketing horizons in general; to help revive and upgrade small-scale infrastructure in towns; and to create more attractive town environments for the enhancement of investments and the social mix. The main objective under these circumstances should not only be survival, but also development that would help incorporate the regions into European and international markets and increase their competitive capacity.

Integration into the European space can be described as a *Europe-oriented priority, permanently present on the agenda of all regions.* Although it includes inter-regional cooperation, the main emphasis is on cross-border cooperation directed at achieving two main results: linking national infrastructure networks to those in neighbouring countries and thus integrating Bulgaria into the European space, and solving the existing problems in border areas, which are among the most unfavourable in the country, the majority of them being typical of the periphery.

The creation of favourable conditions for the development of new economic activities and the diversification of production is strongly linked with the measure of improving the educational structure and the re-qualification of manpower. The link between the skill level of the workforce and its competitiveness is increasingly recognized in all developed countries.

The development of basic and business-related infrastructure in view of the support of economic development and the creation of favourable living conditions for the population should be strongly presented as a main policy objective. The role of infrastructure as a condition for and a motor of business development and the attractiveness of the region is clearly defined.

The ultimate goals of revitalizing and developing all regions requires the sharing of the efforts for their achievement. The limited local financial, and to a certain extent administrative, resources might not be able to produce results that would reverse the observed development trends. To this end, sharing the burden by joining the efforts of the central, regional and local authorities and attracting support and assistance from the EU is indispensable. A special recommendation refers to seeking and locating new funding sources and financing schemes. This means prioritizing the establishment of local investment and guarantee funds, as well as national investment funds that would assist the special regional programs. The focus of spending for the limited financial resources available from the different institutions should be on the implementation of projects that may play the role of development stimuli. The funds allocated under the PHARE programme and other pre-accession funds are another important source of financing. Another issue of no lesser importance is the improvement of the investment image of the regions and cities to raise the level of local entrepreneurship. Local authorities play a significant role in this process, since they may contribute to facilitating the procedures for starting businesses and to abolishing bureaucratic barriers.

The regions and cities should mobilize their endogenous capacity for survival and reconstruction. It is becoming ever clearer that their problems cannot be resolved by national-level efforts alone. In this respect certain regional and local efforts have already been observed, including local development strategies, the application of certain modern institutional arrangements and contributions to endogenous development from regional development agencies, local business centres and business incubators and local development councils and forums.

9 Conclusions

Existing evidence on trade development and foreign direct investment suggests an increasing process of integration of the Bulgarian economy with the European Union in the period from 1990–99. Integration so far seems to have confined Bulgaria to the status of a supplier of low-value-added labour and natural-resource-intensive products; however, in recent years, a decline in the aggregate share of unskilled labour and resource-intensive products has been observed. The commodity groups differentiated by their comparative advantage and strong export potential (based on TCI and RCA indices) are leather and leather products, wood and wood products, fertilizers, oil seeds and oleaginous fruits, tobacco and tobacco processing, non-ferrous metals, beverages, travel goods, handbags and the like, pulp and waste paper, iron and steel, crude rubber (incl. synthetic and reclaimed), textiles and clothing, sanitary, plumbing, heating and lighting fixtures and fittings n.e.c., footwear, crude animal and vegetable materials n.e.c., inorganic chemicals, furniture and parts thereof, and meat and meat preparations.

The analysis of regional disparities (NUTS 2) for Bulgaria and comparisons made with EU member states and CEECs indicate that all Bulgarian regions are 'poor' according to European standards and are relatively 'equally poor'. The analysis also shows that the disparities are much more intra-regional than inter-regional.

At the NUTS 2 level a low and uniform degree of specialization is observed (according to Herfindahl index values, the level of specialization is 0.09 for the southwest region and 0.10 for the other regions). At the NUTS 3 level the values of the absolute and relative measures for specialization are higher. Twenty districts have increased their degree of specialization and eight have retained or reduced their degree of specialization from 1990–99. The following patterns of regional specialization have been identified: regions with the highest and increasing specialization: Blagoevgrad, Varna, Vidin, Kardjali, Lovech, Pernik, Sliven, Stara Zagora and Shumen; regions with the highest specialization and a decreasing trend: Vidin, Gabrovo and Smolyan; and diversified regions with increasing diversification: the Sofia region, Haskovo, Montana and Yambol.

According to the model estimations the annual percentage change of regional specialization over the period 1990–99 is 1.9 per cent for the specialization measured by the Herfindahl index, 1.6 per cent for that measured by the Dissimilarity index and 1.3 per cent as measured by the Gini index.

The present study finds a clear pattern of relocation of manufacturing activity in Bulgaria over the period 1990–99 in terms of manufacturing employment.

The distribution of manufacturing shares compared to population shares is considerably unbalanced at the NUTS 3 level, where the differences in 1999 were almost three times greater. The values are quite high in industrially developed regions (Gabrovo, Lovech, Bourgas, Plovdiv, Blagoevgrad, Pernik, Stara Zagora and Veliko Tarnovo). At the same time, the values of the ratio diminish between 1990–99 in the districts that face a decline in industrial development (Vidin, Silistra, Pazardjik, Montana, Dobrich, Razgrad, Smolyan and Yambol).

The results of our research suggest an increasing geographical concentration of industries. 15 out of 19 industries at the NUTS 2 and NUTS 3 levels demonstrate an increase in geographical concentration. Only four industries experienced a drop in concentration. The four most concentrated industries include: coke, refined petroleum products and nuclear fuel, basic metals except casting of metals, transport equipment and pulp, paper and paper products and publishing and printing. The four most dispersed industries are food, beverages and tobacco, metal products, machinery and equipment, textiles and apparel and other non-metallic products. The annual percentage change of geographical concentration over the period 1990–99 has been estimated to be 1.9 per cent (under the Herfindahl index), 1.4 per cent (under the Dissimilarity index) and 1.6 per cent (under the Gini index).

The impact of economic integration on the regional wage structure has been estimated according to three hypotheses. The first hypothesis is based on the assumption that in a closed economy regional wages will decrease with distance to the industry center, while in an open economy regional wages are determined by access to foreign markets. The estimation results confirm this hypothesis: the distance from the country's capital always has a negative significant coefficient, meaning that the more distant the region is from the country's capital, the less its wages are (relative to those in the country's capital).

From an econometric perspective, we find that nothing changed before and after the entering into force of the EU Association Agreement.

The model estimations of the impact of regional specialization on regional growth show that the specialization coefficient is generally not significant in explaining changes in growth; only when we use the Gini index we do obtain significant results. In this last case, it seems that a higher specialization in year *t* may cause a higher positive change in growth in the same year.

Another important issue in the present chapter is to analyse the relationship between regional specialization, economic growth and unemployment. During the period under review, we found a negative relationship between the level of economic activity in the region, measured by GDP per capita, and the level of unemployment. Hence, poor regions are more likely to suffer from unemployment.

The general conclusion from the descriptive and econometric analyses in this study is that economic integration gains are unevenly distributed across regions and have caused the relocation of manufacturing activities. It seems, therefore, that the spatial effects are related to intensifying polarization and a geographically divided pattern of development.

Winners in the process of economic integration with the European Union seem to be the metropolitan region, central regions and regions with a diversified production base. Losers in the process of European integration are regions with a declining monostructure, rural areas, and certain peripheral regions.

Unlike in the other Eastern and Central European states, the effect of the 'western frontier' does not manifest itself in Bulgaria. One of the reasons for this is the conflicts in the states that Bulgaria borders to the west – the Federal Republic of Yugoslavia and Macedonia. The findings on southern and southwestern border regions show that the process of European integration attaches new qualities to these regions and an important factor here is the border with an EU member state – Greece.

Policy recommendations should take into account changes in regional specialization and the relocation of economic activities that shape, respectively, the development level of a given region and its place among the winner/loser regions. For the winner regions, policy should mainly deal with further restructuring and internationalizing the regional productive system, through fostering SME development, attracting foreign investments and enhancing their innovative and technological potential. In the loser regions, policy measures should be related to enhanced diversification of the local economy; support for local entrepreneurship and attracting external investments; support for human resources development, aimed at building a capacity for dealing with structural weaknesses in performance, in particular, and the expansion of the limited marketing horizons in general; revitalizing and upgrading small-scale infrastructure in towns; and the creation of more attractive town environments for investments.

Notes

* This research was undertaken with support from the European Community's PHARE ACE Programme 1998. The content of the publication is the sole responsibility of the authors and it in no way represents the views of the Commission or its services.

1 I wish to acknowledge the excellent research assistance of Simonetta Longhi in preparing sections 6 and 7 and the estimations of changes in regional specialization and concentration.

2 World Bank, 2000.

3 It is defined by: $TC_i = X_i / M_i$
where:
X_i – Bulgarian export of commodity i;
M_i – Bulgarian import of commodity i.
Values greater than one for a given commodity mean that the country has a comparative advantage in this sector.

4 $$RCA_i = \frac{X_i / \sum_i X_i}{M_i^{EU} / \sum_i M_i^{EU}}$$

where:
Xi – Exports in commodity group i;
M_i^{EU} – Imports to the EU in commodity group i.

5 High values of the RCA index indicate that the country is a preferential supplier of product i to the EU and that the country has a comparative advantage for that product. In comparison with the Trade Coverage index (TCI), which measures 'internal' comparative advantages in the trade, RCA measures 'external' comparative advantages, the position of the product on an external market.

6 $$IIT_i = 1 - \frac{|X_i - M_i|}{(X_i + M_i)}$$

where:
$M(i)$ – countries' import of commodity group i;
$X(i)$ – countries' export of commodity group i.

7 According to BFIA data.

8 According to BFIA data.

9 See Petrakos, 1999.

10 Corresponding to the EU NUTS 4 regions.

11 Corresponding to the EU NUTS 3 regions.

12 Corresponding to the EU NUTS 2 regions.

13 Decree No. 145/27 July 2000 of the Council of Ministers.

14 'Unity, Solidarity, Diversity for Europe, its People and its Territory', 2001.

15 Ibid., pp. 4–10.

16 This is a measure of *absolute specialization*. It indicates how different the distribution of production is from a uniform (national) distribution. Its value is biased towards the largest shares, however. A value close to zero implies a high degree of diversification. A value close to one implies almost complete specialization.

$$H_j^S = \sum_i (S_{ij}^S)^2$$

where:

s_{ij}^S = the share of employment in industry i in region j in total employment of region j.

17 This measure sums up the differences between the shares in a region and a norm (national average) without regard to the signs. A higher index reflects greater specialization.

$$DSR_j = \sum_i \left| S_{ij}^S - S_i \right|$$

where: $S_{ij}^c = \dfrac{Eij}{Ej} = \dfrac{Eij}{\sum_i Eij}$;

where:

E = employment

s = shares

i = industry (sector, branch)

j = region

18 Gini coefficients for regional specialization

$$GINI_i^S = \frac{2}{n^2 \bar{R}} \left[\sum_{i=1}^{n} \lambda_i \left| R_i - \bar{R} \right| \right]$$

where:

S_{ij}^S = the share of employment in industry i in region j in total employment of region j;

S_i = the share of total employment in industry i in total employment;

n = the number of industrial branches.

$R_i = \dfrac{S_{ij}^S}{S_i}$ (for each industry in region j)

$\bar{R}$ = the mean of R_i across industries

$\lambda_i =$ the position of industry i in the ranking of R_i.

19 Footnote missing

20 The *Herfindahl index* for geographic concentration is calculated as follows:

$$H_i^c = \sum_j (s_{ij}^c)^2$$

where:

s_{ij}^c = the share of employment in industry i in region j in total employment of industry i.

21 The *Dissimilarity measure* for geographical concentration is calculated as follows:

$$DCR_i = \sum_j \left| S_{ij}^c - S_j \right|$$

where:

S_j = the share of total employment in region j in total employment.

22 *Gini coefficients* for geographical concentration of industries (Locational Gini coefficient)

$$GINI_i^c = \frac{2}{m^2 \bar{C}} \left[\sum_{i=1}^{m} \lambda_j |C_j - \bar{C}| \right]$$

where:

m = the number of regions;

$$C_j = \frac{S_{ij}^c}{S_j}$$

$\bar{C}$ = the mean of Cj across regions;

λ_j = the position of the region in the ranking of Cj

23 The estimation is based on compliance with the regression model.

24 The estimations were made by Simonetta Longhi.

25 See Hanson, 1996.

26 For the econometric estimations, the western border regions are: the Vidin, Montana, Vratsa, and Sofia regions and the Pernik, Kustendil and Blagoevgrad districts.

27 For the econometric estimations, the south(west)ern border regions are: the Blagoevgrad, Pazardjik, Smolyan, Kardjali and Haskovo districts.

28 Peshel, 1992; Gallup, Sachs and Mellinger, Petrakos, 1999.

References

Agency for SMEs (2000), 'Small and Medium Size Enterprises, 1996–1999', Report, Sofia.

Amiti, M. (1998),'New Trade Theories and Industrial Location in the EU: a Survey of Evidence', *Oxford Review of Economic Policy*, Vol.14 (2).

Berg, A., Borenstsztein, R., Sahay and Zettelmeyer, J. (1999), 'The Evolution of Output of Transition Economies: Explaining the Differences', IMF Working paper, No. 73.

Bradistilov, D. (1974), *Territorial Planning*, Sofia.

Bulgarian Academy of Sciences (1995), 'Bulgaria at the Beginning of XXIc', Sofia.

Bulgarian Common Country Assessment 2000 (2000), UNDP.

Donchev, D. and Karakashev, Hr. (1996), *Physical and Socio-economic Geography of Bulgaria*, 4th edn, Sofia.

Drinov, M. (1997), *Geography of Bulgaria: Physical Geography and Socio-economic Geography*, Academic Publishing House, Sofia.

EC (2001), 'Second Report on Economic and Social Cohesion', pp. 73–6.

Gueshev, G. (1999), 'Problems of Regional Development and Regional Policy of Bulgaria', Bulgarian Academy of Sciences, Sofia.

Hanson, H. (1966), 'Localization Economies, Vertical Organization, and Trade', *The American Economic Review*, Vol. 86 (5).

Human Development Report (2000), 'Bulgaria 2000: The Municipal Mosaic', UNDP.

Landesmann, M. (1995), 'The Patterns of East-Europe Integration: Caching up or Falling Behind', in R. Dobrinsky and M. Landesmann (eds), *Transforming Economies and European Integration*, Edward Elgar, Aldershot.

Mack, R. and Jakobson, D. (1996), 'The Impact of Peripherality upon Trade Patterns in European Union', *European Urban and Regional Studies*, Vol. 3 (4).

Mickiewicz, T. (1999), 'Convergence versus Rapid Deindustrialization: Restructuring of Employment in Central Europe', paper presented at the Workshop on Regional Development and Policy, ZEI, Bonn, 10–11 December 1999.

Ministry of Labour and Social Policy (2001), 'National Plan for Employment Actions in 2001', Final Draft.

National Human Development Report (1999), 'Bulgaria 1999. P.1: Trends and Opportunities for Regional Human Development', UNDP, Sofia, http://www.undp.bg.

National Plan for Economic Development of R.Bulgaria 2000–2006, http://www. government.bg.

National Plan for Regional Development of Bulgaria 2000–2006, http://www. government. mrrb.bg.

National Statistical Institute (2000), 'Bulgaria '98: Socio-economic Development', Sofia.

NCRDHP (1998), 'Setting out the Criteria and Territorial Scope of Regions of Purposeful Impact', Sofia.

Petrakos, G. (1999), 'Patterns of Regional Inequality and Convergence-Divergence Trends in Transition Economies', International Conference on Regional Potentials in an Integrating Europe, Regional Studies Association, Bilbao, Spain.

Petrakos, G. (1999), 'The Spatial Impact of East-West Integration in Europe', in G. Petrakos, G. Maier and G. Gorgelak (eds), *Integration and Transition in Europe: the Economic Geography of Interaction*, Routledge, London.

PHARE Project (1998), 'Bulgaria: Pilot Closure of an Uneconomic Coal Mine', PHARE Project Final Report.

Spiridonova, J. (2000–2001), 'The Future of Old Industrialized Cities and Regions undergoing Structural Changes (FOCUS)', Bulgaria country paper, IOER – EC, INTERREG II C.

Spiridonova, J. (2001),'European Integration, Regional Dynamics and Regional policy in Bulgaria', n Proceedings of the annual meeting of the Austrian Economic Association (NoeG) with the topic 'New Challenges for Europe: Structure, Location and Sustainability', 16–18 May, Graz.

Spiridonova, J. (2001), 'The Impact of East-West Integration and Enlargement on Regional Development in Bulgaria', in Proceedings of Forum Europeen de perspective regionale et locale, 18 and 19 December, Lille.

UNCTAD (2000), *World Investment Report 2000*, Geneva.

Vaknin, Sam (2001), 'Bulgaria's Economy and the Challenge of Accession', CER, Vol. 3 (18), http://www.ce-review.org/01/18/vaknin18.html.

World Bank, Bulgaria (2000), 'Country Economic Memorandum "The Dual Challenge of Transition and Accession"', http://www.government.oficial_docs/index/htmlg.

Yovkov, Ivan (1991), *Wirtshaftsreport Republik Bulgarien*, c/o JUSAUTOR, Berlin.

Appendix

Table 4.A1 Exports and imports by EU countries, 1990–99 (US$m)

	Exports					Imports				
	1990	1991	1995	1999	1999/1990 %	1990	1991	1995	1999	1999/1990 %
Total	3,897.93	3,439.76	5,354.69	3,973.76	101.95	3,807.64	2,706.10	5,657.64	5,470.30	143.67
EU	328.20	598.39	2,013.61	2,060.94	627.95	683.20	705.68	2,065.64	2,582.46	377.99
Austria	16.50	33.78	46.00	67.58	409.59	60.87	126.78	157.37	162.65	267.21
Belgium	9.49	32.69	82.28	174.40	1,838.35	13.14	23.68	74.63	90.60	689.50
United Kingdom	21.89	66.72	168.34	99.85	456.16	62.94	97.79	148.29	131.09	208.28
Germany	165.04	163.61	458.48	390.43	236.56	395.68	188.65	699.49	815.58	206.12
Greece	31.16	74.88	368.63	342.93	1,100.70	12.37	24.10	249.08	308.16	2,491.19
Denmark	2.10	6.05	16.67	22.80	1,083.61	–	–	–	–	–
Ireland	–	–	–	–	–	–	–	–	–	–
Spain	9.52	17.81	134.54	107.53	1,129.02	6.53	13.50	27.08	74.17	1135.83
Italy	30.45	92.89	435.91	552.05	1812.72	72.83	112.87	327.81	459.87	631.43
Luxembourg	0.00	0.00	0.00	0.00	–	–	–	–	–	–
Netherlands	14.43	31.14	102.60	82.49	571.55	19.38	28.49	111.23	109.81	566.62
Portugal	1.07	1.91	16.49	13.17	1,230.13	1.11	4.91	5.55	11.81	1,063.96
Finland	3.51	10.88	12.47	8.05	229.65	4.10	13.08	57.02	60.40	1,473.17
France	20.23	49.61	153.29	179.08	885.28	27.80	57.48	157.76	285.19	1,025.86
Sweden	95.80	57.48	157.76	223.00	232.78	6.46	14.34	50.33	73.13	1,132.04

Table 4.A2 Structure of imports and exports in trade relations with the EU

No.	BRANCH	1995		1996		1997		1998		1999	
		Import %	Export %	Import %	Export %	Import %	Export %	Import %	Export %	Import %	Export %
HS	TOTAL:	100.00	100.00	100.00	100.00	100.00	100.00	100.00	100.00	100.00	100.00
I	Live animals, animal products	1.37	3.20	0.89	2.61	1.73	2.02	1.97	2.05	1.17	2.00
II	Vegetable products	1.32	5.29	0.96	3.45	1.96	2.79	0.83	4.55	0.80	4.77
III	Animal or vegetable fats and oils	0.78	0.42	0.62	0.02	0.68	0.08	0.70	0.02	0.45	0.01
IV	Food, beverages and tobacco	3.37	5.05	3.36	6.12	2.58	4.32	2.73	4.71	2.02	4.75
V	Mineral products	1.73	3.31	1.29	4.26	1.78	6.23	3.33	3.53	2.66	3.83
VI	Chemical or allied products	14.12	12.97	14.54	12.64	12.35	11.68	12.67	6.88	11.14	5.13
VII	Plastics and rubber	5.54	3.84	5.68	4.21	5.00	3.57	5.32	3.73	5.32	3.18
VIII	Leather and leather products	1.47	1.57	2.37	1.55	2.94	1.32	2.08	1.17	1.67	1.26
IX	Woods and articles of wood	0.49	2.36	0.68	2.42	0.77	2.58	0.66	2.70	0.49	2.87
X	Cellulose, paper and articles thereof	5.92	1.62	5.06	1.21	4.50	1.41	4.17	0.99	3.26	0.93
XI	Textiles and textile articles	16.63	14.81	19.13	17.48	22.24	19.38	20.87	23.50	18.97	27.76
XII	Footwear and accessories	1.99	3.97	2.54	4.92	2.43	5.13	1.96	5.00	1.50	5.19

Table 4.A2 cont'd

No.	BRANCH	1995		1996		1997		1998		1999	
		Import %	Export %	Import %	Export %	Import %	Export %	Import %	Export %	Import %	Export %
XIII	Non-metallic minerals	1.69	2.31	1.81	2.06	1.57	1.95	1.42	2.08	1.31	2.27
XV	Base metals and articles thereof	5.84	26.51	6.95	22.58	6.70	26.28	5.49	26.80	5.50	22.12
XVI	Machinery and equipment	23.33	9.25	22.20	9.32	22.59	7.17	23.60	8.73	26.30	9.94
XVII	Vehicles, transport equipment	8.29	0.83	6.26	2.52	5.27	1.65	7.20	0.78	13.16	0.65
XVIII	Precise and optical equipment	4.04	0.51	3.59	0.50	2.90	0.45	2.97	0.59	2.40	0.79
XX	Miscellaneous manufactured articles	2.08	2.18	2.06	2.16	2.00	1.97	2.03	2.17	1.88	2.55
XXI	Art, collections, antiques	0.00	0.01	0.00	0.00	0.01	0.00	0.00	0.00	0.00	0.00
0	Food and live animals	4.84	6.89	3.72	5.52	5.59	4.15	5.00	5.59	3.37	5.37
1	Beverages and tobacco	0.67	4.16	1.23	5.02	0.65	3.55	0.38	3.48	0.43	3.44
2	Crude materials, inedible, except fuels	4.46	8.66	3.84	6.61	4.15	7.97	3.46	6.93	2.77	7.37
3	Mineral fuel, lubricants and related materials	1.06	1.52	0.76	3.21	0.98	4.27	2.25	2.19	1.55	2.70
4	Animal and vegetable oils, fats and waxes	0.90	0.43	0.43	0.03	0.52	0.09	0.62	0.04	0.44	0.03

Table 4.A2 cont'd

No.	BRANCH	1995		1996		1997		1998		1999	
		Import %	Export %	Import %	Export %	Import %	Export %	Import %	Export %	Import %	Export %
5	Chemical and related products, n.e.s.	17.29	14.99	17.93	14.74	15.05	13.50	15.55	8.62	14.04	6.22
6	Manufactured goods classified chiefly by material	27.03	35.96	30.15	31.82	31.48	34.66	28.01	36.07	25.18	29.71
7	Machinery and transport equipment	32.58	9.93	29.04	11.74	28.12	8.80	31.02	9.42	39.20	10.52
8	Miscellaneous manufactured articles	11.34	17.62	12.96	21.47	13.50	23.00	13.68	27.78	12.91	34.61
9	Commodities and transaction not classified elsewhere	0.07	1.09	0.11	0.49	0.08	0.41	0.14	0.39	0.20	0.76

Table 4.A3 Composition of Bulgarian exports to the EU in terms of factor intensities and their share in EU external imports, 1989–98 (in per cent)

A Composition of Bulgaria's EU-oriented exports

	1989	1990	1991	1992	1993	1994	1995	Index 1995 (1989=100)	1996	1997	1998	Index 1998 (1996=100)	Index ROW 1997
Natural resource intensive	46.8	44.3	42.2	41.1	41.8	40.1	36.5	78	32.7	32.0	30.4	93	38.4
Unskilled labour intensive	15.5	18.6	24.0	29.9	31.4	27.0	23.6	152	28.9	30.2	33.7	117	14.6
Capital intensive	21.1	19.4	19.3	16.3	18.2	20.0	19.9	94	22.1	19.8	16.5	75	29.3
Skilled labour intensive	13.7	15.6	13.3	11.9	6.9	11.7	18.5	135	15.5	16.8	18.1	117	14.8
All products (US$m)	657	823	999	1,265	1,210	1,713	2,427	369	2,204	2,410	2,543	115	1,812

B Share in EU-external imports

	1989	1990	1991	1992	1993	1994	1995	Index 1995 (1989=100)	1996	1997	1998	Index 1998 (1996=100)	Index ROW 1997
Natural resource intensive	0.08	0.08	0.10	0.12	0.13	0.16	0.18	225	0.14	0.15	0.17	121	
Unskilled labour intensive	0.07	0.08	0.12	0.17	0.20	0.23	0.25	357	0.27	0.30	0.35	130	
Capital intensive	0.03	0.03	0.04	0.04	0.05	0.06	0.07	233	0.07	0.07	0.05	71	
Skilled labour intensive	0.03	0.03	0.04	0.04	0.03	0.05	0.10	333	0.07	0.09	0.09	129	
All products	0.05	0.05	0.06	0.08	0.09	0.11	0.13	260	0.11	0.12	0.12	109	

Source: World Bank calculations. Data on Bulgaria's exports as reported by the EU to the UN COMTRADE database.

Table 4.A4 Changes in manufacturing employment structure, 1990–99, NUTS 2

Economic activity	Total		Northwest		North central		Northeast		Southeast		South central		Southwest	
	%	%	%	%	%	%	%	%	%	%	%	%	%	%
	1990	1999	1990	1999	1990	1999	1990	1999	1990	1999	1990	1999	1990	1999
Manufacturing (D)	100.00	100.00	100.00	100.00	100.00	100.00	100.00	100.00	100.00	100.00	100.00	100.00	100.00	100.00
Manufacture of food products, beverages and tobacco products (DA)	13.50	17.69	13.51	17.35	14.86	18.66	15.50	19.27	20.70	22.88	13.38	19.11	9.10	13.21
Manufacture of textiles and wearing apparel (DB)	15.50	20.31	16.07	27.95	17.12	19.49	13.13	17.90	17.12	18.43	17.50	20.11	13.07	21.31
Tanning and dressing of leather and manufacture of footwear (DC)	2.81	3.37	1.34	1.08	2.52	3.27	2.86	2.41	1.30	0.73	3.18	3.88	3.52	4.71
Manufacture of wood and furniture and other manufactured goods (DD + DN)	4.79	2.45	3.85	2.20	5.04	2.83	4.85	2.28	5.38	3.17	5.58	2.72	3.89	1.84
Manufacture of paper and paper products; publishing, printing and reprod. of recorded media (DE)	2.83	4.05	2.06	1.83	1.65	2.29	1.03	1.73	0.92	1.03	3.92	4.74	4.54	7.26

Table 4.A4 cont'd

Economic activity	Total % 1990	Total % 1999	Northwest % 1990	Northwest % 1999	North central % 1990	North central % 1999	Northeast % 1990	Northeast % 1999	Southeast % 1990	Southeast % 1999	South central % 1990	South central % 1999	Southwest % 1990	Southwest % 1999
Coke, refined petroleum products and nuclear fuel (DF)	1.13	1.83	0.00	0.02	0.92	1.69	0.03	0.05	10.57	18.40	0.08	0.06	0.10	0.04
Chemicals, chemical products and man-made fibres (DG)	4.59	6.15	4.51	6.25	3.72	5.61	6.47	9.73	3.57	2.66	5.43	6.44	3.74	5.40
Manufacture of rubber and plastic products (DH)	2.60	2.97	9.30	5.18	2.45	2.74	1.66	2.13	2.26	2.18	2.03	3.28	2.11	3.02
Manufacture of other non-metallic mineral products (DI)	4.09	4.67	6.20	7.66	4.03	5.15	8.09	9.52	4.16	4.61	2.60	2.96	2.70	2.88
Manufacture of fabricated metal products and manufacture of machinery and equipment (DJ + DK)	19.56	24.35	20.50	23.75	19.97	24.54	13.61	17.88	11.97	13.11	22.31	28.86	22.26	26.61
Manufacture of electrical machinery and electronics (DL)	14.67	6.04	10.57	4.46	12.27	5.90	9.89	4.36	11.17	6.68	14.30	4.85	21.71	8.38

Table 4.A4 cont'd

Economic activity	Total		Northwest		North central		Northeast		Southeast		South central		Southwest	
	% 1990	% 1999	% 1990	% 1999	% 1990	% 1999	% 1990	% 1999	% 1990	% 1999	% 1990	% 1999	% 1990	% 1999
Manufacture of motor vehicles and transport equipment (DM)	5.41	2.79	1.36	0.34	8.11	3.23	12.95	9.43	3.76	2.94	1.82	0.59	4.18	1.66
Manufacturing, n.e.c.	8.52	3.32	10.73	1.93	7.33	4.59	9.93	3.32	7.12	3.17	7.87	2.39	9.09	3.68

Source: Author's calculations based on REGSTAT.

Table 4.A5 Geographic concentration measures on the NUTS 2 level, 1990–99

(Hic) Herfindahl index – geographic concentration measure for NUTS 2

	1990	1991	1992	1993	1994	1995	1996	1997	1998	1999
Mining of coal; extraction of natural gas	0.35	0.37	0.38	0.38	0.38	0.39	0.39	0.4	0.4	0.4
Mining of ores	0.47	0.47	0.48	0.5	0.49	0.49	0.49	0.48	0.46	0.47
Other mining and quarrying	0.18	0.18	0.18	0.19	0.19	0.19	0.19	0.19	0.19	0.19
Foods, beverages and tobacco	0.19	0.19	0.2	0.2	0.2	0.2	0.2	0.2	0.2	0.2
Textiles	0.21	0.21	0.2	0.2	0.2	0.2	0.2	0.21	0.2	0.2
Apparel	0.2	0.2	0.2	0.2	0.2	0.2	0.21	0.21	0.21	0.22
Leather, leather and fur clothes, footwear and products	0.24	0.25	0.25	0.25	0.26	0.26	0.25	0.27	0.26	0.28
Wood and products of wood and cork, plaiting materials	0.2	0.2	0.19	0.19	0.2	0.2	0.21	0.21	0.21	0.21
Pulp, paper and paper products, publishing and printing	0.31	0.31	0.32	0.32	0.33	0.33	0.33	0.32	0.33	0.34
Coke, refined petroleum products and nuclear fuel	0.67	0.71	0.76	0.77	0.77	0.77	0.79	0.65	0.66	0.68
Chemicals, chemical products and man-made fibres	0.21	0.21	0.21	0.21	0.21	0.21	0.21	0.21	0.21	0.22
Rubber and plastic products	0.19	0.19	0.19	0.19	0.19	0.2	0.2	0.2	0.2	0.21
Other non-metallic products	0.19	0.2	0.2	0.2	0.2	0.2	0.21	0.21	0.2	0.2

Table 4.A5 cont'd

(Hic) Herfindahl index – geographic concentration measure for NUTS 2

	1990	1991	1992	1993	1994	1995	1996	1997	1998	1999
Basic metals except casting of metals	0.43	0.46	0.48	0.49	0.21	0.5	0.51	0.51	0.51	0.51
Metal products, machinery and equipment; casting of metals	0.22	0.23	0.23	0.23	0.23	0.23	0.23	0.24	0.24	0.24
Electrical and optical equipment	0.25	0.24	0.23	0.22	0.21	0.22	0.22	0.22	0.23	0.24
Transport equipment	0.26	0.28	0.3	0.31	0.32	0.32	0.34	0.34	0.35	0.31

(DCRi) Disimilarity index – Geographic concentration measure for NUTS 2

	1990	1991	1992	1993	1994	1995	1996	1997	1998	1999
Mining of coal; extraction of natural gas	0.6	0.65	0.66	0.65	0.67	0.65	0.66	0.71	0.69	0.66
Mining of ores	0.81	0.77	0.79	0.82	0.81	0.8	0.79	0.76	0.7	0.77
Other mining and quarrying	0.22	0.18	0.18	0.18	0.18	0.23	0.22	0.26	0.3	0.29
Foods, beverages and tobacco	0.22	0.22	0.22	0.23	0.21	0.24	0.23	0.21	0.19	0.19
Textiles	0.28	0.26	0.25	0.25	0.25	0.28	0.32	0.32	0.27	0.26
Apparel	0.11	0.13	0.12	0.13	0.11	0.13	0.13	0.12	0.09	0.08
Leather, leather and fur clothes, footwear and products	0.17	0.18	0.19	0.2	0.23	0.25	0.22	0.29	0.23	0.28

Table 4.A5 cont'd

(DCRi) Disimilarity index – Geographic concentration measure for NUTS 2

	1990	1991	1992	1993	1994	1995	1996	1997	1998	1999
Wood and products of wood and cork, plaiting materials	0.16	0.14	0.15	0.15	0.13	0.17	0.21	0.26	0.23	0.21
Pulp, paper and paper products, publishing and printing	0.45	0.45	0.47	0.47	0.5	0.47	0.48	0.46	0.47	0.46
Coke, refined petroleum products and nuclear fuel	1.42	1.49	1.55	1.57	1.56	1.57	1.6	1.45	1.46	1.46
Chemicals, chemical products and man-made fibres	0.21	0.21	0.21	0.21	0.21	0.22	0.21	0.16	0.17	0.2
Rubber and plastic products	0.38	0.38	0.39	0.39	0.4	0.41	0.41	0.39	0.35	0.14
Other non-metallic products	0.41	0.41	0.42	0.43	0.42	0.45	0.46	0.45	0.44	0.44
Basic metals except casting of metals	0.72	0.77	0.8	0.81	0.16	0.84	0.85	0.87	0.86	0.83
Metal products, machinery and equipment; casting of metals	0.26	0.28	0.27	0.28	0.27	0.29	0.29	0.32	0.31	0.33
Electrical and optical equipment	0.21	0.2	0.15	0.16	0.16	0.17	0.19	0.2	0.21	0.21
Transport equipment	0.66	0.75	0.77	0.77	0.74	0.76	0.81	0.79	0.81	0.79

Table 4.A5 cont'd

Gini index – Geographic concentration measure for NUTS 2

	1990	1991	1992	1993	1994	1995	1996	1997	1998	1999
Mining of coal; extraction of natural gas	0.41	0.45	0.46	0.46	0.48	0.47	0.48	0.52	0.51	0.49
Mining of ores	0.53	0.52	0.52	0.53	0.52	0.52	0.53	0.52	0.5	0.57
Other mining and quarrying	0.14	0.14	0.12	0.11	0.1	0.11	0.11	0.12	0.14	0.13
Foods, beverages and tobacco	0.13	0.12	0.11	0.11	0.11	0.12	0.12	0.09	0.1	0.1
Textiles	0.15	0.15	0.18	0.2	0.21	0.23	0.25	0.25	0.24	0.23
Apparel	0.08	0.1	0.09	0.1	0.1	0.1	0.11	0.11	0.09	0.09
Leather, leather and fur clothes, footwear and products	0.18	0.19	0.2	0.22	0.25	0.27	0.22	0.28	0.23	0.29
Wood and products of wood and cork, plaiting materials	0.09	0.09	0.09	0.1	0.08	0.1	0.14	0.16	0.13	0.12
Pulp, paper and paper products, publishing and printing	0.3	0.31	0.32	0.32	0.34	0.33	0.35	0.33	0.35	0.34
Coke, refined petroleum products and nuclear fuel	0.79	0.8	0.8	0.81	0.81	0.81	0.81	0.79	0.79	0.79
Chemicals, chemical products and man-made fibres	0.13	0.14	0.14	0.14	0.14	0.15	0.15	0.14	0.15	0.19
Rubber and plastic products	0.34	0.35	0.37	0.38	0.39	0.39	0.39	0.37	0.36	0.15
Other non-metallic products	0.24	0.24	0.25	0.25	0.26	0.26	0.27	0.27	0.25	0.25

Table 4.A5 cont'd

Gini index – Geographic concentration measure for NUTS 2

	1990	1991	1992	1993	1994	1995	1996	1997	1998	1999
Basic metals except casting of metals	0.42	0.45	0.45	0.46	0.1	0.47	0.48	0.49	0.49	0.46
Metal products, machinery and equipment; casting of metals	0.16	0.17	0.17	0.17	0.17	0.19	0.18	0.18	0.18	0.19
Electrical and optical equipment	0.13	0.12	0.11	0.1	0.1	0.14	0.13	0.13	0.14	0.13
Transport equipment	0.41	0.46	0.5	0.52	0.52	0.53	0.55	0.55	0.55	0.51

Source: Author's calculations based on REGSTAT.

Table 4.A6 Geographic concentration measures on the NUTS 3 level, 1990–99

(Hie) Herfindahl index – geographic concentration measure for NUTS 3

	1990	1991	1992	1993	1994	1995	1996	1997	1998	1999
Mining of coal; extraction of natural gas	0.15	0.18	0.17	0.17	0.17	0.18	0.18	0.19	0.2	0.2
Mining of ores	0.15	0.16	0.19	0.21	0.21	0.2	0.2	0.22	0.23	0.19
Other mining and quarrying	0.07	0.06	0.06	0.07	0.07	0.07	0.07	0.07	0.08	0.08
Foods, beverages and tobacco	0.05	0.05	0.05	0.05	0.05	0.05	0.05	0.05	0.06	0.06
Textiles	0.05	0.05	0.06	0.06	0.06	0.06	0.07	0.07	0.06	0.06
Apparel	0.05	0.05	0.05	0.05	0.05	0.05	0.05	0.05	0.06	0.06
Leather, leather and fur clothes, footwear and products	0.08	0.08	0.08	0.09	0.09	0.09	0.09	0.1	0.09	0.09
Wood and products of wood and cork, plaiting materials	0.05	0.05	0.05	0.05	0.06	0.06	0.07	0.07	0.07	0.07
Pulp, paper and paper products, publishing and printing	0.14	0.14	0.15	0.15	0.15	0.15	0.15	0.14	0.15	0.15
Coke, refined petroleum products and nuclear fuel	0.66	0.71	0.75	0.76	0.76	0.77	0.79	0.64	0.65	0.66
Chemicals, chemical products and man-made fibres	0.07	0.07	0.07	0.07	0.07	0.07	0.07	0.07	0.07	0.07
Rubber and plastic products	0.08	0.09	0.09	0.1	0.1	0.1	0.1	0.1	0.1	0.08
Other non-metallic products	0.06	0.06	0.06	0.06	0.06	0.06	0.06	0.06	0.06	0.06

Table 4.A6 cont'd

(Hic) Herfindahl index – geographic concentration measure for NUTS 3

	1990	1991	1992	1993	1994	1995	1996	1997	1998	1999
Basic metals except casting of metals	0.19	0.21	0.23	0.23	0.06	0.24	0.25	0.25	0.25	0.26
Metal products, machinery and equipment; casting of metals	0.07	0.07	0.07	0.07	0.07	0.07	0.07	0.07	0.08	0.08
Electrical and optical equipment	0.07	0.07	0.06	0.06	0.06	0.07	0.07	0.07	0.07	0.08
Transport equipment	0.08	0.09	0.11	0.12	0.13	0.14	0.15	0.16	0.18	0.15

(DCRi) Disimilarity index – geographic concentration measure for NUTS 3

	1990	1991	1992	1993	1994	1995	1996	1997	1998	1999
Mining of coal; extraction of natural gas	1.27	1.31	1.32	1.32	1.33	1.34	1.34	1.33	1.3	1.29
Mining of ores	1.19	1.19	1.38	1.42	1.43	1.44	1.47	1.47	1.46	1.39
Other mining and quarrying	0.72	0.67	0.71	0.72	0.71	0.75	0.78	0.75	0.76	0.71
Foods, beverages and tobacco	0.34	0.37	0.36	0.37	0.36	0.38	0.37	0.34	0.33	0.32
Textiles	0.49	0.49	0.49	0.52	0.55	0.59	0.6	0.58	0.6	0.56
Apparel	0.41	0.43	0.45	0.46	0.45	0.47	0.48	0.5	0.52	0.54
Leather, leather and fur clothes, footwear and products	0.62	0.64	0.67	0.69	0.7	0.73	0.72	0.78	0.79	0.74

Table 4.A6 cont'd

(DCRi) Disimilarity index – geographic concentration measure for NUTS 3

	1990	1991	1992	1993	1994	1995	1996	1997	1998	1999
Wood and products of wood and cork, plaiting materials	0.44	0.49	0.5	0.52	0.51	0.52	0.63	0.71	0.72	0.7
Pulp, paper and paper products, publishing and printing	0.68	0.69	0.7	0.71	0.73	0.72	0.74	0.72	0.74	0.71
Coke, refined petroleum products and nuclear fuel	1.71	1.74	1.74	1.74	1.74	1.73	1.72	1.72	1.73	1.73
Chemicals, chemical products and man-made fibres	0.66	0.68	0.69	0.68	0.69	0.7	0.69	0.63	0.65	0.6
Rubber and plastic products	0.63	0.66	0.7	0.72	0.72	0.75	0.75	0.74	0.7	0.56
Other non-metallic products	0.59	0.58	0.58	0.59	0.61	0.61	0.62	0.61	0.57	0.6
Basic metals except casting of metals	0.99	1.03	1.05	1.06	0.36	1.07	1.08	1.1	1.1	1.07
Metal products, machinery and equipment; casting of metals	0.38	0.4	0.41	0.41	0.41	0.42	0.41	0.45	0.47	0.48
Electrical and optical equipment	0.34	0.32	0.3	0.33	0.36	0.37	0.4	0.43	0.45	0.43
Transport equipment	0.76	0.86	0.89	0.88	0.91	0.95	1.04	1.01	1.02	0.94

Table 4.A6 cont'd

(Gi) Gini locational coefficients on NUTS 3 level

	1990	1991	1992	1993	1994	1995	1996	1997	1998	1999
Mining of coal; extraction of natural gas	0.96	0.98	0.98	0.98	0.99	0.99	0.99	1.03	1.03	1.03
Mining of ores	0.98	0.99	1.02	1.03	1.04	1.03	1.15	1.16	1.16	1.11
Other mining and quarrying	0.45	0.44	0.45	0.47	0.48	0.49	0.51	0.5	0.5	0.48
Foods, beverages and tobacco	0.21	0.21	0.22	0.22	0.22	0.23	0.23	0.21	0.21	0.2
Textiles	0.36	0.36	0.38	0.4	0.42	0.45	0.48	0.49	0.49	0.48
Apparel	0.28	0.29	0.31	0.31	0.31	0.32	0.33	0.33	0.35	0.36
Leather, leather and fur clothes, footwear and products	0.44	0.45	0.47	0.49	0.5	0.53	0.5	0.54	0.53	0.52
Wood and products of wood and cork, plaiting materials	0.3	0.31	0.31	0.32	0.32	0.35	0.44	0.48	0.49	0.49
Pulp, paper and paper products, publishing and printing	0.44	0.44	0.44	0.45	0.46	0.47	0.48	0.48	0.49	0.47
Coke, refined petroleum products and nuclear fuel	1.33	1.33	1.34	1.34	1.34	1.34	1.34	1.28	1.24	1.24
Chemicals, chemical products and man-made fibres	0.51	0.53	0.54	0.54	0.54	0.55	0.56	0.55	0.54	0.52
Rubber and plastic products	0.57	0.58	0.61	0.62	0.62	0.64	0.64	0.64	0.63	0.5
Other non-metallic products	0.44	0.43	0.44	0.44	0.46	0.46	0.47	0.48	0.47	0.48

Table 4.A6 cont'd

(Gi) Gini locational coefficients on NUTS 3 level

	1990	1991	1992	1993	1994	1995	1996	1997	1998	1999
Basic metals except casting of metals	0.83	0.84	0.84	0.84	0.3	0.85	0.83	0.83	0.86	0.82
Metal products, machinery and equipment; casting of metals	0.24	0.26	0.27	0.27	0.27	0.28	0.27	0.28	0.29	0.29
Electrical and optical equipment	0.25	0.26	0.25	0.28	0.3	0.34	0.32	0.32	0.33	0.31
Transport equipment	0.59	0.62	0.64	0.66	0.68	0.69	0.72	0.71	0.7	0.67

Source: Author's calculations based on REGSTAT.

Chapter 5

The Emerging Economic Geography in Estonia*

Grigory Fainshtein and Natalie Lubenets

1 Introduction

The economy of Estonia is very open and has fairly liberal trade regulations. Since the beginning of the transition, structural changes in the Estonian economy have been primarily determined by shifts in demand from its foreign trading partners, and substantial inflow of foreign direct investment (FDI). These factors have become crucial for the development of a regional structure of industrial manufacturing.

As a result of integration, the European Union (EU) has become the main trading partner and source of FDI for Estonia, and accession into the EU has influenced regional restructuring in the Estonian industrial sector.

The location of manufacturing activities has been a key factor in inconsistency in regional development. To smooth these dissimilarities, a concept for regional policy was approved by the government in 1994. In 1998, the Estonian Regional Development Strategy was introduced, which defined regional policy as an explicit activity of the public authorities with the objective of 'creating premises for development for all the regions of the state and balancing socio-economic development proceeding from the interests of the regions and the state as a whole'.

The implementation of regional policy is financed by the government's budget and has been supported by EU pre-accession structural instruments (ISPA, SAPARD and PHARE). As soon as Estonia becomes an EU member, the amount of structural aid will increase, and by 2008 may account for four per cent of the GDP.[1] However, in order to develop and conduct effective policy it is necessary to understand the processes that occur in the Estonian economy at a regional level. At this level, crucial elements are the development of regional specialization and location of economic activity, particularly of industrial manufacturing, as well as the factors that determine industrial and regional dynamics.

In this chapter, we attempt to analyse the processes of industrial specialization and geographic concentration in the face of the integration of the Estonian economy into the EU. The starting point for our study is the analysis of the structure and dynamics of foreign trade and FDI, and their impact on regional development.

One of the key indicators of regional development is the dynamics of regional income. In this chapter, we analyse how trade liberalization has affected the structure of regional wages. Within the scope of this chapter is the role that transportation costs (captured by distance from the capital city) play in the regional wage differential, as well as the impact of different stages of integration and specific features of border regions.

We also attempt to study how industrial specialization has influenced economic growth in regions, and to determine qualitative characteristics that were relevant for economic growth and regional convergence.

As predicted by the new economic geography theory, the impact of trade liberalization and economic integration is region-specific and varies substantially. Based on this analysis, we can predict potential winner and loser regions on the basis of the location of industrial activity, and can develop some policy implications.

The rest of the chapter is structured as follows. Section 2 describes the data used in this analysis. Section 3 discusses Estonian economic integration into the EU by analysing foreign trade commodity flows and the inflow of FDI. The fourth and fifth sections present a study of regional specialization and geographical concentration patterns, followed by an analysis of the impact of economic integration on the regional wage structure in section 6. In section 7, we study the influence of regional specialization and some regional characteristics on regional growth. In the eighth section regional winners and losers and some policy implications are considered. The final section concludes.

2 The Data

We used employment data provided by the Labor Market Division of the Statistical Office of Estonia for the calculation of industrial specialization and geographical concentration indices. The data was based on the Labor Force Surveys. A wide set of regional parameters was taken from the yearbooks of regional statistics for Estonia for 1992–2000. We also used data on regional wages and regional development indicators for 1992–2000[2] that was published by the Statistical Office of Estonia.

The analysis presented in this chapter was somewhat restricted by a lack of data on regional GDP in Estonia: regional GDP figures have been available for only three years since 1996. For this reason, to study regional specialization and geographic concentration, we have used employment statistics that were collected for each of the NUTS 3 regions of Estonia by sectors of economic activity and by NACE broad manufacturing classifications for 1990–99. Our analysis of the impact of integration with the EU on regional development is based on the data for average wages in industries in regions at the NUTS 4 level (15 administrative units in Estonia). Not only did it give us more panel observations, but it also solved the problem of defining the distance between NUTS 3 regions and the capital city. The distance used in the study was distance from the county's capital to the capital city.

A substantial problem with the industrial employment data was the low degree of confidence for some industries in some regions due to the smallness of the sample of the employment measure. However, as low confidence was likely to appear only in the least significant industries in regions, it would not change the major tendencies in regional development, and might only cause fluctuations of the indicators by years within particular industries.

The data source for our analysis of export and import flows was the database of the Foreign Trade Division of the Statistical Office of Estonia. All commodities were classified according to the Estonian Goods Nomenclature (EGN) issued in 1993. The first six digits in the EGN are equivalent to the Harmonized System (HS) nomenclature that is used in international trade. The nomenclatures divide trade commodities into 21 broad sections and 97 chapters. We analyse development of commodity flows at the level of HS two-digit chapters, since broad NACE classification by manufacturing sectors is an unreasonably high level of aggregation for Estonia. It is easy to link HS two-digit sectors and NACE broad manufacturing sectors.

Estonian foreign trade flows possess some distinctive features that complicate their analysis. The principal obstacle is a large share of re-exports (from a quarter to one-third of total exports in different years). Re-export flows were quite difficult to measure and eliminate in the early years of transition. The foreign trade statistics in Estonia are based on the data of customs statistics. Initially, disaggregated HS two-digit data was available only in the General Trade System format, which includes re-export. Employing such data would substantially influence the outcome of the analysis. However, since 1995, the codes of custom procedures have been added to a declaration form, and the data is now accessible in the Special Trade System format that excludes re-export. The present analysis is based on data with the Special Trade System classification, but for the purposes of analytical comparison, data that includes re-export before 1995 was added to the appropriate tables.

3 Economic Integration with the European Union since 1990

The development of Estonian economic integration with the EU begun just after Estonia gained its independence. This process has been characterized by the re-direction of foreign trade to EU markets and by increased inflow of FDI from the EU. In addition, a legislative base for integration had developed, including free trade agreements and later harmonization of the terms of trade.

3.1 Trade Liberalization and Industrial Specialization

The Estonian economy is characterized by a high degree of openness. As is the case with any small open economy, the main engine of liberalization has been the need to gain a competitive advantage by exploiting economies of scale, thus expanding the tiny domestic market. Domestic consumer demand has also been oriented mostly towards imported goods. A high degree of economic openness can be illustrated by share of Estonian imports and exports in the GDP (83 and 77 per cent respectively in 1999).

An important factor in Estonian economic openness was the competitive advantage of its geographical location and remarkably liberal trade regime. The geographical location of Estonia enables it to have very close trade relations with Western-European, Scandinavian and CIS countries offering transit services for commodity flows from west to east (mainly to Russia) and in the opposite direction.

Estonian foreign trade policy has been based on liberal principles and can be described as having the following features:[3]

- no restrictions on the free movement of goods and capital;
- minimal rules for wage formation, foreign trade and the right of establishment;
- import duties on agricultural products only, with a weighted average of 3.3 per cent;
- liberal price formation.

The ratification of bilateral agreements on free trade formed a legislative basis for the development of trade with the EU (including potential members). In

February 1992, the protocol on temporary measures of economic and trade cooperation with Finland was signed. Ratification of free trade agreements with a number other countries followed, including Sweden (July 1992), Norway (September 1993) and Switzerland (March 1994). The Free Trade Agreement between Estonia and the EU was signed in July 1994 and entered into force in January 1995. As a result, a free trade area has been created between Estonia and the European Union where all trade barriers to industrial products have been abolished. The only restriction on Estonian imports into the EU is applied on Estonian agricultural products.

At present, foreign trade relations between Estonia and the EU are regulated by the Europe Agreement (signed in August 1995, entered into force in February 1998). The Europe Agreement incorporates all provisions and mandates of the Free Trade Agreement.

Estonia has been a full member of the World Trade Organization (WTO) since November 1999.

Change of geographical trade patterns Since the beginning of the transition, a rapid re-orientation of Estonian foreign trade from the markets of CIS countries to western markets took place (see Table 5.1). This reorientation can be explained by the adjustment of the artificial structure of foreign trade with former Soviet states towards territorial structures naturally determined by factors of geopolitical location, comparative advantage and foreign demand.

Among the reasons for trade re-orientation at the beginning of the transition were the deterioration of the terms of trade with Russia since 1992 caused by high inflation in Russia, the collapse of the system of payments, the introduction of import tariffs, the rise in prices of raw materials and the unstable overall economic climate. In many cases, Estonian firms preferred importing more expensive raw materials from the west, since guaranteed deliveries and high quality compensated for the higher prices than those on the more risky eastern market.

Just after monetary reform in Estonia in 1992, exports to western markets became highly profitable and have rapidly increased since, due to the significant difference in prices between the two regions. An initial rise occurred in such commodity groups as textiles, wood, paper and other products that are material- and labor-intensive. The structure of trade with western countries conforms to the principle of geographical closeness: the largest trading partners are Finland, Germany, and Sweden. The share of EU countries in Estonian foreign trade has increased.

Table 5.1   Development of the geographical structure of foreign trade, per cent

	EU	EFTA	Export CIS	ETC*	Other
1991	0.2	2.8	82.9	12.3	1.8
1992	13.6	36.4	35.2	10.0	4.8
1993	17.8	31.5	30.4	14.8	5.5
1994	19.0	30.9	30.3	15.3	4.5
1995	54.0	2.4	25.1	13.8	4.7
1996	51.0	2.8	25.1	15.5	5.6
1997	48.5	3.8	26.4	15.8	5.5
1998	55.0	3.7	20.7	15.1	5.5
1999	62.8	3.0	13.5	13.8	6.9
	EU	EFTA	Import CIS	ETC*	Other
1991	3.1	3.1	73.3	13.2	7.3
1992	15.6	32.0	36.0	5.9	10.5
1993	23.3	38.1	21.6	7.0	10.0
1994	23.9	40.7	20.4	5.8	9.2
1995	66.0	1.7	18.8	5.6	7.9
1996	64.6	2.1	17.0	6.2	10.1
1997	59.2	2.2	17.4	5.7	15.5
1998	60.1	2.2	14.2	6.3	17.2
1999	57.7	2.1	17.0	7.4	15.8

*   European transition countries.

Source: Estonian Statistical Office.

To analyse the geographical reorientation of foreign trade, it is necessary to specify the ways in which it has occurred. One should distinguish growth in trade with western markets at the expense of a decline in trade with eastern markets in respective sectors (trade diversion), and the creation of new commodity flows to the western markets (trade creation). In addition, reasons for re-orientation could also be the growth of re-exports, the reorientation of domestic sales to external markets because of a reduction in domestic demand, a smaller decrease in exports to one market in comparison with another, etc.

No statistical information is available for the analysis of actual trade diversion at the commodity groups level. Therefore, we calculated the Finger-

Kreinin coefficient (FK)[4] to study actual trade diversion at a disaggregated level. This coefficient captures the similarity of trade flows between regions. As a result of trade diversion, similarity may increase. Similarity decreases if no diversion happens.

Table 5.2 presents values of Finger-Kreining coefficients of trade similarity of exports and imports between EC + EFTA and CIS countries in 1993–99. One can observe decreasing similarity of exports between these two areas. It means that no essential trade diversion occurred in export flows in the given period. This can partly be explained by trade creation with western countries based on western investments, which made the structure of exports to the EU and EFTA increasingly diverge from CIS countries.

Import similarity between the two areas has also declined. This indicates that the main import articles from eastern markets (mostly raw materials) have remained unchanged, and growth of imports from the west has occurred due to expansion in other commodity groups.

Table 5.2   Finger-Kreining coefficients of similarity between Estonian foreign trade with EC+EFTA and CIS

	1993	1994	1995	1996	1997	1998	1999
EXPORT	0.4690	0.3779	0.3949	0.3585	0.3232	0.2965	0.2938
IMPORT	0.4114	0.3826	0.3699	0.3371	0.3936	0.2911	0.2929

Source: Estonian Statistical Office, own calculations.

Comparative advantage   One common indicator of the development of comparative advantage is the trade coverage ratio (TC).[5]

Table 5.3 shows commodity groups with the greatest trade coverage ratio for trade with the EU (internal comparative advantage). There are 25 two-digit commodity groups with TC indexes greater than unit. Among the biggest export-oriented groups are wood and articles of wood; furniture; apparel and clothing accessories; textile articles; and fish. Altogether, internal comparative advantage is typical of both capital- and labor-intensive commodities. Over the period considered, the values of the TC index have increased for the following groups: other textile articles, furniture; wood and articles of wood; cotton; apparel and clothing accessories; and textile articles. The TC value for the group 'electrical machinery and equipment' has increased sharply (more than 2.5 times), and in 1999 was close to unit.

Table 5.3   Commodity groups with highest trade coverage ratio in Estonian foreign trade with the EU, 1995–99, ranked by ratio values in 1999

CODE		1993*	1995	1997	1999	1999/1995
14	Vegetable plaiting materials	56.40	428.442	96.350	108.842	0.254
81	Other base metals; cermets; articles thereof	732.00	86.136	18.944	20.683	0.240
44	Wood and articles of wood	18.23	10.615	10.777	15.588	1.468
78	Lead and articles thereof	686.20	7.917	6.078	10.217	1.291
46	Manufactures of plaiting materials	30.45	11.897	5.435	7.244	0.609
63	Other textile articles	1.01	1.674	3.472	4.670	2.790
03	Fish and crustaceans	7.44	2.984	1.037	3.354	1.124
94	Furniture; bedding; mattresses, etc.	1.43	1.371	1.774	3.312	2.416
62	Articles of apparel and clothing accessories, etc.	2.69	2.709	2.592	3.204	1.183
61	Articles of apparel and clothing accessories, knitted or crocheted	2.53	2.505	1.834	2.739	1.093
53	Other textile articles	0.95	1.092	1.243	2.319	2.123
04	Dairy products	12.02	4.445	1.185	1.982	0.446
47	Pulp of wood or of other fibrous cellulose material	3.59	9.761	1.758	1.906	0.195
65	Headgear and parts thereof	2.89	2.319	1.684	1.466	0.632

*   Including re-export.

Source: Database of Foreign Trade Division of the Statistical Office of Estonia, own calculations.

As noted previously, analysing export shares and TC dynamics helps us to evaluate the importance of given commodity groups in Estonian export in total as well as to measure their internal comparative advantage. However, this information is not sufficient for determining international specialization of Estonian exports. Thus, we used the so-called specialization index[6] as an indicator of the international specialization of Estonian exports on the EU market (external comparative advantage).

The largest commodity groups with a fairly high comparative advantage in the EU market are: wood and articles of wood; furniture; dairy products; textile articles, cotton; articles of iron or steel; apparel and clothing accessories; electrical machinery and equipment; footwear; fish; and apparel and clothing accessories.

From 1995–99, the comparative advantage in the EU market increased for the following groups: wood and articles of wood; textile articles, cotton; articles of iron or steel; and electrical machinery and equipment. Comparative advantage in the EU market declined for the following groups: dairy products; fish; apparel and clothing accessories; textiles; apparel and clothing accessories; and footwear.

Intra-industry trade A significant portion of foreign trade flows cannot be explained by the comparative advantage. First of all, it concerns mutual trade flows within one commodity group, so-called intra-industry trade. The main factors determining the level of intra-industry trade are vertical specialization, the location of different stages of production in different countries including subcontracted work, and horizontal intra-industry trade involving finished products.

We used the Grubel-Lloyd (GL) index[7] to measure the level of intra-industry trade for Estonia.

Table 5.5 presents the values of the GL index for the biggest commodity groups. The largest export groups with the most intra-industry trade are cotton; glass and glassware; iron and steel; electrical machinery and equipment; and optical equipment. It is important to note that for the majority of commodity groups the intra-industry trade index has increased, with cotton and articles thereof as the only exception (note that the export share of this article has also decreased).

As already mentioned, the high level of intra-industry trade can be explained by different factors. In developed industrial economies, the largest part of intra-industry trade is trade of the same quality commodities-substitutes (horizontal intra-industry trade). At the same time, countries with a lower

Table 5.4 Commodity groups with the highest specialization index in Estonian foreign trade with the EU

		1995	1997	1999	1999/1995
53	Other vegetable fibres	7.18	1.58	18.30	2.55
44	Wood and articles of wood	11.26	14.11	14.92	1.33
43	Fur skins and artificial fur	5.45	7.91	10.31	1.89
94	Furniture; bedding; mattresses, etc.	6.08	6.18	6.19	1.02
04	Diary products	28.22	18.11	5.80	0.21
63	Other textile articles	4.77	5.93	5.53	1.16
52	Cotton and articles thereof	3.62	5.53	3.96	1.09
56	Wadding, felt and non-wovens	3.83	5.00	3.21	0.84
73	Articles of iron or steel	2.09	3.25	3.20	1.53
62	Articles of apparel and clothing accessories, etc.	2.78	2.57	2.46	0.89
65	Headgear and parts thereof	3.64	2.61	2.27	0.62
85	Electrical machinery and equipment	0.52	1.73	2.01	3.84
14	Vegetable plaiting materials	2.13	1.58	1.94	0.91
64	Footwear, gaiters and the like	1.96	1.66	1.83	0.93
70	Glass and glassware	2.14	2.28	1.80	0.84
72	Iron and steel	1.75	1.70	1.73	0.98
03	Fish and crustaceans	2.92	1.26	1.53	0.53
25	Salt; sulfur; earth and stone, lime and cement	2.23	1.55	1.53	0.68
48	Paper and paperboard	0.84	1.21	1.22	1.45
61	Articles of apparel and clothing accessories; knitted or crocheted	1.88	1.55	1.21	0.65

Source: Database of Foreign Trade Division of the Statistical Office of Estonia, COMEX database, own calculations.

degree of economic development either participate in vertical intra-industry trade of commodities with different quality, or have a large share of sub-contracted work. In both cases, industrial cooperation with other countries may take place.

To analyse the nature of intra-industry trade we assume that the quality of a good is reflected in its price. As suggested by Greenaway (1994), intra-industry trade is horizontal if commodity price shares of exports and imports do not differ more than 15 per cent. Otherwise, intra-industry trade is vertical.

In the present study, we have calculated price shares of exports and imports for 1995 and 1999 according to HS six-digit nomenclature. We then selected the commodity groups with an export and import price share difference of more than 15 per cent and calculated their share in every two-digit commodity group. The results are given in Table 5.6. Obviously, in Estonia, vertical intra-industry trade dominates. This confirms the fact that the trade pattern reflects the transitional character of the Estonian economy.

3.2   Foreign Direct Investment

Since the beginning of the transition, foreign direct investment has been one of the major sources of foreign capital inflow into Estonia. The relatively significant volume of foreign direct investment into Estonia (see Table 5.7) can be explained by sound economic reforms, including macroeconomic stabilization and privatization schemes. The active involvement of foreign capital in privatization was determined by the chosen methods of privatization. Because one of the most important goals of privatization was to create groups of owners with enough resources to restructure companies and to provide effective management, a majority of the formerly state-owned companies were sold to core investors. Since residents did not possess enough financial resources, the market prices of the companies being privatized were not very high. Thus, foreign direct investments into privatized companies were also stimulated by the expected high returns (see Randveer 1999).

Table 5.7 presents the volume of FDI in Estonia in 1993–99. One can see significant fluctuations in volume by years, which can be explained by the small size of the Estonian economy, as well as by some outside economic factors, such as the world financial crisis, the Russian crisis, integration into the EU, etc.

An analysis of major FDI groups shows that the most significant volume of capital was invested, during the course of privatization, into companies that

Table 5.5   Commodity groups with the highest Grubel-Lloyd indexes of intra-industry trade with the EU in 1993–99, ranked by index values in 1999

CODE		1993*	1995	1997	1999	1999/1995
64	Footwear, gaiters and the like	0.86	0.848	0.803	0.997	1.176
28	Inorganic chemicals	0.24	0.786	0.888	0.981	1.249
73	Articles of iron or steel	0.83	0.702	0.770	0.976	1.390
85	Electrical machinery and equipment	0.402	0.495	0.884	0.956	1.929
95	Toys, games and sports equipment	0.83	0.773	0.979	0.947	1.225
75	Nickel and articles thereof	0.014	0.047	0.747	0.947	20.291
80	Tin and articles thereof	0.53	0.678	0.434	0.934	1.378
72	Iron and steel	0.23	0.875	0.784	0.927	1.060
42	Articles of leather	0.71	0.618	0.681	0.901	1.458
89	Ships, boats and floating structures	0.32	0.539	0.682	0.885	1.640
56	Wadding, felt and non-wovens	0.47	0.691	0.965	0.883	1.278
12	Oil seeds and oleaginous fruits	0.55	0.766	0.830	0.882	1.152
16	Preparations of meat, fish, or crustaceans	0.52	0.999	0.855	0.854	0.855
71	Natural or cultured pearls, precious or semi-precious stones	0.56	0.864	0.935	0.854	0.988
43	Fur skins and artificial fur	0.22	0.764	0.993	0.845	1.106
52	Cotton and articles thereof	0.87	0.991	0.638	0.834	0.842

* Including re-export.

Source: Database of Foreign Trade Division of the Statistical Office of Estonia, own calculations.

Table 5.6   Shares of vertical intra-industry trade with the EU (as a percentage of total intra-industry trade) for commodity groups with the largest GL values

CODE		1995	1999
64	Footwear, gaiters and the like	61.76	92.16
28	Inorganic chemicals	100.00	100.00
73	Articles of iron or steel	99.98	98.06
85	Electrical machinery and equipment	62.47	42.94
95	Toys, games and sports equipment	71.24	99.19
75	Nickel and articles thereof	100.00	100.00
80	Tin and articles thereof	81.23	100.00
72	Iron and steel	99.98	98.03
42	Articles of leather	100.00	100.00
89	Ships, boats and floating structures	100.00	100.00
56	Wadding, felt and non-wovens	100.00	100.00
12	Oil seeds and oleaginous fruits	100.00	90.25
16	Preparations of meat, fish, or crustaceans	100.00	100.00
71	Natural or cultured pearls, precious or semi-precious stones	99.41	98.06
43	Fur skins and artificial fur	99.81	81.21
52	Cotton and articles thereof	90.92	99.68
31	Fertilizers	100.00	100.00
88	Aircraft, spacecraft and parts thereof	10.46	100.00
25	Salt; sulfur; earth and stone, lime and cement	82.35	100.00
65	Headgear and parts thereof	100.00	99.28
90	Optical, photographic, cinematographic, measuring, checking, etc.	99.37	96.55
70	Glass and glassware	61.01	68.90
86	Locomotives, rolling stock and parts thereof	18.50	100.00
57	Carpets and other textile floor coverings	100.00	100.00

Source: Database of Foreign Trade Division of the Statistical Office of Estonia, own calculations.

already existed on the market. The volume of re-invested income has been growing in recent years (the temporary decrease in 1997–98 was influenced by a world financial crisis). Altogether, this indicates the crucial role FDI has played in industrial restructuring in Estonia.

Table 5.8 shows the distribution of FDI by sectors of economic activity. Its present structure was determined primarily by the structure of the Estonian

Table 5.7   Foreign direct investments into Estonia (US$ million)

	1993	1994	1995	1996	1997	1998	1999
Total FDI	162.7	217.4	201.8	150.8	266.1	573.9	302.7
Into capital stock	–	145.8	101.7	17.9	98.0	402.6	173.7
O/w into new enterprises	–	–	–	4.1	3.8	3.0	3.1
Additional investments into existing enterprises	–	–	–	12.9	93.9	396.7	166.4
Reinvested income	–	42.4	15.6	18.0	93.9	27.7	49.1
Other direct investment capital	–	29.1	84.5	114.9	74.2	143.6	79.9
Share of inward FDI in GDP (%)	9.96	9.52	5.68	3.46	5.74	11.01	5.91
FDI per capita (USD)	107.3	145.0	136.0	102.7	182.5	395.8	209.8

Source: Bank of Estonia, own calculations.

Table 5.8   Foreign direct investment stock by fields of activity (as of 31 December 2000)

	US$ million	%
Finance	654.48	25.0
Transport, storage, communication	571.91	21.8
Manufacturing	562.30	21.5
Wholesale, retail trade	408.55	15.6
Real estate, renting and business activities	181.99	6.9
Electricity, gas and water supply	61.90	2.4
Hotels, restaurants	51.60	2.0
Construction	38.16	1.5
Agriculture, hunting, forestry	33.60	1.3
Other community, social and personal service activities	29.84	1.1
Mining, quarrying	10.89	0.4
Other	15.01	0.5
TOTAL	2620.29	100.0

Source: Bank of Estonia, own calculations.

economy as a whole and by the rapid development of the financial sector and transit, transport and storage services and trade (a large share of trade is attributed to re-export). The share of investment into the manufacturing sector is relatively small compared to that in the majority of East-European economies, though it is still significant in volume.

An analysis of the geographical structure of FDI in Estonia (see Table 5.9) reveals the dominant role of EU countries. This can be explained by geopolitical factors as well as by the overall integration process of the Estonian economy into the EU.

Table 5.9   Direct investment stock by countries (as of 31 December 2000)

Country	USD million	%
EU countries	2,373.3	90.6
Sweden	1,060.2	40.5
Finland	781.0	29.8
Norway	111.7	4.3
Denmark	105.9	4.0
Germany	68.3	2.6
Great Britain	62.0	2.4
Netherlands	57.9	2.2
East-European and CIS countries	44.3	1.7
Other	202.6	7.7
Total	2,620.3	100.0

Source: Bank of Estonia, own calculations.

4 Regional Specialization Patterns

4.1 *Regional Structure and Disparities*

Due to the small size of Estonia's territory, the largest regional divisions are at the NUTS 3 level only (the second aggregated regional level is Estonia as a whole: as a regional unit it is represented at the NUTS 2 level, and as an administrative entity at NUTS 1. Table 5.10 presents five regions of Estonia[8] at the NUTS 3 level, the administrative units (counties) that each of the regions contains, and some general economic characteristics.

Table 5.10   Regional structure of Estonia

Regions (NUTS 3)	Included counties (NUTS 4)	Area in km² (% of total)	Population of region (% of total)	Regional GDP per capita, 1998 (% of country average)
Northern Region	Harju county (Tallinn included)	4,331.6	37.02	162.99
West Estonia	Lääne, Pärnu, Hiiu and Saare counties	1,1134.68	12.94	55.90
Southern Estonia	Tartu, Põlva, Võru and Valga counties	9506.48	18.77	60.14
Central Estonia	Viljandi, Jõgeva, Järva and Rapla counties	11,629.22	12.72	69.21
Northeast Estonia	Lääne-Viru and Ida-Viru counties	6,828.63	18.56	66.51

Source: Statistical Office of Estonia.

Fairly noticeable differences between the regions exist for most of the indicators. Below we present a brief description of each of the NUTS 3 regions.

The northern region, including Tallinn, is by far the largest economic region in Estonia, with more than a third of Estonia's inhabitants. The region's share of industrial employment in the country's total industrial employment has varied slightly around 37 per cent; the share of employment in industry to total regional employment is c.30 per cent. In 1998 the northern region produced more than half (52 per cent) of the total industrial output. As one might expect, the unemployment rate in this region is one of the lowest in Estonia. In 2000 it was 5.6 per cent. The labor force in the capital region has the greatest number of people with higher education.

In recent years the key industries in the region were (calculated for the employment figures and sorted according to their importance) manufacturing of wood, food products, beverages and tobacco products, textiles and wearing apparel, furniture, and fabricated metal products.

The northeastern region is a big industrial region. The region's share of the total Estonian population is 19 per cent. Although the region's share of industrial employment in the country's total industrial employment has decreased in recent years, it constitutes an impressive 23 per cent; industrial employment accounts for roughly half (47 per cent) of all employment in the region. In 1998 the region produced c.19 per cent of the total industrial output. The region's industrial structure is determined by its location close to the Russian border and its reliance on close economic connections with the Russian market. The share of unemployment in this region is the highest in Estonia (11.2 per cent). The structure of employment has not changed significantly during the transition years, and the most important industries remain the manufacture of food products, beverages and tobacco products, textiles and wearing apparel, wood and furniture, and the manufacture of rubber and plastic products.

By size of manufacturing sector, the next region is southern Estonia, including the second largest city of Tartu. This region includes c.19 per cent of the total population. The region's share of industrial employment in the country's total industrial employment is 14.8 per cent; manufacturing accounts for c.30 per cent of employment in the region. In 1998, roughly 12 per cent of the total industrial output was produced in the region. The unemployment rate is 6.6 per cent. In recent years, among the most important industries were the manufacture of food products, beverages and tobacco products, textiles and apparel, wood and furniture, and the manufacture of electrical machinery.

The central region is the biggest agricultural region in Estonia. About 13 per cent of the total population lives in this region. The region's share of industrial employment in the country's total industrial employment is 11 per cent; industrial employment accounts for c.31 per cent of the region's total employment. In 1998, the region's share in industrial production was 7 per cent. The unemployment rate in the region in 2000 was 7.9 per cent. In recent years, among the most important industries were the manufacture of textiles and apparel, food products, beverages and tobacco products, wood and furniture, and the manufacture of fuels, chemicals and chemical products.

The western region's share of the total population is c.13 per cent. The region's share of industrial employment in the country's total industrial employment is 12 per cent; the share of industrial employment in the total regional employment is c.12 per cent. In 1998, the region produced 6.7 per cent of the total industrial output. This is the second agricultural region of Estonia. The unemployment rate is 5.8 per cent. In recent years, the most important industries were the manufacture of wood, food products, beverages and tobacco products, furniture, textiles and apparel, and the manufacture of fabricated metal products.

4.2   Specialized and Diversified Regions

To analyse the development of regional specialization, we calculated three indices of regional specialization for Estonian regions at the NUTS 3 level for 1990–2000. As a measure of absolute specialization in regions, the Herfindahl index[9] was chosen. The Dissimilarity index[10] and GINI coefficients[11] were taken as measures of relative specialization in the regions. The values of the indices are presented in the Appendix (Tables 5.A1–5.A3). .

As expected, the least specialized regions are the most industrialized: northern Estonia, northeastern Estonia and southern Estonia. Favorable market conditions (demand and logistics networks) attract companies in many industries to locate in these regions. Accordingly, as one might expect, the most specialized regions are the least industrialized regions of central Estonia and western Estonia.

To analyse the dynamics of specialization in regions, we calculated the growth rates of the absolute and relative specialization indices at a regional level (see Table 5.11. for results).

A general increase in the level of specialization indices can be observed in northern and southern Estonia. This allows for the assumption that the optimization of industrial structures has occurred in these regions. In northern Estonia, stable growth can be observed prior to 1997, followed by a decrease from 1998–2000. This pattern may be explained by the theory of agglomeration, which states that a large amount of investment in the region, as well as the region's having a dominant share of the total FDI, provides for rapid development of infrastructure. These factors induce the largest enterprises to move into the region. As the costs of production rise due to increasing demand for less mobile factors of production (primarily labor and mortgage) along with the development of infrastructure in other parts of a country, an increasing number of companies move their activity to peripheral regions. In Southern Estonia, it is possible to observe an overall increase in absolute specialization. However, this tendency cannot be observed in relative specialization due to significant variation in the index values. The dynamics of the relative specialization index in the third industrial region, northeastern Estonia, show a stable increase in diversification of manufacturing. This indicates a change in the regional industrial structure.

In less developed regions the level of specialization has decreased. Specialization has continuously declined in central Estonia. In western Estonia, specialization has fallen since the beginning of the industrial recovery. Such developments are consistent with our previous explanation of the regional dynamics of manufacturing, as additional evidence of the relocation of manufacturing activity to peripheral regions.

Overall, specialization dynamics reveal some tendencies towards the homogenization of industrial specialization in Estonia across the regions, and there are some signs that industrial activity is starting to relocate from fairly developed central regions to the relatively less industrialized periphery.

To evaluate specialization dynamics in Estonia as a whole, we calculated the change in specialization indices as a weighted average of specialization indices' changes at the regional level, using employment shares of regions as weights (see Table 5.12. for index values).

The dynamics of all three specialization indices follow similar inverse U-shaped patterns. Before 1995, an overall increase in specialization can be observed. However, since 1997 these dynamics have been reversed, and industrial specialization in Estonia has started to decrease. Such dynamics reveal similarities with general economic development trends (especially for industrial growth) in Estonia. Before 1995, the Estonian GDP declined. Increasing specialization during that period not only reflected an optimization of the industrial structure, but also accounted for a decline in the number of industrial branches and enterprises in the regions. Accordingly, a tendency towards decreasing specialization in the later years coincided with stable economic growth. Consequently, new industrial enterprises are emerging in more uniform regional patterns.

An important contribution to the present analysis is a study of general trends in specialization levels during the observed period. One of the ways to analyse these trends is to calculate the average changes for every index for the given period.[12] If we divide this change by the number of years, we obtain the average growth rate of the level of specialization. For the Herfindahl index of absolute specialization, it is 3.3 per cent; for regional dissimilarity, it is 5.2 per cent, and for relative specialization measured by the GINI index, it is 2.2 per cent. In conclusion, for the observed period, the level of regional specialization in Estonia has increased on average by two to five per cent a year. In our case, time is a good proxy for the economic integration of Estonia into the EU. Therefore, we can also conclude that integration processes are an important factor in increasing regional specialization, as predicted by the new economic geography hypotheses.

Table 5.11 The dynamics of specialization indices at the regional level (1990 = 100)

Years	1991	1992	1993	1994	1995	1996	1997	1998	1999	2000
Northern Estonia										
Herfindahl	101.0	99.0	103.1	106.1	116.3	114.3	118.4	106.1	102.0	106.1
Dissimilarity	97.5	96.3	102.0	103.3	177.5	166.4	157.0	148.4	148.0	150.8
GINI	100.0	95.4	101.8	98.6	139.2	138.2	143.8	121.7	99.5	110.6
Central Estonia										
Herfindahl	90.4	86.5	75.5	75.0	71.2	72.6	83.2	88.0	89.4	82.7
Dissimilarity	96.8	101.0	88.0	84.1	68.4	57.3	70.3	68.4	66.5	56.1
GINI	99.8	94.8	85.3	88.0	76.3	86.7	87.4	76.3	72.9	70.9
Northeastern Estonia										
Herfindahl	99.3	99.3	103.4	105.5	93.1	96.6	102.1	100.0	102.1	104.1
Dissimilarity	95.5	89.0	80.7	73.0	78.9	77.1	78.2	67.9	56.8	64.1
GINI	96.4	92.5	94.4	90.8	100.0	98.3	88.1	83.1	76.8	79.7
Western Estonia										
Herfindahl	99.5	96.6	94.1	104.4	131.9	129.9	119.1	103.9	100.5	92.6
Dissimilarity	94.9	96.9	89.0	92.3	112.7	109.6	94.9	80.9	90.3	73.0
GINI	94.8	96.7	84.9	76.9	113.4	117.9	105.4	97.6	100.9	88.0
Southern Estonia										
Herfindahl	100.7	94.8	94.8	100.7	111.9	117.8	123.7	120.0	117.8	118.5
Dissimilarity	97.8	92.6	101.3	105.8	81.3	95.3	101.1	85.3	92.2	104.9
GINI	100.3	95.9	110.8	121.2	86.1	105.1	110.8	93.0	95.6	108.2

Source: Authors' calculations based on REGSTAT.

Table 5.12 The dynamics of specialization indices* at the country level (1990 = 100)

Years	1991	1992	1993	1994	1995	1996	1997	1998	1999	2000
Herfindahl	99.3	96.7	97.5	101.2	107.8	108.6	112.3	104.8	103.0	103.4
Dissimilarity	96.8	94.9	94.7	94.4	122.4	118.7	115.5	105.5	106.2	107.3
GINI	98.7	95.0	98.0	97.0	112.5	116.9	116.8	101.6	91.9	97.4

*   Calculated as weighted average from regional changes using employment shares of regions as weights.

Source: Authors' calculations based on REGSTAT.

5 Location and Concentration of Industrial Activity

5.1 Structure of Manufacturing, 1991–99

The transition to a market economy in Estonia has been characterized by significant structural changes in industry. The greatest decline in industrial production in Estonia occurred in 1990–91. During these years price and demand shocks occurred simultaneously. The former was caused by inflating prices in Russia of the raw materials on which Estonian industry was initially fully dependent. One of the main determinants of the demand shock in these years was the loss of the traditional, mainly Russian, market. Table 5.14 presents cumulated indices of industrial growth in comparative prices.

Following the approach of Repkine and Walsh (1999), we analyse the general structural change in sectors of manufacturing based on sectoral rates of growth.

A discrete measure of growth g over a period t-1 to t in sector i is the following:

$$g_{it} = \left(\frac{y_{it} - y_{it-1}}{(y_{it} + y_{it-1})/2} \right)$$

The contributions of increasing and declining industries are calculated separately as a sum of growth rates of rising sectors weighted by sector size (POS) and a sum of absolute values of growth rates of declining sectors weighted by sector size (NEG), respectively.

$$POS_{it} = \sum_{i-1}^{n} S_{it}\, g_{it} \text{ if } g_{it} > 0$$

$$NEG_{it} = \sum_{i-1}^{n} S_{it} \left| g_{it} \right| \text{ if } g_{it} < 0$$

Net change (NET) is a net outcome that is induced by output growth in rising sectors offset by output fall in declining sectors:

$$NET_{it} = POS_{it} - NEG_{it}$$

The reallocation of output between sectors is captured by the EXCESS index:

$$EXCESS_{it} = NET_{it} + POS_{it} - \left| NEG_{it} \right|$$

Indicators described above are shown in Table 5.13:

Table 5.13   Indicators of general structural change in Estonian industry

	1992	1993	1994	1995	1996	1997	1998
POS	0.007	0.083	0.036	0.026	0.044	0.089	0.054
NEG	0.197	0.088	0.029	0.014	0.029	0.001	0.009
NET	−0.190	−0.005	0.007	0.012	0.014	0.087	0.045
EXCESS	0.014	0.166	0.058	0.028	0.059	0.003	0.018

Sources: *Statistical Yearbook 1999* (1999), pp. 252–3; *Statistical Yearbook 2000* (2000); authors' calculations.

One can observe an initial decrease in 1992, followed by the simultaneous growth and contraction of sectors and the reallocation of output between sectors. The largest structural change in the sectors occurred in 1993. Since 1995, a clear growth trend has been observed, with only some inter-sectoral structural change occurring before 1997. The financial crisis in Russia in 1998 induced another wave of structural change.

Next, we consider structural changes across manufacturing industries. From Table 5.14 we can conclude that, during the transition years, the most significant fall in shares of total output has occurred in food products, leather products and chemicals. The most rapid growth is associated with industries such as wood products, paper, printing and publishing, rubber and plastic products, metals, and production of furniture. In general, one can conclude that the greatest increase in real output occurred in resource-intensive, mainly export-oriented industries.

5.2   Concentrated and Dispersed Industries

We start our analysis of geographical concentration by considering regional employment shares[13] by industries and geographic concentration rates[14] according to the NACE broad industrial classifications. Next, we calculate three generalized indices for 1990–2000. Absolute geographical concentration of industries is captured by the Herfindahl index.[15] The Dissimilarity index[16] and

Table 5.14 Cumulative output indices for Estonian industry (in constant 1991 prices)

NACE	Category	1992	1993	1994	1995	1996	1997	1998	1999
D	Total manufacturing	62.00	50.50	48.90	50.30	51.40	60.9	64.40	62.8
DA	Food, beverage and tobacco	73.60	57.90	51.90	50.20	41.60	53.9	51.30	41.2
DB	Textiles and textile products	64.35	49.74	46.29	50.25	57.43	63.62	65.79	67.56
DC	Leather and leather products	63.30	36.70	32.70	31.10	29.00	33.1	38.90	39.9
DD	Wood products	70.50	84.50	122.40	149.50	207.40	283.8	347.30	427.6
DE	Paper, printing and publishing	170.90	628.40	637.57	598.00	481.32	501.20	587.58	698.6
DG	Chemicals, products, fibres	54.90	36.00	40.80	44.20	43.10	43.3	36.20	34.2
DH	Rubber and plastic products	37.40	34.20	59.10	64.10	79.40	120	137.70	127.6
DI	Mineral materials and products	52.70	44.70	46.10	41.90	40.80	54.2	61.00	51.9
DJ	Basic metals and fab. products	57.00	87.70	109.30	123.70	141.60	179.1	223.9	212.7
DK	Machinery, excluding electrical	65.10	75.30	80.80	89.90	86.60	87.5	91.90	87.8
DL	Electrical and optical equipment	37.83	26.54	19.30	18.57	16.83	19.74	23.79	30.83
DM	Transport equipment	67.00	91.70	64.80	53.20	52.30	61.8	96.25	98.04
DN	Other manufactured products	59.80	55.20	68.40	76.50	87.40	109.7	117.10	120.2

Source: *Statistical Yearbook 1999* (1999), pp. 252–3.

GINI coefficients[17] measure relative geographical concentration. The values of the indices obtained are presented in the Appendix (Tables 5.A4–5.A6).

One can observe that the most concentrated industries are the manufacture of paper, publishing, printing, chemicals and chemical products, vehicles, electrical machinery, and optical instruments. In most cases, concentration was driven by industry-specific production needs such as economies of scale and the demand for trained and educated labor, which induced companies to locate close to industrial centers.

The least concentrated industries are manufacturing of food products, beverages and tobacco products, textiles and apparel, mineral products, and furniture. These industries traditionally locate close to production resources and need relatively cheap labor with average skills.

We base our analysis of geographical concentration dynamics on the percentage change of the geographic concentration index. The calculated results are presented in Tables 5.15–5.17. Next, we consider the development of concentration by industries.

Manufacture of food products, beverages and tobacco products (DA) is one of the biggest export-oriented manufacturing industries (it accounts for 28.7 per cent of total industrial output; 33.4 per cent of output in this industry was exported in 1999). Its employment's share in total industrial employment has risen remarkably, with only a slight correction occurring in 1998–99 due to the financial crisis in Russia (the biggest export market for this commodity group). The regional location pattern of this industry has noticeably changed. The share of employment in the northeastern region has skyrocketed as the region became a gateway to the Russian market. The share in the southern region has somewhat increased, and the concentration of this industry in other regions has declined altogether. An analysis of absolute concentration indicates an increase in industry concentration in the northeastern region, and declining rates of concentration in the rest of the country. The largest shares of employment are in the northern and northeastern regions (the northern region is a gateway to western markets and has a powerful demand structure).

The general cumulative dynamics of the concentration indices show decreases in both absolute and relative concentration of this industry, evidently as a result of companies' decisions to locate production closer to production resources.

Manufacture of textiles and wearing apparel (DB) is also one of the biggest export-oriented industries (it accounts for 11.6 per cent of manufacturing output; 48.6 per cent of the industry's output was exported in 1999). Its employment share in total industrial employment has not varied remarkably. Historically,

Table 5.15 The dynamics of the Herfindahl geographic concentration index (1990=100)

NACE	Category	1991	1992	1993	1994	1995	1996	1997	1998	1999	2000
DA	Food, beverages and tobacco	104.6	100.4	95.8	97.1	89.5	90.8	95.0	91.2	93.7	101.7
DB	Textiles and textile products	101.1	101.1	104.6	101.4	107.8	102.5	100.7	97.5	93.6	84.1
DC	Leather and leather products	100.6	94.1	90.1	90.4	103.7	105.9	107.1	119.3	128.3	97.5
DD	Wood products	87.0	77.4	71.4	72.1	70.1	73.4	70.1	70.4	72.8	69.4
DE	Paper, printing and publishing	96.3	85.4	79.7	72.4	91.6	102.3	98.9	97.0	102.3	115.1
DF+ DG	Chemicals, products, fibres	102.8	100.4	102.5	104.3	108.9	112.6	118.1	96.8	74.5	92.5
DH	Rubber and plastic products	100.7	93.6	97.4	77.2	57.1	55.2	50.3	44.8	47.1	42.2
DI	Mineral materials and products	102.5	107.6	103.5	110.5	82.8	76.8	104.5	77.4	100.3	104.8
DJ	Basic metals and fab. products	100.0	93.8	99.6	99.1	110.7	99.4	55.2	53.5	58.0	55.2
DK	Machinery excluding electrical	98.6	97.6	102.0	102.7	110.5	112.8	113.2	112.5	123.3	104.7
DL	Electrical and optical equipment	98.1	98.9	100.0	98.3	75.7	91.4	119.0	117.1	124.5	119.4
DM	Transport equipment	96.2	99.1	106.2	111.0	106.5	111.2	93.3	86.3	79.9	70.0
DN	Other manufact. products	94.4	87.5	86.6	88.2	95.3	87.2	83.8	77.9	72.6	75.4
Average Manufacturing		99.8	97.1	96.8	95.9	94.0	93.8	92.9	87.6	87.8	86.7

this industry has been concentrated in the northern and northeastern regions. There has been an increase in relative concentration of this industry in the northern and northeastern regions (only until 1996 in the latter).

The generalized indices of relative concentration show that relative concentration declined until 1996, followed by a stable increase. An important role in this can be assigned to investment into the industry, which boosted a rise in productivity.

Tanning and dressing of leather and manufacture of footwear (DC) is another export-oriented industry. The industry's share in total manufacturing employment has declined. The largest employment shares are in the northern, southern and northeastern regions. An analysis of concentration rates shows a decline in relative concentration in the northeastern region; no clear trend is evident in the two other regions.

General indices show increases in both absolute and relative concentration over the period considered. This reflects the development of a network of sub-contract work with western partners in this industry.

Manufacture of wood (DD) is the biggest export-oriented industry and one of the biggest industries overall, with a constantly growing share of employment (it accounts for 10.7 per cent of manufacturing output; 54.1 per cent of this industry's output was exported in 1999). Its geographical structure has changed remarkably during the transition period. At first, the largest employment share was in northeastern, southern and northern Estonia. However, the northeastern region's share has more than halved. At the same time, central and western Estonia's shares have significantly increased, which indicates a move closer to production resources. Geographic concentration rates by region follow the same dynamics. The largest decline in relative concentration occurred in the northeastern region; the maximum growth occurred in the central region.

Generalized indices show declines in both absolute and relative concentration in the period considered.

The manufacture of paper and paper products (DE) industry is highly concentrated, with the lion's share of its employment in the northern and southern regions. Over the years, the share of employment in the central region has increased, which also reflects a tendency to locate closer to production resources. The dynamics of geographic concentration rates show an increase in the relative concentration of the industry in northern and central Estonia.

General indices show that both absolute and relative concentration has risen overall, although until 1995 both had been declining due to a notable fall in production in this industry.

The share of employment in the manufacture of fuels, chemicals and chemical products (DF+DG) has steadily declined, indicating a decrease in its relevance in general. This is another industry with very high concentration rates. The largest share of employment is in northeastern and northern Estonia. The structure of employment in this industry has been relatively stable. An analysis of the rates of geographic concentration reveals growth in concentration in northeastern Estonia, despite a general decline in concentration over last three years. This trend is captured by the dynamics of the Herfindahl, GINI and dissimilarity indices. Obviously, it can be attributed to the decrease in exports to Russia for this commodity group.

No remarkable change has occurred in employment shares in the manufacture of rubber and plastic products (DH). Initially, the industry's largest share was in the northern and northeastern regions. However, since 1995, the share of employment has grown in the southern region and has declined in the northern region. The development of geographic concentration rates has followed exactly the same pattern: concentration has increased in the southern region and decreased in northern Estonia. In general, one can observe a decline in both absolute and relative concentration of this industry, which is reflected in the dynamics of Herfindahl, GINI and dissimilarity indices.

Since the beginning of the 1990s, the share of employment in the manufacture of other non-metallic mineral products (DI) has halved. The largest share of employment was in northeastern and northern Estonia, although the share of employment in other regions has grown lately. Geographic concentration rates, as well as the Herfindahl, GINI and dissimilarity indices, have fluctuated loosely.

There has been a decline in employment in the manufacture of fabricated metal products (DJ) industry in the northern region and an increase in employment in the northeastern region; these regions account for most of the employment in the industry. These dynamics coincide with the development of relative concentration. General indices of both absolute and relative concentration show an increase until 1996, followed by a subsequent decrease.

The share of employment in the manufacture of machinery and equipment (DK) industry has decreased substantially, although this could be attributed to productivity growth due to vast investment in the industry. The largest employment share is in the northern region, and this share has steadily increased. The share of employment in the northeastern region has risen as well. This is reflected in an increase in the relative concentration of the industry in these regions (measured by geographic concentration rates). General indices of both absolute and relative concentration show a growth trend.

The manufacture of electrical machinery (DL) industry has the largest share in manufacturing and in Estonian exports (it accounts for 5.9 per cent of manufacturing output; 66.8 per cent of the industry's output was exported in 1999). Due to its high productivity, the industry's employment share is relatively small. The industry is highly concentrated. The largest share of employment is in the northern and northeastern regions. The share in the northern region has grown due to voluminous foreign investment; the level of relative concentration of manufacturing in this region has also grown. Measured by general indices, the levels of both absolute and relative concentration of the industry have increased, mostly due to relocation into the northern region.

The manufacture of motor vehicles and transport equipment (DM) is highly concentrated, with the largest employment share in the northern region. However, during the last two years, the southern region's share has grown somewhat, and the northern region's share has decreased respectively. This is reflected by the dynamics of relative concentration in those regions. Until 1997, both Herfindahl and dissimilarity indices showed a tendency towards increase in absolute and relative concentration, but this was followed by a decline in concentration due to the increased share in the southern region.

The manufacture of furniture and other manufactured goods (DN) industry is one of the most important in Estonian exports. It is quite dispersed geographically, since companies tend to locate close to production resources. The largest shares of employment are in the northern, southern and northeastern regions, though in recent years, the shares in other regions have grown as well. Relative concentration has increased somewhat in the western region; no remarkable change can be noticed for the rest of the country. General indices show a decrease in absolute concentration. Indices of relative concentration fluctuate, so no clear tendency can therefore be distinguished.

Overall, the most rapid increase in geographical concentration of industries has occurred in the manufacturing of electrical machinery and optics; paper and paper products; publishing and printing; machinery and equipment; tanning and dressing of leather; and the manufacture of footwear. This can be attributed to the investments into these industries that were directed primarily to the northern region. Noticeable geographic diversification relative to other industries has occurred in the manufacturing of food, beverages and tobacco products; wood, fabricated metal products, and in the manufacture of rubber and plastic products. The dynamics of the first two industries can be explained

Table 5.16 The dynamics of the geographic dissimilarity index (1990 = 100)

NACE	Category	1991	1992	1993	1994	1995	1996	1997	1998	1999	2000
DA	Food, beverages and tobacco	86.9	91.7	86.4	77.6	116.8	101.3	73.3	51.5	47.2	52.0
DB	Textiles and textile products	106.2	102.3	69.4	68.6	76.7	72.9	88.8	97.3	104.3	122.1
DC	Leather and leather products	92.7	104.2	103.9	96.9	197.7	242.5	194.2	196.9	179.5	170.7
DD	Wood products	103.8	83.9	90.7	83.9	65.4	74.7	82.5	85.3	89.4	75.3
DE	Paper, printing and publishing	95.0	81.5	79.0	73.9	127.8	135.0	145.7	132.6	128.8	153.5
DF+											
DG	Chemicals, products, fibres	104.4	100.4	105.3	107.1	118.8	123.3	123.3	108.9	70.4	97.6
DH	Rubber and plastic products	100.9	92.5	102.0	78.4	62.3	60.5	64.4	56.7	58.1	81.2
DI	Mineral materials and products	99.0	80.1	54.9	80.4	154.2	90.5	246.7	100.7	56.9	80.1
DJ	Basic metals and fab. products	102.1	97.4	111.8	117.9	157.7	138.0	51.1	33.3	42.3	45.7
DK	Machinery, excluding electrical	96.9	101.3	101.3	102.8	87.8	96.9	84.7	98.1	118.8	72.5
DL	Electrical and optical equipment	93.5	94.6	104.0	109.9	86.8	133.6	201.3	191.9	186.8	186.0
DM	Transport equipment	96.8	103.3	111.8	120.3	126.9	131.9	120.4	108.8	91.8	81.6
DN	Other manufactured products	81.7	91.7	105.0	85.6	127.2	91.7	138.3	93.9	111.7	83.9
Average											
Manufacturing		96.3	95.3	90.1	86.9	106.7	100.5	107.0	92.4	92.1	93.7

Source: Authors' calculations based on REGSTAT.

Table 5.17 The dynamics of the Gini concentration index (1990 = 100)

NACE	Category	1991	1992	1993	1994	1995	1996	1997	1998	1999	2000
DA	Food, beverages and tobacco	93.7	97.9	195.8	82.7	113.5	107.2	82.3	62.0	59.5	107.2
DB	Textiles and textile products	104.3	101.9	338.5	84.5	84.5	87.6	88.2	112.4	109.9	87.6
DC	Leather and leather products	103.4	106.5	245.3	120.7	222.8	240.9	186.2	157.3	149.6	240.9
DD	Wood products	91.6	66.6	122.5	84.1	71.9	75.0	84.7	88.1	89.4	28.4
DE	Paper, printing and publishing	100.7	87.6	215.6	74.3	125.4	131.9	157.3	147.9	139.4	131.9
DF+											
DG	Chemical products, fibres	97.9	96.1	97.2	100.4	80.6	88.3	84.7	78.4	73.3	88.3
DH	Rubber and plastic products	106.1	110.5	205.2	71.4	57.1	99.4	67.1	77.0	80.8	99.4
DI	Mineral materials and products	106.3	101.6	300.5	104.2	134.4	103.2	201.6	105.3	78.3	103.2
DJ	Basic metals and fab. products	100.5	69.3	177.0	83.6	105.6	89.5	50.4	30.4	42.2	89.5
DK	Machinery, excluding electrical	98.5	102.3	217.2	99.6	100.0	115.7	83.1	81.6	112.3	115.7
DL	Electrical and optical equipment	98.0	99.5	166.3	88.5	54.0	84.6	116.6	103.9	104.6	84.6
DM	Transport equipment	88.2	99.5	131.9	114.1	118.0	120.3	100.2	94.1	88.7	120.3
DN	Other manufact. products	83.1	86.5	374.3	98.0	97.3	61.5	89.9	73.6	79.7	61.5
Average Manufacturing		97.9	96.3	236.2	90.6	98.6	95.9	94.7	87.1	87.3	87.1

Source: Authors' calculations based on REGSTAT.

by the trend towards relocation closer to production resources, among other factors.

To analyse the dynamics of industrial concentration in general, we have calculated the average changes in the three indices considered.[18] The results are presented in Tables 5.15–5.17. As can be observed, each of the indices captures the tendency towards decline in concentration. To evaluate this trend, we have calculated the average changes in every index.[19] Dividing these parameters by the number of years in the period considered, we obtain an average rate of change in industrial concentration. For the Herfindahl index, it is −0.69 per cent; the regional dissimilarity index shows a decrease of −0.63 per cent, the locational GINI coefficient is −0.73 per cent (we do not take the year 1993 into account when systematic deviation can be observed since this deviation is obviously a result of error in the initial data). Altogether, the level of geographical concentration of manufacturing in Estonia has decreased annually, on average by 0.63 to 0.73 per cent.

6 The Impact of Economic Integration on the Regional Wage Structure

6.1 Model Specification

The starting point for an analysis of the impact of economic integration on the regional structure is the assumption that, due to integration, the share of the costs of transportation to industry centers in the total production costs becomes less significant. In an empirical analysis, such transportation costs can be captured by the distance between the regional capital and the central capital city. This proxy is quite fair in case of Estonia, where the quality of the infrastructure is relatively uniform and does not vary with distance from the central region. The relocation of production is reflected in the regional wage structure. In our model, we use industrial wage differentials (regional wages related to wages in Tallinn) as a proxy for industrial relocation. Wages are constructed from data from the Estonian Statistical Office. Estonia includes 15 counties, including the capital county. The wage variable we use in calculations is the average annual remuneration per employee in industry in county j in year t. Complete data is available for the years 1990–2000. By employing relative wages, we operate with a complete data set of 126 observations (nine years x 14 counties – omitting the capital county and adjusting the endpoints).

The second factor we study within this model is the effect of integration and distance on western border regions.[20] We introduce border dummies to capture time and distance-invariant factors that are specific to western border regions and that may explain the dynamics of wage differentials. One such factor might be the possibility of cross-border co-operation and work.

Within the framework of this model, we test several hypotheses, following Hanson (1994).

The main proposition is formulated as follows: before trade liberalization, *regional relative wages decrease with distance from the capital city.* We specify the model below:

$$\log (WAGE_{jt}/WAGE_{ct}) = \alpha + \beta_t \log(DIST_j) + \lambda_t (\log DISTj \times BORDj)$$
$$+ \varepsilon_{jt} \tag{6.1}$$

where:
$WAGE_j$ = wage in county j;
$WAGE_c$ = wage in the capital city;
$DISTj$ = distance between county i (county capital) and capital city;
$BORD$ = dummy variable for western border regions. $BORD_j$ is one if county j is a western border region, and zero otherwise.

Next, the hypothesis to be tested is the following: *trade liberalization eliminates distance effects.* Using the notation shown above, we re-specify the model:

$$\log (WAGE_{jt}/WAGE_{ct}) = \alpha + \beta_t \log(DIST_j) + \lambda t (\log DIST_j \times BORD_j)$$
$$+$$

$$+ \mu_t (\log DISTj \times YEAR) + \varepsilon_{jt} \tag{6.2}$$

where:
YEAR is a dummy variable for the years after the entering into force of the trade liberalization agreements.

In Estonia, the first significant step towards liberalization of trade with the EU was made in 1995 when the Free Trade Agreements with the EU entered into force. In 1998, the EU Association Agreements entered into force. We take both events into account and introduce impact YEAR dummies for 1995 and 1998.

Finally, we test the hypothesis that: *after the entering into force of the EU agreements, distance effects in western border regions and in other regions converge to similar levels.* Our basic model is specified as follows:

$$\log (WAGE_{jt}/WAGE_{ct}) = \alpha + \beta_t \log(DIST_j) + \lambda t \ (\log DISTj \times BORDj) +$$

$$+ \mu_t \ (\log DIST \times YEAR) + \nu_t \ (\log DIST \times BORD \times YEAR) + \varepsilon_{jt} \ (6.3)$$

If this hypothesis is confirmed, the following relation for the regression coefficients should hold:

$$\mu_t = \lambda_t + \nu_t$$

This model was also estimated for 1995 and for 1998.

6.2   Econometric Issues

In our data set, the dependent variable to be explained is relative wages across counties and across time. Our basic hypothesis attempts to explain variation in relative wages by variation in transport costs measured as distance from the center. That is, in all regressions, we expect the term β_t to be negative in the pre-liberalization period. One would also like to know if easier access to foreign industry centers (in border counties) eliminates the dependency of relative wage variation on the distance from the center. Following Hanson (1994), we test this hypothesis by allowing distance effects for border counties to differ from those for interior counties; that is, in regression (6.1) we expect $\lambda_t = 0$ for border counties.

In our basic regressions, the distance variable will not only capture transportation costs, but will also pick up any other county-specific effects that are correlated with distance from the centre. Therefore, basic regressions contain an error term that captures some regional characteristic components:

$$\varepsilon_{jt} = \omega_i + \zeta_{i,}$$

where ω_i is the effect for county i, and ζ_i is a random variable with mean zero and variance σ^2. If county effects are random, we can use OLS for estimation. In our case, independent variables in the panel regression (distance) do not vary across groups of observations (time). Therefore, the OLS estimates are efficient even if ε_{jt} is correlated across counties.

If county effects are fixed instead of random, we need to look for another solution to stationarity. If we use the general method of first differencing the data, this will eliminate the distance variable. Introducing fixed effects into the model would lead to perfect multicollinearity.

To determine what proportion of fixed county-specific effects are related to distance, we regress wage differentials on county-specific dummies. The estimated county effects represent the mean of regional characteristics on relative wages. We then regress the estimated dummies on distance to determine what proportion of the variance in relative wages that is related to county characteristics is in fact accounted for by distance (a share of relative wage variance across the counties that is determined by the distance).

6.3   Empirical Results

Estimation results are presented in Table 5.18.

As can be observed from the table, the estimation results confirm the hypothesis that relative wages decrease with distance from the capital city. In all regressions, the log distance to Tallinn is negative ($\beta_t = -0.03$) and significant at the 0.01 level. However, the quantitative effect is small, which is reasonable due to the small size of Estonia. We also find strong evidence that trade liberalization eliminates the distance effect: in all regressions except one we fail to reject the null hypothesis that $\log(DIST_j)$ has been zero since 1995.

The distance variable remains significant even if controlled for the border states, though its quantitative impact is rather small and close to zero. One should notice that due to the specific definition of border counties (counties with marine access), not all trade effects could be captured. Therefore, counties that are relatively distant from Tallinn (Pärnu on the southwest and Ida-Virumaa on the east) also have good access to the Baltic market and the Russian market, respectively. This explains the sign of the coefficient (0.007).

Trade liberalization had a positive effect on border regions. In the regressions for 1995 and 1998 we fail to reject the null hypothesis that logDIST x BORD x YEAR is zero at the 0.01 confidence level. There is also evidence to support the hypothesis that, due to trade liberalization, distance effects in border regions and in the other counties converge to the same level. Distance effects prior to liberalization have the coefficient of -0.03, whereas after the entering into force of the Free Trade Agreements, the distance effect fell close to zero both in border counties and in the other regions.

The models described above explain 20–25 per cent of the variation of the dependent variable. To determine county-specific effects, we regress relative wages on county dummies. Table 5.19 shows results from regressing log relative wages on county dummies. The estimated county effects for all border regions are positive; four are significant at the 0.05 level. Replacing distance variables with county dummies increases the adjusted R-squared to 0.78. Distance appears to account for about a third of the explained variance in regional relative wages that is attributable to county characteristics. To verify this, we regress the estimated county effects on the distance variables, again controlling for border counties. The resulting regression is:

$$\tau_i = 0.15 - 0.029\log(DIST) + 0.008\log(DIST)*BORD, \quad N=154, \quad \text{R-squared}=0.42$$
$$(0.027)* \quad (0.05)* \qquad\qquad (0.001)*$$

where standard errors are in parentheses and τ_i is the estimated county effect for county i. Distance accounts for 42 per cent of the variance in regional relative wages that is county-specific. Therefore, distance is relatively important. To ensure border counties do not affect the results, we regress county effects for non-border counties on distance:

$$\psi i, = 0.86 - 0.04\log(DIST), \quad N=72, \quad \text{R-squared} = 0.22$$
$$(0.048)* \quad (0.009)*$$

where $\psi_{i,}$ indicates county effects for non-border counties. The coefficient estimates are consistent with those above, but the explained variance is lower.

7 Regional Specialization and Growth

7.1 Model Specification

In this section we attempt to determine the regional factors that account for regional diversification and economic growth. Until now, the process of integration in Europe has been characterized by a twofold effect on geographical distribution of income. Amiti (1998) and Hallet (1997) notice that while incomes in the EU tend to converge on a national level, regional inequalities within the member countries are widening.

Conventional growth theory has provided an analytical framework that underlined the differences in productivity across regions and predicted convergence of regional income levels in the long run. However, a number of

Table 5.18 Estimation of the relationship between regional relative wages and distance (proxy of transportation cost; values of standard errors are given in parentheses)

Variable	Model 1(a)	Model 1(b)	Model 2 (for year 1995)	Model 2 (for year 1998)	Model 3 (for year 1995)	Model 3 (for year 1998)
Intercept	0.8 (0.05)	0.81 (0.05)	0.8 (0.05)	0.8 (0.05)	0.8 (0.05)	0.8 (0.05)
LogDIST	-0.03* (0.01)	-0.03* (0.01)	-0.03* (0.01)	-0.03* (0.01)	-0.03* (0.01)	-0.03* (0.01)
LogDISTxBORD	0.007* (0.002)	0.008* (0.002)	0.008* (0.002)	0.008* (0.002)	0.009* (0.003)	0.008* (0.002)
LogDISTxYEAR			0.003 (0.002)	-0.005* (0.002)	0.004 (0.002)	-0.002 (0.002)
LogDIST x BORD x YEAR					-0.002 (0.004)	-0.007** (0.004)
Adjusted R-squared	0.19	0.25	0.20	0.23	0.19	0.25
F-statistic	15.54	11.42	11.19	13.64	8.46	11.53
Prob(F-statistic)	0.00	0.00	0.00	0.00	0.00	0.00
Number of observations	126	126	126	126	126	126
Year dummies	No	Yes	No	No	No	No

Table 5.19   Fixed effects estimation of county regression

County	Coefficient	Standard error	County	Coefficient	Standard error
Harju^	0.327**	0.023	Pärnu^	0.032	0.023
Hiiu^	0.054***	0.023	Rapla	0.056	0.023
Iida–Viru^	0.097**	0.023	Saare^	0.030	0.023
Jõgeva	−0.027	0.023	Tartu	0.049***	0.024
Järva	0.022	0.023	Valga	0.008	0.024
(Lääne^	0.013	0.023	Viljandi	−0.007	0.024
Lääne–Viru^	0.056**	0.023	Võru	−0.043**	0.023
Põlva	−0.004	0.023	Constant	0.641	0.018

** (***) Indicates significance at the 0.05 (0.01) level. ^ Indicates border county.

empirical studies conducted in Europe at various times confirmed the opposite result, and demand for a new explanation arose (see, for example, Karsten, 1996; Aiginger, 1999; Haaland et al., 1999).

A different approach was suggested by the new economic geography theorists. Their models were based on the explicit assumptions of monopolistic competition and increasing returns to scale. Under these assumptions, new theoretical predictions about the development of industrial production patterns have been proposed.

According to the new explanations, the main sources of regional variation in growth are infrastructure that minimizes transportation costs and good quality human capital, along with the resulting high labor productivity in a region. However, various researchers have frequently pointed out the difficulty of verifying empirically the assumption outlined above (e.g., Venables, 1998; Krugman, 1998).

Nonetheless, some attempts have been made. Following one of them, proposed in Hanson (1994), we look at a wide range of regional indicators that are proxies for the quality of infrastructure and labor. One would also like to know if any regional convergence has occurred during the process of transition. The main analytical model is specified below:

$$\log (y_{j,t+1}/y_{jt}) = \alpha + \beta_t \log(SPEC_{jt}) + \sum_{i-1}^{10} \gamma_{jt} Xi_{jt} + \varepsilon_{jt} \qquad (7.1)$$

where:

$y_{j,t}$ = regional GDP in year t in region j;
SPEC = specialization measure in year t in region j;
X_{ij} = structural variables for qualitative regional characteristics;
$X1j$ = the number of firms with foreign capital per 100,000 inhabitants;
$X2j$ = the number of self-employed per 100,000 inhabitants;
$X3j$ = the number of students per 100,000 inhabitants;
$X4j$ = the number of telephone lines per 100,000 inhabitants;
$X5j$ = the density of the network of roads;
$X6j$ = public expenditure per capita;
$X7j$ = the percentage of the population in the age group;
x_{8j} = the share of employment in the secondary sector;
x_{9j} = the share of employment in services.

Although the specifications presented above would provide a fairly good explanation of the process of regional growth, we had to adjust it for the data available. Statistics for regional GDP are available only for 1996–98. Using these data would not allow us to take advantage of the fairly wide set of regional indicators that have been provided since 1992. Therefore, we calculated a proxy for regional GDP as an aggregated index including major GDP components measured by the outflows method:

$$g_{j,t} = 0.5 * \text{Icons}_{j,t} + 0.2 * \text{Iinv}_{j,t} + 0.2 * \text{Igov}_{j,t} + 0.1 * \text{Iexp}_{j,t}$$

where:
I indicates growth indices, $g_{j,t} = y_{j,t}/y_{jt-1}$;
$\text{Icons}_{j,t}$ = index of regional disposable income in the year t;
$\text{Iinv}_{j,t}$ = index of investments regional investment in fixed assets in year t;
$\text{Igov}_{j,t}$ = index of expenditures in local budgets in year t;
$\text{Iexp}_{j,t}$ = index of special export without re-export in year t;

Weights were taken from the national account tables as 1995–2000 averages. Export weight was estimated by the authors so that all estimation coefficients were normalized. The index gives a proxy of regional GDP growth from 1995–2000. As can be seen from Table 5.20, these regional indicators are fairly consistent with the annual growth rates of the total Estonian GDP. Table 5.20 presents the calculated indexes of regional economic growth.

Table 5.20 Composition and dynamics of the index of regional GDP growth

Region	1995	1996	1997	1998	1999	2000
Northern Estonia	1.144	1.059	1.154	1.116	0.998	1.103
Central Estonia	1.168	1.048	1.120	1.089	1.031	1.063
Northeastern Estonia	1.045	1.027	1.087	1.057	0.989	1.062
Western Estonia	1.128	1.090	1.127	1.081	0.975	1.091
Southern Estonia	1.110	1.007	1.143	1.117	1.031	1.088
Estonia in total	1.119	1.046	1.126	1.092	1.005	1.081
GDP average growth rate	1.04	1.04	1.1	1.05	0.993	1.06

Source: Authors' calculations based on REGSTAT.

The measures of specialization in the basic model are the Herfindahl, Krugman (dissimilarity) and GINI specialization indices at the NUTS 3 level for Estonia from 1995–2000. Therefore, we had 25 observations in the panel data altogether.

7.2   Econometric Issues

In the basic model, all regional variables are taken in value per capita. One might expect an initially high level of multicollinearity of data. Therefore, we omitted those variables that were highly correlated. Next, we tested the series for autocorrelation. Correlogram specification determined that most of the variables were integrated. Eventually, we chose to estimate the model in first differences.

We tested the models both with fixed effects (region-specific) and with common intercepts (with regional weights). Our preference between the common intercept and fixed effects models was determined on the basis of F-statistics according to the hypothesis that all intercepts are equal.

7.3   Estimation Results

Table 5.21 presents estimation results for regional growth regressed on specialization measures and the group of regional indicators. It is important to note that, out of whole set of regional parameters, only a few of them were statistically significant in all three specifications.

Most of the models present three statistically significant variables that can be considered as proxies for factors of economic growth. The number of telephone lines in a region may be an indicator of the state of regional infrastructure. This measure is statistically significant in five out of the six models. The share of employment in industry in a region can stand for the level of regional industrial development. This indicator is significant in all of the specified models. Absolute (Herfindahl) specialization and dissimilarity indices are also highly significant.

As can be seen in the table, for the Herfindahl and dissimilarity indices, regression results show a statistically significant positive relationship between regional specialization and regional growth. In terms of regional convergence, these results can be interpreted as follows. First, during the observed period positive specialization dynamics have occurred in the Northern and Southern regions, indicating a relocation of production into the regions with lower costs for immobile production factors. At the same time, specialization in poorer agricultural regions has decreased (see Appendix). As empirical estimation shows, the increase in specialization has been accompanied by higher growth rates. Therefore, a positive relationship between specialization and growth in the more advanced regions has been dominant. In the specific case of Estonia, this may indicate widening regional disparities in recent years.

Positive changes in the variables that indicate levels of industrial development (share of industrial employment) and improvement in infrastructure (increase in number of telephone lines) have shown that there has been a generally positive impact on regional growth. Quantitative effects are displayed in Table 5.21.

8 Regional Winners and Losers and Policy Implications

In this section we define the winning and losing regions in the process of Estonia's integration with the EU. This exercise is quite subjective, since the variables with which we operate and the level and growth rate of the GDP do not capture the entire pattern of regional development. Due to the size of the Estonian economy, even one significant investment into a peripheral region can lead to greater increases in GDP than in living standards.

The scarcity of regional databases requires that we find proxies for the level of living standards. Next, we evaluate the dynamics of the level of personal income and the unemployment rate and determine their relationship with the level of industrial specialization in a given region.

Table 5.21   **Estimation results of the model of mutual dependence of regional specialization and growth (model is estimated in first differences, values of standard errors are given in parentheses)**

Model	Model 1 (Herfindahl)		Model 2 (GINI)		Model 3 (Dissimilarity)	
	GLS	FEM	GLS	FEM	GLS	FEM
Method of estimation	GLS	FEM	GLS	FEM	GLS	FEM
Constant	−0.04* (0.02)		−0.05 (0.02)		−0.05* (0.02)	
Specialization index	1.46*** (0.49)	2.81* (1.51)	2.81* (0.27)	0.33 (0.41)	0.34** (0.15)	0.41 (0.28)
TEL/POP	0.63** (0.32	1.06* (0.54)	0.71* (0.38)	1.18* (0.58)	0.7*** (0.39)	0.7*** (0.56)
EMP/POP	0.87*** (0.13)	0.86*** (0.31)	0.86*** (0.14)	0.77** (0.32)	0.92*** (0.17)	0.77** (0.31)
Adjusted R-squared	0.82	0.41	0.81	0.31	0.76	0.37
F-statistic Probability	37.1	11.78	35.9	9.0	26.2	10.5
F-statistic	0.000	0.0006	0.000	0.002	0.000	0.001
Observations	25	25	25	25	25	25

*, ** and *** denote coefficient estimates significant at 10, 5 and 1 per cent levels.

The results of a previous analysis show that the level of industrial specialization and the structure of industrial employment have changed significantly during the years of transition. Relocation of manufacturing activity between regions can be noted. Table 5.22 illustrates the evolution of the regional shares in industrial employment by region. There has been a decrease in industrial employment in the biggest industrial region (northern) as well as a significant increase in the industrial employment share in the peripheral regions (central and western Estonia). The most significant changes in regional structure of industrial employment have occurred since 1995. In general, this signifies that some relocation of industrial activity into the peripheral regions has occurred, driven by the development of infrastructure and FDI inflow. The analysis displayed in Table 5.23 also supports these conclusions. We look at the ratios of manufacturing shares to overall population shares. Higher values indicate a very homogenous spread of manufacturing across the population.

Table 5.22 Evolution of the regional shares of manufacturing employment by regions

Year	1991	1992	1993	1994	1995	1996	1997	1998	1999
Northern Estonia	0.99	0.96	0.96	0.93	0.82	0.82	0.75	0.78	0.86
Central Estonia	0.97	1.01	1.14	1.12	1.36	1.35	1.48	1.55	1.57
Northeastern Estonia	1.02	1.03	1.01	1.01	1.06	1.04	1.13	1.01	0.88
Western Estonia	1.04	1.14	1.25	1.32	1.48	1.47	1.69	1.69	1.48
Southern Estonia	0.99	0.99	0.94	1.00	1.05	1.1	0.99	1.07	1.12

Source: Authors' calculations based on REGSTAT.

Table 5.23 Ratio of manufacturing share to overall population share

Region	1992	1993	1994	1995	1996	1997	1998	1999
Northern Estonia	1.20	1.20	1.17	1.04	1.05	0.96	1.00	1.09
Central Estonia	0.58	0.64	0.63	0.76	0.75	0.82	0.86	0.87
Northeastern Estonia	1.28	1.25	1.26	1.33	1.30	1.42	1.27	1.12
Western Estonia	0.70	0.76	0.80	0.89	0.88	1.00	1.00	0.87
Southern Estonia	0.78	0.74	0.78	0.82	0.86	0.78	0.83	0.87

Source: Authors' calculations based on REGSTAT.

It can be noticed that this coefficient is quite high in industrially developed regions (northern and northeastern Estonia), but has decreased, reflecting the tendency for employment to spread into other sectors of the economy. At the same time, in the less developed regions of central and western Estonia the values of the index have increased.

To analyse the impact of regional specialization and the growth rate on unemployment, we have estimated the following model:

$$\log (u_{j,t}) = \alpha + \beta_t \log(SPEC_{jt}) + \gamma_t \log(growth_{jt}) + \varepsilon_{jt} \qquad (8.1)$$

where:
u is unemployment rate in region j

The model was estimated for both the fixed effects (region-specific) and the common intercept (with regional weights) specifications. Results are presented in Table 5.24.

Estimation results show a statistically significant relationship between the level of specialization and the unemployment rate. Economic growth depends negatively on the level of unemployment; this supports the theoretical predictions.

Next, the model evaluates the impact of specialization and economic growth on regional wages. The model is specified as follows:

$$\log (w_{j,t}) = \alpha + \beta_t \log(SPEC_{jt}) + \gamma_t \log(growth_{jt}) + \varepsilon_{jt}$$

where:
w = real wage (deflated by PPI) in region *j*.

This model was also tested both with fixed effects (region-specific) and with common intercept (with regional weights). Table 5.25 shows the estimation results.

The regions that are the most industrially diverse have the highest wage levels, and the most specialized regions, with small number of industries, have the lowest wages. The model does not estimate a statistically significant relationship between economic growth rates and regional real wage levels.

In spite of the relocation of industrial activity and a slight tendency towards equalization of the regional specialization levels, no noticeable regional convergence is measured by the growth rates of the regional GDP. Therefore, one can distinguish between winning and losing regions in Estonia.

Table 5.24   Estimation of the relationship between real wages and specialization and growth rates (model is estimated in first differences, values of t-statistics are given in parentheses)

Model	Model 1 (Herfindahl)		Model 2 (GINI)		Model 3 (Dissimilarity)	
Method of estimation	OLS	FEM	OLS	FEM	OLS	FEM
Constant	−0.003* (−1.66)	(−2.16)	−0.003**		−0.006 (−5.64)	
Specialization index	0.01** (3.8)	−1.01* (−1.92)	0.007** (4.53)	−0.002 (−0.45)	0.008** (4.48)	−0.006 (−0.95)
Growth	−0.034 (−1.98)	−0.04** (−2.25)	−0.046** (−3.35)	−0.04* (−1.93)	−0.037** (−2.46)	−0.039* (−1.87)
Adjusted R-squared	0.81	0.84	0.59	0.74	0.54	0.75
F-statistic Probability	37.99	70.57	13.67	53.15	11.12	56.63
F-statistic	0.00	0.00	0.00	0.00	0.00	0.00
Observations	30	30	30	30	30	30

*, ** and *** denote coefficient estimates significant at the 10, 5 and 1 per cent levels.

An evident winner in the process of economic integration with the EU is the Northern region, which has attracted the lion's share of the total FDI and consequently has the best infrastructure. Industry in the region has developed more rapidly than that in the rest of Estonia. Major factors in the growth are the fairly good infrastructure, access to seaports and location close to the neighboring countries with strongest market potential, as well as good quality of labor.

The region that lost its leading position over the course of transition is the North-Eastern region. Since the breakdown of economic connections to Russia after 1998, the regional industrial structure has undergone drastic changes. As a result, unemployment in the region is the highest in Estonia, personal incomes are low, and no competitive industries are initiating regional growth.

Initial stagnation in the peripheral agricultural regions ended in the mid–90s. In subsequent years, however, regions have experienced a boost in economic activity that was caused primarily by improving infrastructure and the competitive costs of factors of production (mostly labor) compared to the central region. As economic integration with EU speeds up, these regions are acquiring strong growth potential.

Table 5.25 **Estimation of the relationship between real wages and specialization and growth rates (model is estimated in first differences, values of t-statistics are given in parentheses)**

Model	Model 1 (Herfindahl)		Model 2 (GINI)		Model 3 (Dissimilarity)	
Method of estimation	OLS	FEM	OLS	FEM	OLS	FEM
Constant	7.47**		7.48**		7.6***	
	(60.1)		(224.3)		(142)	
Specialization index	−0.41***	−0.06	−0.29	−0.21***	−0.31***	−0.15
	(−3.19)	(−0.31)	(−11.05)	(−4.04)	(−6.97)	(−1.17)
Growth	−0.71	−0.5	−0.45***	−0.37	−0.7	−0.38
	(−0.77)	(−0.4)	(−1.69)	(−1.65)	(−1.24)	(−0.97)
Adjusted R-squared	0.33	0.88	0.98	0.99	0.73	0.89
F-statistic	5.23	129.3	190	133	24.6	143.1
Probability F-statistic	0.01	0.00	0.00	0.00	0.00	0.00
Observations	30	30	30	30	30	30

*, ** and *** denote coefficient estimates significant at 10, 5 and 1 per cent levels.

Overall, Estonian manufacturing has become more flexible in terms of regional allocation. An inverse U-shaped specialization pattern in the northern (capital) region, along with declining specialization levels in the periphery, suggests that producers have been aware of falling transportation costs and have been sensitive to the cost of immobile factors of production. Therefore, our findings show an increase in the interregional allocation of industries, which has developed as a consequence of regional dynamics of main production factors such as wages and infrastructure.

Below, we present some implications for regional policy implementation, which are based on the results obtained in the course of this study. Our conclusions are general, but significant nonetheless:

1) Due to the agglomeration process associated with early transition, the most effective instruments of regional industrial policy are regional transfers. Our research results show that infrastructure has been one of the main determinants of regional growth. Therefore, allocation of funds to finance

the development of regional infrastructure will help to boost growth in peripheral regions.

2) Stimulation of regional investment (and, first of all, of FDI) can be an effective regional policy instrument if applied according to regional specialization. First, such strategies would increase the economic effectiveness of the investments thanks to the various effects that can be considered 'new growth' factors. Among these are learning-by-doing and other concepts related to labor productivity. Also, regional specialization is often accompanied by an extended logistical network and improved infrastructure (when similar businesses are concentrated in one region). In addition, it can improve regional growth, as shown by our estimations of regional growth and specialization.

3) Development of regional infrastructure is the most effective stimulus for industrial companies to relocate into a particular region.

4) The development of regional integration with the EU in recent years has been an important factor for the relocation of manufacturing activity and for regional income convergence.

5) The results of this study suggest that wage dynamics vary in the smaller administrative units. Therefore, regional policy has an impact at the county level, and crucial productive investment has to be directed to the peripheral administrative units.

9 Conclusion

The process of integration with the EU has resulted in the reorientation of Estonian foreign trade flows from Eastern to Western markets and in substantial structural changes. The analysis of the comparative advantage of Estonian goods on the EU market and of share of intra-industry trade enables us to conclude that, as in most transition economies, a core of Estonian exports is constituted by labor- and resource-intensive commodities. As our study shows, trade liberalization, as a part of the general integration process, has significantly influenced the regional development of Estonia in the past decade.

The analysis of industrial specialization in Estonian NUTS 3 level regions has shown that the level of specialization has increased, on average, by 2 to 5 per cent per year. Because for a transition economy, time is a fair proxy to integration, we may conclude that the initial stages of establishing closer economic relations with the EU and voluminous target investments into the regions have stimulated specialization. The overall increase in specialization was supported by the recent shift of economic activity from the northern (central) region to the periphery as a result of improved infrastructure and the persisting wage differential.

However, specialization varied by region. In developed regions (northern and southern Estonia), industrial activity developed in an inverse U-shape, as predicted by the new economic geography hypothesis. The level of specialization has decreased in the agricultural regions (central and western). Therefore, our study reveals a tendency for industrial specialization in Estonia to homogenize across the regions, and suggests that industrial activity has started to relocate from fairly developed regions to the poorly industrialized periphery.

The level of geographical concentration of manufacturing in Estonia has decreased annually, on average by 0.63 to 0.73 per cent. However, dynamics across industries varied greatly. The most rapid increase in geographical concentration has occurred in the manufacturing of electrical machinery and optics; paper and paper products; publishing and printing; machinery and equipment; tanning and dressing of leather; and the manufacturing of footwear. This can be attributed to investments into these industries because they are labor- or resource intensive, and production is relatively cheaper than in the EU. Those investments were directed primarily to the Northern region. Noticeable geographic diversification has occurred in the manufacturing of food, beverages and tobacco products, wood, fabricated metal products, and the manufacturing of rubber and plastic products. The dynamics of the first two industries can be explained by relocation closer to production resources, among the other factors.

An econometric analysis of the relationship between relative regional wages and distance to the capital suggests an explanation consistent with the new economic geography hypothesis. Surprisingly, in spite of the small size of Estonian territory, distance as a proxy of transportation costs has been a significant factor behind variations in regional wages. Our estimates show that integration with the EU and trade liberalization minimize the negative impact of distance. It is also possible to make a distinction between border and internal regions in these terms, since in border regions distance as a proxy for transportation costs is less important.

An econometric analysis of the impact of specialization on economic growth has revealed a strong direct relationship between these two variables on the regional level. Regional growth is also positively influenced by indicators of regional industrial development (the share of industrial employment) and by improvements in infrastructure (an increase in the number of telephone lines).

Northern Estonia (a central region) has undoubtedly benefited from liberalization. Nonetheless, since 1995, growth in peripheral agricultural regions could be partly explained by the positive effects of integration. The northeastern region, which is situated close to the eastern border, has experienced a drastic reduction in economic activity and has undergone the most substantial structural change. This has resulted in deep social problems, but solutions can be found by means of consistent and efficient regional policy.

Notes

*   This research was undertaken with support from the European Community's PHARE ACE Programme 1998. The content of the publication is the sole responsibility of the authors and it in no way represents the views of the Commission or its services.

1   See the draft version of the Estonian National Development Plan provided by the Ministry of Finance, www.fin.ee.

2   The content of the database is described in Chapter 3.

3   See http://www.mineco.ee for details.

4    $FK\ (i) = \sum_i [\min s(i,k), s(i,l)]$

where:

s(i,k) and s(i,l) are export and import shares of sector i in export (import) to (from) markets k and l respectively.

5   It is defined as a ratio of a country's exports of a given commodity group to a country's imports of the same commodity group: $TC_i = X_i/M_i$ where M(i) is the import of commodity i, X(i) is the export of commodity i. If the ratio is greater than one, a country specializes in a given sector and has a comparative advantage in this sector. Since the trade coverage ratio reflects a proportion of exports to imports of the same country, it describes 'internal' comparative advantage versus 'external' comparative advantage in the export markets.

6   Specialization index calculated as follows:

$$SI = \frac{X_i / \sum_i X_i}{M_i^{EU} / \sum_i M_i^{EU}}$$

where:

X(i)   = Estonian export to the EU in commodity group;
M(i)$_{eu}$ = Imports to the EU in commodity group i.

7    $GL_i = 1 - \dfrac{|X_i - M_i|}{(X_i + M_i)}$

where:

M(i) = countries' import of commodity group i;
X(i) = countries' export of commodity group i.

8 Since 2001, the new regional breakdown laid down by Regulation No. 126 of the Government of Estonia has been used. Counties are aggregated as follows:
northern Estonia – Harju county (incl. capital city Tallinn);
central Estonia – Järva, Lääne-Viru, Rapla counties;
northeastern Estonia – Ida-Viru county;
western Estonia – Hiiu, Lääne, Pärnu, Saare counties;
southern Estonia – Jõgeva, Põlva, Tartu, Valga, Viljandi, Võru counties.

9 The index was calculated according to the following formula:

$$H_j^s = \sum_i (S_{ij}^s)^2$$

where s_{ij}^s is the share of employment in industry i in region j in total employment of region j

$$S_{ij}^s = \frac{Eij}{Ej} = \frac{Eij}{\sum_i Eij} \quad ; Eij \quad \text{is the employment in industry } i \text{ in region } j.$$

10 The dissimilarity index for regional specialization is calculated as follows:

$$DSR_j = \sum_i \left| S_{ij}^s - S_i \right|,$$

where:

s_i is the share of total employment in industry i in total employment

$$s_i = \frac{Ei}{E} = \frac{\sum_j Eij}{\sum_i \sum_j Eij}$$

11 Gini coefficients for regional specialization are calculated following Devereux et al. (1999)

$$GINI_j^s = \frac{2}{n^2 R} \left[\sum_{i=1}^n \lambda_i (R_i - \bar{R}) \right]$$

where n is the number of industrial branches;

$$R_i = \frac{S_{ij}^s}{S_i} \quad \text{(for each industry in region j)};$$

$\bar{R}$ is the mean of R_i across industries;

λ_i is the position of the industry i in the ranking of R_i.

12 This indicator was calculated as a geometrical average of changes in indices by years.

13 S_{ij}^c is the share of employment in industry i in region j in total employment of industry i,

$$S_{ij}^c = \frac{Eij}{Ei} = \frac{Eij}{\sum_j Eij} \quad ; Eij \quad \text{is the employment in industry } i \text{ in region } j.$$

14 Rates of geographical concentration are a measure of the relative concentration of a given industry in a region. They were calculated as follows:

$$CR_{ij} = \frac{S_{ij}^c}{S_j}, \text{ where } s_j = \frac{Ej}{E} = \frac{\sum_i Eij}{\sum_i \sum_j Eij}$$

15 The index was calculated according to the following formula:

$$H_i^c = \sum_i (S_{ij}^c)^2$$

16   The dissimilarity index for geographical concentration is calculated as follows:

$$DCR_i = \sum_{j} | s_{ij}^c - s_j |$$

where s_j = the share of total employment in region j in total employment.

17   Gini coefficients for geographical concentration are calculated as follows:

$$GINI_i^c = \frac{2}{m^2 \bar{C}} \left[\sum_{j=1}^{m} \lambda_j (C_j - \bar{C}) \right]$$

where:
m is the number of regions; $C_j \cdot \dfrac{s_{ij}^c}{s_j^c}$;

$\bar{C}$ is the mean of C_j across regions, λ_j is the position of the region in the ranking of C_j.

18   Calculated as a weighted average of changes across the industries using employment shares as weights.

19   This indicator has been calculated as a geometric average of changes of the indices by years.

20   In the case of Estonia, the main border with the EU is marine. Another important factor to mention is that the biggest border region is Harju county, which includes the capital city.

References

Aiginger, K. (1999), 'Trends in the Specialisation of Countries and the Regional Concentration of Industries: a Survey on Empirical Literature', WIFO-Working Papers.

Amiti, M. (1998), 'New Trade Theories and Industrial Location in the EU: A Survey of Evidence', *Oxford Review of Economic Policy*, Vol. 14 (2), pp. 45–53.

Devereux, M., Griffith, R. and Simpson, H. (1999), 'The Geographic Distribution of Productive Activity in the UK', Institute for Fiscal Studies, Working Paper, W99/26.

Fainshtein, G. (1998), 'Development of Estonian Foreign Trade Flows in 1992–96', Preprint 52, Estonian Institute of Economics, Tallinn.

Fainstein, G. and Lubenets, N. (2001), 'Development of Estonian Trade Flows, Comparative Advantage and Intra-industry Trade in the EU Markets', in U. Ennuste (ed.), *Factors of Convergence*, EMI, Tallinn, pp. 83–105.

Greenaway, D., Hine, R.C. and Milner, C.R. (1994), 'Country-specific Factors and the Pattern of Horizontal and Vertical Intra-industry Trade in the UK', *Weltwirtschaftliches Archiv*, Vol. 130, pp. 77–100.

Grubel, H.G. and Lloyd, P.J. (1975), *Intra-industry Trade. The Theory and Measurement of International Trade in Different Products*, Macmillan, London.

Haaland, J.I., Kind, H.J., Knarvik, K.H. and Torstensson, J. (1999), 'What Determines the Economic Geography of Europe', CEPR Discussion Paper, no 2072.

Hanson, G. (1994), 'Localization Economies, Vertical Organization, and Trade', *NBER Working Paper No. 141*.

Hansson, Ardo H. (1995), 'Macroeconomic Stabilization in the Baltic States', Stockholm Institute of East European Economies, Working Paper No. 108, Stockholm.

Karsten, J. (1996), 'Economic Development and Industrial Concentration; An Inverted U-curve', Kiel Working Paper no. 770, Kiel Institute for World Economics, Kiel.

Krugman, P. (1998), 'What's New About the New Economic Geography?', *Oxford Review of Economic Policy*, Vol. 14, No. 2, pp. 7–17.

Repkine, A and Walsh, P. (1999), 'Evidence of European Trade and Investment U-Shaping Industrial Output in Bulgaria, Hungary, Poland and Romania', LICOS Centre for Transition Economics, Katholieke Universiteit Leuven, Discussion Paper 83.

Venables, Anthony J. (1998), 'The Assessment: Trade and Location', *Oxford Review of Economic Policy*, Vol. 14(2).

Appendix

Indexes of Regional Specialization

Table 5.A1 Herfindahl regional specialization index

	Northern Estonia	Central Estonia	Northeastern Estonia	Western Estonia	Southern Estonia
1990	0.098	0.208	0.145	0.204	0.135
1991	0.099	0.188	0.144	0.203	0.136
1992	0.097	0.180	0.144	0.197	0.128
1993	0.101	0.157	0.150	0.192	0.128
1994	0.104	0.156	0.153	0.213	0.136
1995	0.114	0.148	0.135	0.269	0.151
1996	0.112	0.151	0.140	0.265	0.159
1997	0.116	0.173	0.148	0.243	0.167
1998	0.104	0.183	0.145	0.212	0.162
1999	0.100	0.186	0.148	0.205	0.159
2000	0.104	0.172	0.151	0.189	0.160

Source: Authors' calculations based on REGSTAT.

Table 5.A2 Regional dissimilarity index

	Northern Estonia	Central Estonia	Northeastern Estonia	Western Estonia	Southern Estonia
1990	0.244	0.585	0.555	0.544	0.448
1991	0.238	0.566	0.530	0.516	0.438
1992	0.235	0.591	0.494	0.527	0.415
1993	0.249	0.515	0.448	0.484	0.454
1994	0.252	0.492	0.405	0.502	0.474
1995	0.433	0.400	0.438	0.613	0.364
1996	0.406	0.335	0.428	0.596	0.427
1997	0.383	0.411	0.434	0.516	0.453
1998	0.362	0.400	0.377	0.440	0.382
1999	0.361	0.389	0.315	0.491	0.413
2000	0.368	0.328	0.356	0.397	0.470

Source: Authors' calculations based on REGSTAT.

Table 5.A3 Specialization GINI

	Northern Estonia	Central Estonia	Northeastern Estonia	Western Estonia	Southern Estonia
1990	0.211	0.443	0.413	0.425	0.316
1991	0.217	0.442	0.398	0.403	0.317
1992	0.207	0.420	0.382	0.411	0.303
1993	0.221	0.378	0.390	0.361	0.350
1994	0.214	0.390	0.375	0.327	0.383
1995	0.302	0.338	0.413	0.482	0.272
1996	0.300	0.384	0.406	0.501	0.332
1997	0.312	0.387	0.364	0.448	0.350
1998	0.264	0.338	0.343	0.415	0.294
1999	0.216	0.323	0.317	0.429	0.302
2000	0.240	0.314	0.329	0.374	0.342

Source: Authors' calculations based on REGSTAT.

Indexes of Geographical Concentration

Table 5.A4 Herfindahl geographic concentration index

Year	1990	1991	1992	1993	1994	1995	1996	1997	1998	1999	2000
DA	0.239	0.250	0.240	0.229	0.232	0.214	0.217	0.227	0.218	0.224	0.243
DB	0.283	0.286	0.286	0.296	0.287	0.305	0.290	0.285	0.276	0.265	0.238
DC	0.322	0.324	0.303	0.290	0.291	0.334	0.341	0.345	0.384	0.413	0.314
DD	0.301	0.262	0.233	0.215	0.217	0.211	0.221	0.211	0.212	0.219	0.209
DE	0.562	0.541	0.480	0.448	0.407	0.515	0.575	0.556	0.545	0.575	0.647
DF+DG	0.530	0.545	0.532	0.543	0.553	0.577	0.597	0.626	0.513	0.395	0.490
DH	0.578	0.582	0.541	0.563	0.446	0.330	0.319	0.291	0.259	0.272	0.244
DI	0.314	0.322	0.338	0.325	0.347	0.260	0.241	0.328	0.243	0.315	0.329
DJ	0.533	0.533	0.500	0.531	0.528	0.590	0.530	0.294	0.285	0.309	0.294
DK	0.296	0.292	0.289	0.302	0.304	0.327	0.334	0.335	0.333	0.365	0.310
DL	0.474	0.465	0.469	0.474	0.466	0.359	0.433	0.564	0.555	0.590	0.566
DM	0.811	0.780	0.804	0.861	0.900	0.864	0.902	0.757	0.700	0.648	0.568
DN	0.321	0.303	0.281	0.278	0.283	0.306	0.280	0.269	0.250	0.233	0.242

Source: Authors' calculations based on REGSTAT.

Table 5.A5 Geographic dissimilarity index

Year	1990	1991	1992	1993	1994	1995	1996	1997	1998	1999	2000
DA	0.375	0.326	0.344	0.324	0.291	0.438	0.380	0.275	0.193	0.177	0.195
DB	0.258	0.274	0.264	0.179	0.177	0.198	0.188	0.229	0.251	0.269	0.315
DC	0.259	0.240	0.270	0.269	0.251	0.512	0.628	0.503	0.510	0.465	0.442
DD	0.653	0.678	0.548	0.592	0.548	0.427	0.488	0.539	0.557	0.584	0.492
DE	0.525	0.499	0.428	0.415	0.388	0.671	0.709	0.765	0.696	0.676	0.806
DF+DG	0.832	0.869	0.835	0.876	0.891	0.988	1.026	1.026	0.906	0.586	0.812
DH	0.559	0.564	0.517	0.570	0.438	0.348	0.338	0.360	0.317	0.325	0.454
DI	0.306	0.303	0.245	0.168	0.246	0.472	0.277	0.755	0.308	0.174	0.245
DJ	0.468	0.478	0.456	0.523	0.552	0.738	0.646	0.239	0.156	0.198	0.214
DK	0.320	0.310	0.324	0.324	0.329	0.281	0.310	0.271	0.314	0.380	0.232
DL	0.372	0.348	0.352	0.387	0.409	0.323	0.497	0.749	0.714	0.695	0.692
DM	0.849	0.822	0.877	0.949	1.021	1.077	1.120	1.022	0.924	0.779	0.693
DN	0.180	0.147	0.165	0.189	0.154	0.229	0.165	0.249	0.169	0.201	0.151

Source: Authors' calculations based on REGSTAT.

Table 5.A6 GINI coefficients

Year	1990	1991	1992	1993	1994	1995	1996	1997	1998	1999	2000
DA	0.237	0.222	0.232	0.464	0.196	0.269	0.254	0.195	0.147	0.141	0.254
DB	0.161	0.168	0.164	0.545	0.136	0.136	0.141	0.142	0.181	0.177	0.141
DC	0.232	0.240	0.247	0.569	0.280	0.517	0.559	0.432	0.365	0.347	0.559
DD	0.320	0.293	0.213	0.392	0.269	0.230	0.240	0.271	0.282	0.286	0.091
DE	0.307	0.309	0.269	0.662	0.228	0.385	0.405	0.483	0.454	0.428	0.405
DF+DG	0.675	0.661	0.649	0.656	0.678	0.544	0.596	0.572	0.529	0.495	0.596
DH	0.343	0.364	0.379	0.704	0.245	0.196	0.341	0.230	0.264	0.277	0.341
DI	0.189	0.201	0.192	0.568	0.197	0.254	0.195	0.381	0.199	0.148	0.195
DJ	0.391	0.393	0.271	0.692	0.327	0.413	0.350	0.197	0.119	0.165	0.350
DK	0.261	0.257	0.267	0.567	0.260	0.261	0.302	0.217	0.213	0.293	0.302
DL	0.409	0.401	0.407	0.680	0.362	0.221	0.346	0.477	0.425	0.428	0.346
DM	0.595	0.525	0.592	0.785	0.679	0.702	0.716	0.596	0.560	0.528	0.716
DN	0.148	0.123	0.128	0.554	0.145	0.144	0.091	0.133	0.109	0.118	0.091

Source: Authors' calculations based on REGSTAT.

Chapter 6

The Emerging Economic Geography in Hungary

Alessandro Maffioli*

1 Introduction

During the 1990s, Hungary had almost completed the political transition it began in 1989. As a result of this process and of the increasing trade liberalization of the Eastern European Area, the country achieved significant results in both production and trade restructuring. This structural reorientation of production and commerce was followed by a deep change in the regional patterns of specialization within the country. The ongoing process of economic integration with the European Union (EU) common market can be considered one of the main forces that have driven the restructuring of Hungarian trade and industrial patterns. In 1989 the EU granted Hungary the General System of Preference (GSP). In 1992 the preferential trade agreement with the European Free Trade Association came into effect; in 1993 the Central European Free Agreement (CEFTA) was established; in 1997 the Pan-European Cumulation Agreement was signed. Following these achievements, Hungary has entered a wide multilateral free trade area encompassing the EU, EFTA and nine other Central and Eastern European countries. In 2001 all tariffs on manufacturing products were removed, giving Hungarian products the opportunity to access a market of 487 million consumers.

A cursory analysis of the evolution of Hungary's reintegration into the international economy in the last ten years allows us to identify two different phases of production and trade reorientation (Kaminski, 1999): the first, which occurred between 1989 and 1993, was characterized by an increase in the exports of firms that had already established links abroad, and by a significant redirection of manufacturing exports from the collapsed Council for Mutual Economic Assistance (CMEA) markets to the EU market.[1] During 1993 and 1994 total exports fell, but from 1994 onwards the value of EU-oriented exports witnessed a new expansion. However, this period of continuous export increases was led by different agents, namely Transnational Corporations

(TNCs). These operate in global networks of production that had undertaken significant Foreign Direct Investments (FDIs) in Central and Eastern Europe. These foreign-owned and export-oriented firms were not only one of the main factors of enforcement of the geographical patterns of exports (from CMEA to EU), but were also at the origin of the Hungarian industrial system's reconversion process towards a manufacturing specialization.

According to the theoretical and empirical literature on the effect of trade liberalization on the location and concentration of the economic activity (Hanson, 1996; Krugman and Venables, 1997), one may expect a significant relocation within Hungary, especially towards EU borders, and a significant change in regional manufacturing wages (given the assumption of perfect regional mobility and absolute international immobility of the labour force).

The aim of the present chapter is to analyse patterns of regional manufacturing relocation in Hungary after the trade and capital liberalization towards the EU. I will first provide a brief analysis of the impact of the EU on trade specialization. Then, I shall analyse in depth the evolution of regional specialization and geographical concentration of manufacturing activity, focusing mainly on the relocation patterns that have emerged in the last ten years. Finally, I shall test some of the predictions of the literature on regional adjustment for trade liberalization in Hungary's case. The econometric analyses mainly follow Hanson (1996).

2   Evidence of Increasing Integration with the EU

2.1   Trade Liberalization and Trade Performance

The evolution of Hungary's external trade during the 1990s can be divided into two phases. During the 1989–93 period, the main determinant of the geographical orientation of Hungarian trade was the collapse of the Former Soviet Union (FSU) and the dissolution of the CMEA. Macroeconomic variables such as the fall of oil prices and of FSU exports led many former CMEA members to relocate their exports towards the EU, a market that was both close and stable. Hungary's share of trade with ex-CMEA members decreased from 60 per cent in 1986 to 38 per cent in 1990, and to 20 per cent in 1997 (Kaminski, 1999).

As a result, the export expansion between 1990 and 1991, shown in Table 6.1, was mainly due to the reorientation of export products to new markets. Since 1993 the expansion rhythms have slowed, as the values of both total

Table 6.1   The role of the EU in Hungary's foreign trade, 1990–2001 (€ and percentages)

	Ex-total	Ex-EU 15	EU share (%)	Im-total	Im-EU 15	EU share (%)	EU Ex/Im (%)
1988	–	2,158,372	–	–	2,354,581	–	91.7
1989	–	2,587,278	–	–	2,988,305	–	86.6
1990	5,016,744	2,934,049	58.5	4,717,457	2,876,155	61.0	102.0
1991	7,463,900	3,625,111	48.6	7,321,864	3,486,087	47.6	104.0
1992	7,761,368	3,987,852	51.4	7,800,077	4,060,886	52.1	98.2
1993	6,931,457	3,952,945	57.0	9,717,397	4,967,004	51.1	79.6
1994	6,409,966	4,923,072	76.8	9,463,636	6,152,789	65.0	80.0
1995	9,894,118	7,610,396	76.9	11,787,241	8,730,773	74.1	87.2
1996	11,326,775	8,846,917	78.1	13,437,105	10,027,617	74.6	88.2
1997	17,388,324	11,684,207	67.2	19,120,945	13,596,336	71.1	85.9
1998	18,447,012	14,655,420	79.4	20,526,557	16,863,194	82.2	86.9
1999	20,520,523	17,416,628	84.9	22,574,075	18,572,983	82.3	93.8
2000	27,987,754	21,900,000	78.2	29,903,860	23,000,000	76.9	95.2
2001	31,345,841	23,650,725	75.5	33,610,870	24,204,186	72.0	97.7

Source: Author's calculations on the basis of data from the National Bank of Hungary (2002) and Eurostat (2000).

and EU-oriented exports have decreased significantly. This contraction is due to a reduction in agricultural exports.[2]

Table 6.2 shows the geographical destinations of EU-oriented exports in detail, and reveals the pivotal role of the German markets (Germany and Austria), the Netherlands' increasing relevance, and Italy's decreasing but still appreciable contribution. Note that in 1999 almost 70 per cent of total EU-oriented exports were directed to the three nearest European regions (Northern Italy, Southern Germany and Austria).

After 1993, the strong enforcement of commercial ties with the EU (from around 50 per cent to more than 70 per cent of total exports) started to have a discernible impact on export composition and on Hungary's production specialization. Increasing integration with developed markets and the collapse of CMEA markets required a deep restructuring of the industrial system in order to develop a sustainable foreign trade pattern. As part of the literature has already pointed out (Kaminski, 1999; Nemes-Nagy, 2000; Szanyi, 2002), this restructuring was mainly driven by TNCs. FDI both allowed the emergence of a 'second generation' of firms, and had an important impact on the organization of industrial linkages and on local value chains.

With reference to the sectoral composition of Hungary's exports, Table 6.3 reveals the increasing presence of manufactured goods, which in 1999 represented 90 per cent of EU-oriented exports.

An in-depth analysis of the composition of manufacturing exports provides evidence for the assumption that a structural change occurred during the second half of the 1990s. As shown in Table 6.3, since 1993 the EU-oriented share of exports of several basic manufactured goods (furniture, clothing and footwear) has been constantly decreasing, though the total share of manufactured goods has significantly increased. New manufacturing sectors have emerged as the main pillars of Hungary's foreign trade: *non-electric machinery* and *electrical machinery*, which have provided more than 50 per cent of the total exports to the EU since 1996.

In order to further analyse this structural change in Hungary's position within the international arena, I take into account two indices of specialization: the Balassa (1967) index of Revealed Comparative Advantage (RCA) and the Gruber-Lloyd (1975) index of Intra-Industry Trade (IIT). The first allows us to analyse the competitiveness of a country in the global market or in a specific target market. In the case of Hungary, I consider the RCA relative to EU-oriented exports. Table 6.4 indicates, first of all, that since 1996 manufactured goods are the only ones in which Hungary has maintained a comparative advantage with respect to EU markets. Secondly, among the manufactured

Table 6.2   Exports to the EU by country (percentages)

	1988	1989	1990	1991	1992	1993	1994	1995	1996	1997	1998	1999
Austria								12.4	17.0	16.2	14.9	13.4
Belgium and Luxembourg	3.3	3.7	3.6	3.5	3.9	4.0	4.5	4.0	3.7	3.3	4.3	4.0
Denmark	1.7	1.3	1.4	1.1	1.0	0.9	0.9	0.6	0.6	0.6	0.7	0.6
Finland								0.7	0.7	0.6	0.6	0.5
France	9.9	10.5	9.4	8.1	7.2	7.2	6.6	5.6	5.5	5.8	5.9	5.8
Germany	50.4	50.1	53.7	57.1	57.1	59.0	56.8	47.7	45.5	46.3	48.6	46.7
Greece	1.6	2.4	1.6	2.4	1.4	1.0	0.8	0.7	0.9	0.7	0.6	0.4
Ireland	0.2	0.2	0.2	0.1	0.1	0.1	0.1	0.2	0.1	0.3	0.7	1.4
Italy	18.9	18.5	18.3	18.0	18.7	16.5	16.5	12.8	11.2	9.8	8.5	7.8
Netherlands	4.9	4.8	4.7	4.1	4.4	4.9	4.5	4.1	4.4	6.2	5.9	6.5
Portugal	0.2	0.1	0.1	0.1	0.2	0.1	0.2	0.1	0.1	0.1	0.4	0.4
Spain	1.5	1.8	1.4	1.4	1.9	1.3	2.7	3.8	2.7	2.6	2.1	1.9
Sweden								1.6	1.6	1.4	1.3	1.2
United Kingdom	7.4	6.6	5.5	4.3	4.2	5.0	6.3	5.8	6.0	6.0	5.4	5.4

Source: Author's calculations on the basis of data from the National Bank of Hungary (2002) and Eurostat (2000).

Table 6.3   Composition of exports to the EU over 1988–99 (percentages)

	1988	1989	1990	1991	1992	1993	1994	1995	1996	1997	1998	1999
All food products	23.5	25.1	20.9	20.8	14.8	13.4	11.9	9.2	8.4	6.1	4.7	4.4
Agricultural materials	6.1	5.2	4.5	4.5	3.8	3.5	2.6	2.2	2.1	1.9	1.5	1.5
Textile fibres	1.0	0.9	0.8	0.8	0.5	0.2	0.4	0.2	0.2	0.3	0.1	0.1
Ores, minerals and metals	5.2	5.4	5.4	3.9	3.5	4.2	5.5	6.4	4.3	3.6	3.0	2.6
Energy	2.7	2.9	2.8	2.1	1.3	1.6	0.9	2.2	2.3	1.4	1.1	1.2
All manufactured goods	59.2	57.6	63.4	65.9	74.1	76.5	78.1	79.3	82.1	86.7	89.4	90.1
Other:	2.2	2.9	2.3	2.0	2.0	0.6	0.6	0.5	0.5	0.1	0.2	0.1
Chemical elements	4.4	4.6	4.0	3.5	3.5	3.9	3.4	2.7	2.3	1.7	1.4	1.3
Leather and leather goods	0.5	0.7	0.6	0.5	0.7	0.4	0.5	0.3	0.4	0.4	0.3	0.2
Wood and manufacturing	1.5	1.2	1.7	1.4	1.3	0.7	1.0	1.3	1.3	1.2	1.0	1.0
Textiles yarn and fabric	2.2	1.7	2.0	2.2	1.8	1.6	1.2	1.4	1.3	1.1	1.4	1.4
Iron and steel	3.7	3.1	3.3	2.5	2.7	1.8	3.5	3.2	2.9	1.8	1.7	1.3
Metal manufacturers	2.4	2.5	3.3	3.9	4.4	5.0	4.1	4.3	4.1	3.4	3.1	3.1
Non-electric machinery	6.2	7.1	8.3	8.6	8.2	11.0	14.0	19.1	19.5	22.4	25.3	24.5
Electrical machinery	6.0	5.5	6.3	7.3	9.6	12.7	14.5	15.2	21.0	28.3	30.6	33.5
Transport equipment	0.5	0.9	1.1	1.8	2.6	1.0	1.4	1.1	1.5	1.6	1.7	1.8
Furniture	2.9	2.6	2.6	3.1	3.3	3.2	2.7	2.5	2.3	1.7	2.0	2.2
Clothing	13.2	10.7	12.3	11.9	15.2	15.5	14.6	10.9	10.3	9.1	8.0	7.2
Footwear	2.9	3.7	4.3	3.7	5.2	3.3	2.4	2.1	2.1	1.9	1.6	1.5
Scientific instruments	2.9	3.7	4.3	3.7	5.2	3.3	2.4	2.1	2.1	1.9	1.6	1.5

Source: Author's calculations on the basis of data from Eurostat (2000).

goods, non-electrical and electrical machinery are the only ones that showed a significant improvement between 1988 and 1999.

The Intra-Industry Trade (ITT) index adds some important information about the competitive position of the country. In fact, it allows us to analyse how a specific sector is inserted into the international production network and value chain. A value of the ITT index close to zero means that the country is almost a pure exporter or a pure importer of the good, while a value close to one means that the country is significantly participating in international production networks. The results reported in Table 6.6 confirm that there was a significant change in Hungary's international position and, thus, a relevant change in its production system. In fact, while sectors like food products and textile fibres have seen a reduction in their indices, an important increase has occurred in several manufacturing sectors. The index's significant increase in the two sectors where Hungary has had better performance in terms of RCA (non-electric and electrical machinery) is also noteworthy. In contrast, the index's decrease in the transport equipment sector is quite surprising, since one might expect an increase in this sector's participation in the international value chain due to relevant investments carried out by TNCs.

In conclusion, figures relative to the Hungarian economy's international setting confirm that a structural change occurred during the 1990s as a consequence of increasing trade liberalization and integration with EU markets. In this period, non-electrical and electrical machinery has emerged as the new core sector with reference to EU-oriented trade, due to its revealed advantages and its increasing participation in the European production process.

2.2   Foreign Direct Investment

FDI has played a crucial role in the definition of Hungary's new specialization patterns. In particular, foreign investments have compensated for the decline in domestic savings, preventing a fall in aggregate investment activity. During the last decade, this phenomenon has demonstrated impressive growth: the number of multinational and joint-ventures operating in the country increased from 1,800 units in 1989 to 19,000 in 1993, and to 76,000 in 2001. Estimates show that around 30 per cent of the Hungarian labour force was employed in multinational enterprises in 2001 (ICE, 2002).

Despite these impressive figures, FDI inflows decreased during the 1990s (Table 6.5). After they reached a peak in 1995, FDI inflows showed a constantly decreasing trend during the second half of the decade. This effect is probably due to a natural reduction in investment opportunities.[3]

Table 6.4   Revealed Comparative Advantages and Intra-industry trade, 1988–99

Industry	RCA*						ITT+					
	1988	1991	1994	1997	1998	1999	1988	1991	1994	1997	1998	1999
All food products	2.42	2.49	1.31	0.74	0.58	0.61	0.93	0.91	0.66	0.57	0.53	0.55
Agricultural materials	1.42	1.43	0.75	0.77	0.64	0.69	0.62	0.56	0.74	0.73	0.78	0.71
Textile fibres	0.81	0.94	0.45	0.45	0.21	0.22	0.80	0.82	0.58	0.97	0.65	0.56
Ores, minerals and metals	0.85	0.84	1.22	0.74	0.64	0.62	0.45	0.45	0.47	0.78	0.79	0.77
Energy	0.22	0.15	0.07	0.10	0.11	0.11	0.34	0.78	0.77	0.88	0.79	0.68
All manufactured goods	1.00	1.04	1.16	1.27	1.24	1.24	0.75	0.84	0.81	0.85	0.89	0.88
Other	0.30	0.39	0.25	0.06	0.06	0.05	0.33	0.62	0.36	0.97	0.89	0.81
Chemical elements	2.44	2.05	1.47	0.75	0.62	0.57	0.81	0.96	0.80	0.93	0.85	0.90
Leather and leather goods	0.90	1.10	1.04	0.77	0.54	0.27	0.30	0.24	0.24	0.25	0.21	0.12
Wood and manufacturing	0.51	0.49	0.38	1.28	1.02	1.07	0.32	0.33	0.64	0.72	0.78	0.78
Textiles, yarn and fabric	1.33	1.32	0.46	0.39	0.54	0.51	0.37	0.37	0.20	0.27	0.35	0.42
Iron and steel	1.90	1.48	1.83	1.29	0.92	0.92	0.83	0.96	0.86	0.80	0.78	0.69
Metal manufacturing	1.61	2.34	2.06	1.82	1.56	1.54	0.92	0.93	0.89	0.77	0.73	0.72
Non-electric machinery	0.78	1.06	1.63	2.60	2.63	2.52	0.39	0.55	0.75	0.97	0.95	0.91
Electrical machinery	0.45	0.53	1.03	1.73	1.75	1.86	0.68	0.80	0.90	0.95	0.90	0.85
Transport equipment	0.08	0.27	0.20	0.21	0.19	0.20	0.31	0.76	0.36	0.22	0.23	0.17
Furniture	5.17	4.83	3.33	1.81	1.90	1.92	0.14	0.50	0.94	0.89	0.82	0.88
Clothing	2.98	2.14	2.60	1.51	1.32	1.21	0.25	0.45	0.49	0.52	0.53	0.57
Footwear	4.04	4.04	2.74	2.01	1.78	1.70	0.43	0.42	0.68	0.48	0.51	0.45
Scientific instruments	4.04	4.04	2.74	2.01	1.78	1.70	0.43	0.42	0.68	0.48	0.51	0.45

*   Balassa Index of Revealed Comparative Advantage (RCA): $RCA_i = (x_{iHUN}/X_{HUN})/(m_{iEU}/M_{iEU})$, i.e., share of exports of product i in total Hungarian exports relative to the share of the import of product i in total extra-EU imports.

+   Gruber-Lloyd index of intra-industry trade (ITT): $ITT_i = |x_i - m_i|/(x_i + m_i)$, where x_i and m_i refer to exports to the EU and imports from the EU of product i.

Source: Author's calculations on the basis of data from Eurostat (2000).

Table 6.5 FDI inflows in Central and Eastern Europe, 1997–2000 (US$m)

	1990	1991	1992	1993	1994	1995	1996	1997	1998	1999	2000
E. Europe	568	2,523	4,436	6,748	5,931	14,268	12,730	19,188	21,008	23,221	25,419
Albania	=	–0.7	20	58	52.9	70	90.1	47.5	45	41	92
Belarus	...	...	7	18	11	15	105	352	203	444	90
Bulgaria	4	60	42	40	105	90	109	505	537	819	1,002
Czech R.	...	...	...	654	869	2,562	1,428	1,300	3,718	6,324	4,595
Estonia	...	...	82	162	215	202	151	267	581	305	399
Hungary	311	1,459	1,471	2,339	1,146	4,453	2,275	2,173	2,036	1,944	1,957
Latvia	...	...	27	45	214	180	382	521	357	348	407
Lithuania	...	...	10	30	31	73	152	355	926	487	379
Moldova	...	...	17	14	28	67	24	79	74	39	128
Poland	88	359	678	1715	1,875	3,659	4,498	4,908	6,365	7,270	10,000
Romania	–	42	80	94	342	420	265	1215	2,031	1,041	998
Russian F.	...	...	700	1211	640	2,016	2,479	6,638	2,761	3,309	2,704
Slovakia	...	...	...	168	245	195	251	206	631	356	2,075
Ukraine	...	...	200	200	159	267	521	624	743	496	595

Missing values:
... = not applicable; _ = not applicable or not separately reported.

Source: Author's calculations on the basis of data from UNCTAD (2001).

Between 1994 and 1995, several large foreign manufacturing firms almost completed the construction or restructuring of their Hungarian production plants, thus reducing capital imports, both in terms of financial flows and equipment (Kaminski, 1999; Hunya, 1997). Most of them are strongly export-oriented. In order to further analyse the increasing outward orientation of Hungarian subsidiaries, I calculate the FDI stocks/export outflows ratio for some Central and Eastern European countries in the 1990s. Table 6.6 clearly shows that in Hungary this ratio peaked in 1995 and then started to decrease due to the high export performance of the foreign-owned plants that began operations in the second half of the decade.

The timing of this process is clearly consistent with the hypothesis that the emergence of Hungary's new revealed advantages vis-à-vis the EU was led by the competitiveness of direct investments made by EU TNCs (Szanyi, 2002). In fact, I have previously shown that the greatest increase in the RCA index for the new leading sector occurred between 1994 and 1997, while, in many cases, the ITT index started to grow in 1991. All these elements allow the definition of a specific replacement process in the Hungarian industrial system: first of all, trade liberalization increased domestic and international competition; then, foreign-owned and domestic firms started to replace domestic suppliers with higher-quality foreign suppliers in order to improve their competitiveness (this can explain the ITT trend in the first half of the 1990s); meanwhile, the gradual integration into the EU attracted new greenfield FDI from industrialized countries, reinforcing the import of capital goods (this explains the inflow of capital and the growth of the FDI/exports ratio until 1995); finally, the export performance of new TNC subsidiaries determined the emergence of a new revealed advantage (this explains the figures for RCA and the contraction of the FDI/exports ratio).

Looking at the geographical sources of FDI, the leading role of EU investors, and particularly German-speaking investors (see Table 6.7) is not surprising. Among European investors, another key player is the Netherlands, which has confirmed its role as a dynamic partner, as shown by the data on exports' destinations (see Table 6.2). These results are consistent with the process of internationalization of EU firms towards EU Candidate Countries. In fact, until 1998, the German share in total assets in the region was close to the total amount of investment made by France, the United Kingdom, the Netherlands and Austria together (Eurostat, 2001). Among extra-EU investors, Japan and the United States are emerging as the main investors. According to information from the Hungarian Ministry of Economic Affairs, TNCs from these two countries have often identified Hungary as a strategic platform

Table 6.6   Ratio between FDI stocks and exports in Central and Eastern Europe, 1990–2000

	1990	1991	1992	1993	1994	1995	1996	1997	1998	1999	2000
E. Europe	1.4	5.9	3.8	6.1	4.1	7.7	6.4	9.2	10.3	11.5	10.1
Albania	–	-0.7	27.8	47.4	37.9	34.6	43.4	34.2	22.0	15.5	35.4
Belarus	...	...	0.2	0.9	0.4	0.3	1.8	4.8	2.9	7.5	1.2
Bulgaria	0.1	1.9	1.1	1.1	2.7	1.7	1.7	9.5	12.5	20.9	20.8
Czech R.	...	...	...	4.5	5.4	11.8	6.5	5.7	14.1	23.6	15.9
Estonia	...	...	18.4	20.2	16.3	11.0	7.2	9.1	18.5	10.4	12.7
Hungary	3.2	14.3	13.8	26.4	10.7	35.8	18.0	11.7	8.9	7.8	7.0
Latvia	...	...	3.5	4.5	21.6	13.8	26.4	31.2	19.7	20.2	21.8
Lithuania	...	...	1.2	1.5	1.5	2.7	4.5	9.2	24.9	16.2	9.9
Moldova	...	...	3.6	2.9	4.5	9.1	3.0	8.9	11.5	6.0	16.1
Poland	0.6	2.4	5.1	12.1	11.0	16.0	18.4	19.1	23.4	26.5	31.6
Romania	–	1.0	1.8	1.9	5.6	5.3	3.3	14.4	24.5	12.2	9.6
Russian F.	...	...	1.7	2.7	0.9	2.5	2.8	7.5	3.7	4.4	2.6
Slovakia	...	...	...	3.1	3.7	2.3	2.8	2.5	5.9	3.5	17.6
Ukraine	...	...	2.5	2.6	1.5	2.0	3.6	4.4	5.9	4.3	4.0

Missing values:

... = not applicable; _ = not applicable or not separately reported.

Source: Author's calculations on the basis of data from UNCTAD (2001).

useful for penetrating EU markets (Hungarian Ministry of Economic Affairs, 2001).

A look at Hungarian Statistical Office data on the sectoral composition of FDI stocks between 1990 and 2000 shows a relevant concentration in manufacturing sectors (36.5 per cent of total FDI stocks). Other large shares were mainly determined by the privatization of financial services and many public utilities, as revealed by banking and finance (12.5 per cent), energy (9.6 per cent) and communication (7.7 per cent).

Finally, I take into account the localization of foreign-owned firms in the Hungarian regions. I refer to a NUTS 3 level classification: in the case of Hungary this means counties. Table 6.8 shows the ratio between firms with foreign participation and domestic firms in each county, normalized by the national average. Figures reveal the capital city and the county of Pest's expected (as the traditional center of the Hungarian economic system) leading role as a favourite destination for foreign firms, though this is now declining. The only location characterized by an above-average and increasing presence of foreign-owned firms is the Western Transdanubiana region (namely the counties of Győr-Moson-Sopron, Vas and Zala), while a below-average and declining share characterizes almost all of the eastern part of the country (North Hungary and the North Great Plain and South Great Plain regions), with Heves, Jász-Nagykun-Szolnok and Szabolcs-Szatmár-Bereg as the only significant exceptions. These regions have reduced the gap between their performance and the national average (the latter thanks to the recovery in 1999).

In summary, FDI played a crucial role in Hungary during the 1990s. In Central and Eastern Europe, Hungary has been one of the main destinations for manufacturing FDI from the EU (mainly from Germany, Austria, and the Netherlands), Japan and the United States. Figures show some preliminary evidence of the important role of TNCs in the process of restructuring and relocation of the Hungarian industrial system towards a model driven by the catalyzing force of the integrated European markets.

3 Regional Specialization Patterns

This section focuses on the evolution of regional specialization in Hungary during the 1990s. I will refer to an unofficial database on economic activities at a county level (REGSTAT) constructed for this purpose.[4] The dataset covers the period 1990–99, but since for the earlier part of the period some data may be missing, I will sometimes refer to the 1992–98 period, which, in

Table 6.7   FDI inflows by home countries (% of net value)

Investors' country/ region	Net flows			% of total (inflows + outflows)		
	1999	2000	2001	1999	2000	2001
Austria	65.33	20.92	219.69	4.1	1.4	9.2
Belgium	−12.17	201.51	39.5	(0.8)	13.2	1.7
Finland	10.72	112.09	2.67	0.7	7.4	0.1
France	57.42	70.45	45.98	3.6	4.6	1.9
Netherlands	453.27	515.37	−646.47	28.6	33.9	(27.1)
Ireland	13.42	51.85	−5.5	0.8	3.4	(0.2)
Luxemburg	43.24	15.53	61.04	2.7	1.0	2.6
United Kingdom	95.9	111.2	19.24	6.0	7.3	0.8
Germany	410.41	16.08	633.29	25.9	1.1	26.5
Italy	40.04	−2.41	58.64	2.5	(0.2)	2.5
Sweden	4.66	45.73	36.38	0.3	3.0	1.5
Total EU	*1,182.24*	*1,158.32*	*464.46*	*74.5*	*76.1*	*19.5*
Switzerland	15.69	31.08	37.60	1.0	2.0	1.6
Other European countries	90.49	−26.51	53.31	5.7	1.7	2.2
United States	186.86	238.41	134.63	11.8	15.7	5.6
Canada	24.44	11.41	59.02	1.5	0.7	2.5
Japan	21.91	16.96	291.74	1.4	1.1	12.2
South Korea	4.04	20.36	24.78	0.3	1.3	1.0
Other Countries	−4.88	6.39	17.2	(0.3)	(0.4)	0.7
Not allocated	31.56	8.22	0.10	2.0	0.5	0.0
Total	**1,552.35**	**1,464.65**	**1,082.84**			

Source: Author's calculations on the basis of data from the National Bank of Hungary (2001).

Table 6.8 Geographical distribution of foreign-owned firms (1992–99)

Reg*	Counties	1992	1993	1994	1995	1996	1997	1998	1999	1999/92
CH	Budapest	1.79	1.89	1.82	1.77	1.75	1.73	1.74	1.65	0.92
CH	Pest	0.72	0.77	0.75	0.76	0.75	0.76	0.78	0.75	1.05
CT	Fejér	0.57	0.55	0.56	0.56	0.55	0.55	0.51	0.47	0.83
CT	Komárom-Esztergom	0.74	0.79	0.74	0.75	0.85	0.96	0.98	0.62	0.84
CT	Veszprém	0.71	0.65	0.74	0.76	0.74	0.77	0.78	0.89	1.27
WT	Győr-Moson-Sopron	1.14	1.17	1.34	1.38	1.42	1.40	1.34	1.26	1.11
WT	Vas	1.14	1.08	1.25	1.30	1.29	1.34	1.31	1.27	1.11
WT	Zala	0.81	0.67	0.74	0.83	0.77	0.77	0.81	1.09	1.35
ST	Baranya	0.78	0.72	0.75	0.74	0.76	0.73	0.75	0.88	1.12
ST	Somogy	0.41	0.39	0.41	0.45	0.50	0.49	0.51	0.61	1.49
ST	Tolna	0.52	0.56	0.54	0.52	0.49	0.52	0.51	0.49	0.95
NH	Borsod-Abaúj-Zemplén	0.37	0.35	0.33	0.33	0.32	0.35	0.35	0.33	0.88
NH	Heves	0.33	0.41	0.46	0.52	0.52	0.56	0.56	0.49	1.49
NH	Nógrád	0.48	0.49	0.48	0.45	0.48	0.45	0.48	0.43	0.90
NGP	Hajdú-Bihar	0.45	0.43	0.38	0.38	0.36	0.34	0.32	0.31	0.68
NGP	Jász-Nagykun-Szolnok	0.33	0.39	0.39	0.37	0.40	0.42	0.39	0.36	1.08
NGP	Szabolcs-Szatmár-Bereg	0.36	0.37	0.34	0.35	0.37	0.39	0.36	0.61	1.69
SGP	Bács-Kiskun	1.07	0.88	0.79	0.76	0.74	0.70	0.64	0.68	0.64
SGP	Békés	0.46	0.39	0.34	0.35	0.34	0.33	0.34	0.35	0.76
SGP	Csongrád	1.18	0.68	0.79	0.69	0.63	0.58	0.57	1.06	0.90
	National average	1.00	1.00	1.00	1.00	1.00	1.00	1.00	1.00	

The indices have been computed as follows: $(F_{jt} / D_{jt}) / \left(\sum_j F_{jt} / \sum_j D_{jt} \right)$, where F (D) is the number of foreign (domestic) owned firms in region j at time t.

Source: Author's calculations on the basis of the REGSTAT database.

any case, allows me to take into account a considerable part of the transition and integration process.

3.1   Patterns of Regional Disparities

Analysing population data, Table 6.9 reveals a large concentration of population in the capital city (more than 18 per cent in 1999). The only other county with a share exceeding 10 per cent is Pest, whose size has increased significantly with the decrease in the capital's size. Given that the two counties belong to the same region (Central Hungary), these shifts have not significantly affected the size distribution of counties.

There were no noteworthy variations in the figures between 1992 and 1999 at the county or regional level. So I can conclude that the population distribution of counties and regions remained stable during the 1990s.

The analysis of GDP per capita highlights a quite different picture. Relevant changes occurred between 1975 (the socialist period) and 1998. In particular, Budapest maintained a very high level of development throughout the entire period, increasing its advantage over the rest of the country. It is worth noting that two out of the four counties that had higher per capita GDPs than the national average in 1998 belong to the Western Trasdanubian region[5] and that most of the regions with per capita GDPs below the national average in 1994 have experienced slower growth than the national rate, thus further increasing the gap between their GDPs and those of the richest counties. The only exceptions are the counties of Pest, Veszéprem and Fejér, all belonging to regions in the Central-eastern part of the country.[6] Fejér, Győr-Moson-Sopron and Vas have also been the most dynamic counties in terms of economic growth, with growth rates of 10.1 per cent, 7.6 per cent and 6.6 per cent, respectively. Vas, furthermore, has been the only county that moved from a below-average GDP per capita in 1975 to an above-average GDP per capita in 1998.

There may be two reasons for the success of the Western Trasdanubian region: first of all, a relatively good transportation system, with several links to the adjacent Austrian regions; secondly, a geographical location along the western border. Because of this location, the region has always been at the margin of socialist industrialization, which was strongly oriented toward heavy industry. As a consequence, it arrived at the threshold of the transition with a less obsolete and more flexible production system. In addition, the region's proximity to the EU border meant that it was relatively attractive for EU-oriented firms seeking to establish plants outside the EU. This covers all

Table 6.9   Distribution of NUTS 3 regions by population share, 1992–99 (million)

Reg*	Counties	1992	1993	1994	1995	1996	1997	1998	1999	Δ 92–99+
CH	Budapest	19.50	19.48	19.42	18.84	18.67	18.54	18.37	18.22	-0.94
CH	Pest	9.23	9.29	9.39	9.50	9.65	9.77	9.93	10.09	1.34
NH	Borsod-Abaúj-Zemplén	7.28	7.27	7.24	7.32	7.30	7.30	7.29	7.28	-0.02
NGP	Szabolcs-Szatmár-Bereg	5.48	5.47	5.46	5.59	5.60	5.63	5.65	5.67	0.49
NGP	Hajdú-Bihar	5.32	5.33	5.34	5.37	5.38	5.38	5.38	5.39	0.19
SGP	Bács-Kiskun	5.25	5.25	5.24	5.28	5.30	5.30	5.30	5.29	0.13
CT	Fejér	4.09	4.10	4.11	4.15	4.17	4.19	4.21	4.22	0.48
WT	Győr-Moson-Sopron	4.13	4.14	4.15	4.15	4.17	4.18	4.19	4.20	0.25
SGP	Csongrád	4.24	4.24	4.25	4.18	4.18	4.17	4.17	4.16	-0.26
NGP	Jász-Nagykun-Szolnok	4.09	4.08	4.08	4.13	4.12	4.12	4.11	4.11	0.05
ST	Baranya	4.04	4.05	4.05	4.02	4.01	4.00	3.99	3.99	-0.18
SGP	Békés	3.94	3.92	3.91	3.95	3.94	3.94	3.93	3.92	-0.07
CT	Veszprém	3.66	3.67	3.67	3.70	3.71	3.71	3.71	3.71	0.21
ST	Somogy	3.31	3.30	3.29	3.30	3.30	3.30	3.30	3.29	-0.07
NH	Heves	3.21	3.20	3.20	3.22	3.22	3.22	3.22	3.21	0.00
CT	Komárom-Esztergom	3.03	3.04	3.04	3.05	3.06	3.06	3.06	3.07	0.18
WT	Zala	2.94	2.93	2.93	2.95	2.95	2.94	2.94	2.93	-0.06
WT	Vas	2.66	2.66	2.65	2.66	2.66	2.66	2.66	2.66	-0.01
ST	Tolna	2.44	2.43	2.43	2.44	2.44	2.43	2.43	2.43	-0.04
NH	Nógrád	2.17	2.16	2.15	2.19	2.18	2.17	2.16	2.16	-0.08
	Total	100.00	100.00	100.00	100.00	100.00	100.00	100.00	100.00	

*   CH = Central Hungary; CT = Central Transdanubia; WT = West Transdanubia; NH = North Hungary; NGP = North Great Plain; SGP = South Great Plain; ST = South Transdanubia.

+   Annual growth rate.

Source: Author's calculations on the basis of the REGSTAT database.

the emerging sectors in the reconversion of the Hungarian industrial system (Nemes-Nagy, 2000). In this process, several instruments of the industrial policy adopted by the Hungarian central and local governments (particularly industrial parks and the Customs-free Zone) have played a significant role.[7] In particular, the 11 industrial Customs-free Zones located in the two EU border counties (five in Győr-Moson-Sopron and six in Vas) have been shown to be very effective in attracting FDI in the automotive industry (among others, Audi in Győr and Opel in Szentgotthard, not to mention their suppliers), in electronics (e.g., the Philips Manufacturing Hungary Customs Free Zone Ltd, which has been producing color PC monitors since 1995, reaching a production level of around two million units in 2000) and in the textile industry.[8]

The impressive performance of western border regions is confirmed by Table 6.11, which compares percentages of the national GDP per capita for different groups of regions classified according to their geographical location along borders or within the country. Such further analysis allows us to underline the outstanding performance of the two EU border counties. In fact, not only are they the only border counties to show GDP per capita annual growth rates higher than the national average, but also, during the second half of the decade, they have grown more than twice the national average and even faster than the capital city.

In addition to the EU border counties, another winner in terms of GDP per capita is Fejér. In fact, this county, with its outstanding 10.1 per cent average annual growth rate, was the only county in the second half of the 1990s that moved from having a GDP per capita below the national average to having a GDP per capita higher than the national average. It is worth noticing that, unlike the two previously analysed counties, Fejér has traditionally been more involved in the heavy industrialization of the Socialist period and has passed through a more difficult reconversion process. Its impressive performance in the second half of the decade was mainly due to the location in the cities of Székesfehérvár and Dunaújváros of some large multinationals such as Parmalat, Philips, IBM, Aikawa, Ford/Visteon, Denso and Alcoa.[9] Initially many TNC investments were attracted by the privatization of many domestic firms located in Dunaújváros.[10] Afterwards many TNCs located their new greenfield investments within the local industrial parks which, in their initial phase, were designed mainly to attract large multinational investments, and frequently took advantage of Custom-free Zones.[11]

With regard to the other counties, the analysis confirms the results of previous studies (Nemes Nagy, 2000). Table 6.11 shows North Hungarian counties' radical decline. During the second part of the 1990s, the gap

Table 6.10　Distribution of NUTS 3 regions by GDP per capita, 1975 and 1994–99 (percentages)

Reg*	Counties	1975	1994	1995	1996	1997	1998	Diff. 75–98+	Δ 94–98++
CH	Budapest	139	177	180	185	186	186	47	4.7
CT	Fejér	106	97	99	104	117	124	18	10.1
WT	Győr-Moson-Sopron	111	104	109	111	110	121	10	7.6
WT	Vas	82	104	106	109	114	116	34	6.6
WT	Zala	88	95	92	93	91	90	2	2.3
SGP	Csongrád	109	94	94	93	90	89	-20	2.2
ST	Tolna	77	94	93	91	84	86	9	1.3
CT	Komárom-Esztergom	131	81	86	89	86	84	-47	4.6
CT	Veszprém	116	80	84	81	80	81	-35	3.8
ST	Baranya	108	83	80	78	80	79	-29	2.1
CH	Pest	61	77	73	73	78	78	17	4.0
NGP	Hajdú-Bihar	83	83	78	78	76	76	-7	1.1
NH	Heves	100	73	74	74	72	73	-27	3.4
NGP	Jász-Nagykun-Szolnok	93	79	77	76	75	72	-21	1.1
SGP	Bács-Kiskun	79	77	79	76	73	71	-8	1.5
ST	Somogy	71	77	76	75	70	69	-2	0.8
NH	Borsod-Abaúj-Zemplén	111	71	76	71	69	69	-42	2.8
SGP	Békés	89	80	78	76	72	69	-20	-0.1
NH	Nógrád	77	62	59	57	53	57	-20	1.3
NGP	Szabolcs-Szatmár-Bereg	59	62	61	59	58	57	-2	1.3
	National average	100	100	100	100	100	100		3.5

* 　CH = Central Hungary; CT = Central Transdanubia; WT = West Transdanubia; ST = South Transdanubia; NH = North Hungary; NGP = North Great Plain; SGP = South Great Plain.

+ 　Difference between the national percentages in 1975 and 1998; ++ GDP per capita annual growth rate.

Source: 　Author's calculations on the basis of Nemes-Nagy (2000) and the REGSTAT database.

Table 6.11 Distribution of NUTS 3 regions by GDP per capita, 1994–99 (percentages)

Regions*	1994	1995	1996	1997	1998	Diff. 94–98*	Δ 94–98[+]
Budapest	177	180	185	186	186	8	4.7
BEU	104	108	110	111	119	15	7.2
INT	82	81	81	84	85	3	4.4
BAC	78	77	76	74	73	–5	1.9
BEX	79	78	76	74	73	–6	1.5
Total border	81	81	80	79	79	–2	2.7
National average	100	100	100	100	100	0	3.5

* INT = Pest, Fejér, Veszprém, Tolna, Heves, Jász-Nagykun-Szolnok; BEU = Győr-Moson-Sopron, Vas; BAC = Komárom-Esztergom, Zala, Borsod-Abaúj-Zemplén, Nógrád, Hajdú-Bihar, Szabolcs-Szatmár-Bereg, Békés, Csongrád; BEX = Baranya, Somogy, Bács-Kiskun.

+ Difference between the national percentages in 1974 and in 1998.

++ GDP per capita annual growth rate.

Source: Author's calculations on the basis of the REGSTAT database.

between these formerly highly industrialized counties and the 'winners' in the transition process kept growing, while their difference from the less-developed agrarian Great Plain was significantly reduced. This slow but inexorable decay is primarily due to a crisis in obsolete heavy industrial plants and to the difficulties involved in reconversion while facing competition from a developed market. In the Great Plain region, the only county that has a relatively better position is Csongrád. This is due to the city of Szeged's capacity to attract the labour force and capital escaping from former Yugoslavia (Nemes Nagy, 2000). Besides, after the cancellation of the embargo with Yugoslavia, trade with the Balkans has recovered greatly. The eastern counties (Szabolcs-Szatmár-Bereg, Hajdú-Bihar, Békés and Jász-Nagykun-Szolnok) obviously were strongly affected by the crisis in Soviet-oriented industrial and agricultural mass production. Moreover, as a consequence of the very low level of infrastructure and of the obsolescence of existing industrial plants, this part of the country could not benefit from inflows of capital from abroad, which remained in the western part of the country or in the capital. Over the last part of the decade, the central and local governments have fostered the creation of new industrial parks in the eastern counties, in order to support the process of industrial reconversion.

In order to summarize the evidence previously pointed out, I can conclude that during the last decade, in terms of economic growth, the winner counties have been the two EU-border regions and Fejér. In the first case, proximity to new major export markets and a good system of infrastructure have played a pivotal role in determining successful performance. The second case is more evidently a success of Hungarian industrial reconversion and FDI promotion programs. However, both Industrial Parks and Customs-free Trade Zones have significantly affected the results of the 'winners'.

3.2   Relocation Patterns of Manufacturing Activities

Are these emerging regional disparities related to the relocation of manufacturing activities? In order to answer this question, I analyse some indices of regional specialization. The available dataset takes into account eight manufacturing sectors that can be interpreted as a synthetic version of NACE two digit sectors: *food, beverages and tobacco* (DA); *textiles, leather and related products* (DB + DC); *wood, paper and related products* (DD + DE); *non-metallic mineral products* (DF + DI); *chemicals* (DG); *machinery and equipment* (DK + DL + DM); *metallurgy and the manufacture of metal products* (DJ); and *other manufacturing* (DH + DN).

Table 6.12 Distribution of manufacturing activity across regions, 1992–98 (percentages)

Reg*	Counties	Employment			Output			Export		
		1992	1998	Diff.	1992	1998	Diff.	1992	1998	Diff.
CH	Budapest	31.5	23.3	-8.2	38.7	24.6	-14.1	5.1	16.8	11.6
CH	Pest	6.3	6.3	0.0	4.5	5.2	0.7	3.5	3.8	0.3
CT	Fejér	3.7	6.3	2.6	6.1	16.0	9.9	3.7	22.4	18.7
CT	Komárom-Esztergom	2.9	3.8	0.8	2.5	3.7	1.2	3.5	3.4	-0.1
CT	Veszprém	3.8	4.3	0.5	3.8	3.0	-0.8	6.8	2.4	-4.3
WT	Győr-Moson-Sopron	5.4	7.0	1.7	5.2	12.3	7.1	5.7	18.4	12.7
WT	Vas	3.1	5.1	2.0	2.8	6.3	3.5	5.5	8.9	3.4
WT	Zala	2.5	2.9	0.4	2.1	1.5	-0.6	5.3	1.0	-4.3
ST	Baranya	3.2	2.9	-0.3	2.4	2.3	-0.1	5.9	2.1	-3.8
ST	Somogy	1.8	2.5	0.7	1.2	1.9	0.7	5.2	2.2	-2.9
ST	Tolna	1.8	2.1	0.3	1.3	1.0	-0.3	5.4	0.8	-4.5
NH	Borsod-Abaúj-Zemplén	7.1	5.5	-1.6	7.3	5.1	-2.2	4.3	4.1	-0.2
NH	Heves	2.2	2.6	0.4	1.9	1.9	0.0	4.5	1.5	-3.0
NH	Nógrád	2.0	2.1	0.1	1.2	1.0	-0.2	4.5	0.7	-3.8
NGP	Hajdú-Bihar	4.5	4.6	0.2	3.4	3.1	-0.3	5.2	2.2	-3.0
NGP	Jász-Nagykun-Szolnok	3.6	3.8	0.2	3.7	2.5	-1.2	5.1	2.1	-3.0
NGP	Szabolcs-Szatmár-Bereg	2.9	3.3	0.4	2.2	2.1	-0.1	4.8	1.8	-3.0
SGP	Bács-Kiskun	4.3	4.4	0.1	3.2	2.2	-1.0	5.3	2.1	-3.2
SGP	Békés	3.4	3.3	-0.1	3.1	1.9	-1.2	5.7	1.5	-4.2
SGP	Csongrád	4.0	3.9	-0.1	3.3	2.4	-0.9	4.9	1.7	-3.3
	National total	100.0	100.0		100.0	100.0		100.0	100.0	0.0

Source: Author's calculations on the basis of the REGSTAT database.

First of all, I analyse the evolution of the distribution of manufacturing activities as whole, using three different units of measure: employment, output and export. Table 6.12 indicates that, during the 1990s, Hungary undertook a redistribution of these activities. In fact, Budapest has significantly decreased its role, to the benefit of such counties as Fejér, Győr-Moson-Sopron and Vas, which are emerging as newly industrialized regions. In the same direction, Borsod-Abaúj-Zemplén's crisis in both employment and output confirms the decline of the formerly industrialized northern regions. It is noteworthy that all the counties that have increased their share of output do not show a proportional growth in employment. The main reason for this effect is that, especially in the second half of the decade, the reorientation of manufacturing industries has been determined by the emergence of so-called 'second generation' firms. It should not be surprising if these new firms, which are mainly foreign owned,[12] were characterized by a definitively higher level of productivity.

In order to prove this hypothesis, I analyse the relationship between the productivity of labour and the weight of foreign-owned firms at the county level, both normalized by their respective national averages. To this purpose, the following model has been estimated:

$$\log (PROD_{jt}) = \alpha + \beta FDI_{jt} + \varepsilon_{jt} \tag{1}$$

where:
j is equal to the 20 Hungarian counties;
t is equal to the years between 1992 and 1998;
$PROD_{jt}$ is the employment/output ratio in region j normalized by the employment/output ratio at a national level;
FDI_{jt} is the foreign/domestic firm ratio in region j normalized by the foreign/domestic firm ratio at a national level (see Table 6.8).

I use both a regression with robust standard errors and regional dummy variables and a panel data regression with regional fixed effects. Table 6.13 shows the results of this econometric exercise, according to which the hypothesis of a positive correlation between the localization of FDI and productivity is verified. County dummy variables also allow us to identify significant county-specific effects on productivity. Dummies relative to the counties of Fejér, Vas, Komárom-Esztergom and Borsod-Abaúj-Zemplén are significant at 1 per cent, and show a positive coefficient.

These results can be interpreted with the help of a more in-depth analysis of the characteristics of the economic and industrial system of each county

Table 6.13   Does FDI affect productivity?

Counties	OLS[+]	FE
Constant	**-0.646	***-0.482
	(0.258)	(0.113)
FDI	***0.449	***0.449
	(0.145)	(0.161)
Adj. R^2	0.837	0.06
Prob. F-stat	0.000	0.0061
Observations	140	140
Test F on restrictions		***26.08

+ = OLS estimation with counties' dummy variable; standard errors in parentheses.
** and *** denote coefficient estimates significant at 5 and 1 per cent.

Source: Author's calculations on the basis of the REGSTAT dataset.

(Hungarian Ministry of Economic Affairs, 2000). In particular, the county of Komárom-Esztergom has experienced a process of industrialization similar to that of Fejér. It was a relatively well-industrialized county that has had to pass through a serious process of reconversion. Afterwards, it has benefited from some multinationals' investments, which have taken advantage of local industrial parks and Customs-free Zone incentives. As in the case of Fejér, it is possible that this FDI's impact on productivity is greater than the impact that can be explained by the increase in the number of foreign-owned firms relative to domestic ones. Finally, the case of Borsod-Abaúj-Zemplén is quite different. This county used to be Hungary's most remarkable industrial zone, thanks to the significant availability of ores and minerals and the related chemical industry. During the 1990s, this county's entire industrial system experienced a serious crisis and many large firms dramatically reduced their labour force. In this case, the positive coefficient of the dummy is probably due to local firms' reorganization process, which might have slightly increased the output/employment ratio.

In order to take into account possible region size effects, I have also measured patterns of location and relocation of manufacturing activities over population. Given that, in the past decade, there have been no significant shifts in population (see Section 3.1), this approach allows us to identify regions with over- (under-) proportional manufacturing intensity and shifts of manufacturing intensity across the counties. I consider the output/population ratio: when this index equals one, it suggests an even spread of manufacturing across the regions. When it changes over the course of the decade, it means that manufacturing activities have been reorganized and have moved across regions. When the change brings the index towards one, it means that regions were homogenized, while the opposite result means they have specialized.

Figure 6.1 shows the results of this approach. The bars higher than one represent the counties with over-proportional manufacturing intensity. In 1998 these were the counties of Budapest (1), Fejér (3), Komárom-Esztergom (4), Veszprém (5), Győr-Moson-Sopron (6), Vas (7) and Borsod-Abaúj-Zemplén (12). As expected, both of the EU border counties are in this group and have registered appreciable rates of intensification (the intensification of manufacturing activities in Vas is the most impressive).

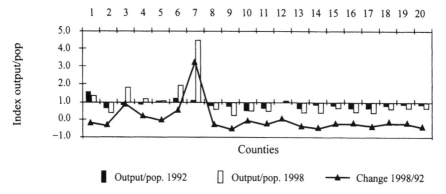

Figure 6.1   Relocation of manufacturing activity between regions (1992–98)*

*   Counties are ranked as in Table 6.12.

Source: Author's calculations on the basis of the REGSTAT dataset.

All the counties located in the Great Plain region have significantly reduced their concentration of manufacturing activities. The same has occurred in the eastern part of the country (with the exception of Borsod-Abaúj-Zemplén), and in the South Transdanudiana region (with the sole exception of Tolna).[13] It is interesting to note that Budapest is the only county that has reduced intensity among those that had an over-proportional manufacturing/population share in 1992, and that only Frejer and Borsod-Abaúj-Zemplén have changed their status from under- to over-proportional intensity of manufacturing activities.

Generally speaking, the relocation process of manufacturing activities is driven by three forces: first of all, a shift of manufacturing intensity from the capital (Budapest and Pest) counties to emerging counties such as Fejér, Győr-Moson-Sopron and Vas has occurred, due to the attractiveness of the EU border region and to the impact of FDI. Secondly, a trend of de-industrialization in the already less manufacturing-intensive regions of the Great Plain has occurred; thirdly, a process of reconversion of the formerly industrialized northern regions has occurred, which has involved a significant reduction of their manufacturing intensity.

In order to identify a clear trend in the concentration of manufacturing activities in Hungary, I take into consideration the 1992–98 shifts of the Lorenz curves measured in terms of employment, output and exports. Figure 6.2 indicates that, during the considered period, the concentration of manufacturing activities shows the most appreciable changes when we are considering export data.

In terms of employment (the first quadrant), one can observe a decrease in concentration in the largest manufacturing counties (Budapest has decreased its share from 31.5 per cent to 22.7 per cent and Borsod-Abaúj-Zemplén, from 7.1 per cent to 5.4 per cent) and a redistribution across a group of newly industrialized regions (Fejér has increased from a share of 3.7 per cent to a share of 6.5 per cent, Győr-Moson-Sopron from 5.4 per cent to 6.9 per cent and Vas from 3.1 per cent to 4.9 per cent). Part of the capital's weight has probably moved towards the larger county of Pest (from 6.3 per cent to 7.4 per cent), in which Budapest county is actually included.

In terms of output (the second quadrant), we can observe an even stronger decrease in concentration in Budapest (from 38.7 per cent to 24.6 per cent) and a weak increase in concentration in a larger group of winners in comparison with the country's losing regions. These results are consistent with the previous conclusion about the existence of a trend of relative de-industrialization of the capital and real de-industrialization of the northeastern part of the country to the benefit of the western part. Really spectacular examples of this are the increases occurring in Fejér (from 6.1 per cent to 16 per cent) and in the two EU border counties (Győr-Moson-Sopron, from 5.2 per cent to 12.3 per cent, and Vas from 2.9 per cent to 6.3 per cent).

Finally, export data (the third quadrant) shows a strong process of concentration in the northwestern part of the country. In fact, starting from a situation where there was substantially equal export distribution across the counties, in 1998, 67 per cent of all exports were concentrated in four counties (Budapest, with a share of 22.3 per cent, Győr-Moson-Sopron, with

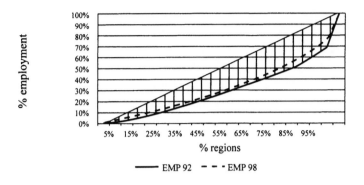

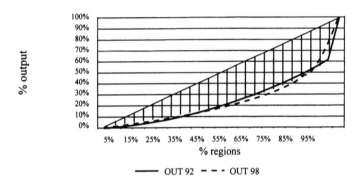

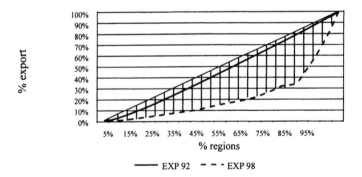

Figure 6.2   Evolution of concentration of manufacturing activity at a county level (1992–98)

Source: Author's calculations on the basis of the REGSTAT dataset.

18.4 per cent, Fejér, with 16.8 per cent, and Vas,with 8.9 per cent). All the other counties have lost shares. This effect is due to the high concentration of export-oriented FDI promoted by the Customs-free trade Zone incentives that characterize these counties.

The analysis carried out in this section allows me to conclude that during the 1990s, manufacturing activities in Hungary experienced not only a structural change in terms of export specialization, but also a relocation process within the country. The 'winner' counties in terms of economic growth are also the counties where manufacturing activities have increased outstandingly in terms of both output and export and where productivity has gained from the presence of TNCs' subsidiaries. The emergence of newly industrialized counties and the relocation of part of the economic activity outside of the capital city have produced a change in the concentration of the manufacturing sector within the country.

3.3 Changes in Regional Manufacturing Specialization

The previous section pointed out a shift in manufacturing activities and underlined both the relevance of proximity to the EU border and the presence of subsidiaries of foreign-owned enterprises. In this section, I analyse whether these shifts might be associated with processes of regional specialization.[14] To this purpose, I analyse the evolution of three indices of regional specialization: the Herfindahl index of absolute specialization, which ranges between zero and one, where one indicates complete specialization; the Balassa (1967) index of relative specialization, which ranges between zero and infinity (the higher the index, the more specialized the region); the Krugman (1991) dissimilarity index, which ranges between zero and infinity (the higher the index, the more differentiated the region).

Table 6.14 shows that, on average, the Herfindahl index remained virtually constant, around 0.17 (which is relatively low), during the 1990s. Looking at the county data, we can observe that, in 1999, the most specialized counties were: Fejér, where 52 per cent of employment has been concentrated in the production of *machinery and equipment* since 1997; Tolna, where more than 40 per cent of manufacturing employment is concentrated in the production of *textiles, leather and related products*; Somogoy, specialized in *foods, beverages and tobacco*; and Vas, where the index of specialization has decreased over time.

As a matter of fact, specialization seems to have moved from *textiles, leather and related products* (from 41 per cent of employment in 1992 to 35

Table 6.14 Herfindahl indices of absolute regional specialization (1992–99)

Reg*	Counties	1992	1993	1994	1995	1996	1997	1998	1999	1992/99
CH	Budapest	0.19	0.19	0.19	0.19	0.19	0.20	0.20	0.18	0.9
CH	Pest	0.21	0.19	0.19	0.19	0.19	0.19	0.23	0.22	1.1
CT	Fejér	0.26	0.22	0.23	0.25	0.29	0.33	0.30	0.33	1.3
CT	Komárom-Esztergom	0.21	0.17	0.17	0.19	0.18	0.18	0.22	0.21	1.0
CT	Veszprém	0.15	0.15	0.15	0.15	0.15	0.15	0.15	0.15	1.0
WT	Győr-Moson-Sopron	0.21	0.21	0.21	0.20	0.20	0.21	0.22	0.23	1.1
WT	Vas	0.26	0.27	0.28	0.27	0.28	0.27	0.27	0.25	0.9
WT	Zala	0.19	0.19	0.18	0.18	0.18	0.17	0.18	0.17	0.9
ST	Baranya	0.21	0.21	0.21	0.23	0.23	0.23	0.20	0.18	0.8
ST	Somogy	0.22	0.23	0.21	0.21	0.21	0.22	0.24	0.25	1.1
ST	Tolna	0.25	0.25	0.26	0.25	0.26	0.29	0.30	0.28	1.2
NH	Borsod-Abaúj-Zemplén	0.19	0.19	0.20	0.19	0.18	0.17	0.18	0.17	0.9
NH	Heves	0.22	0.22	0.22	0.22	0.23	0.24	0.24	0.22	1.0
NH	Nógrád	0.18	0.17	0.19	0.19	0.19	0.21	0.22	0.19	1.0
NGP	Hajdú-Bihar	0.21	0.20	0.21	0.20	0.21	0.21	0.24	0.23	1.1
NGP	Jász-Nagykun-Szolnok	0.23	0.22	0.21	0.21	0.21	0.21	0.24	0.23	1.0
NGP	Szabolcs-Szatmár-Bereg	0.19	0.18	0.19	0.18	0.18	0.19	0.20	0.18	1.0
SGP	Bács-Kiskun	0.22	0.22	0.23	0.22	0.21	0.21	0.21	0.19	0.9
SGP	Békés	0.25	0.24	0.24	0.21	0.21	0.21	0.19	0.19	0.8
SGP	Csongrád	0.20	0.19	0.19	0.18	0.19	0.19	0.19	0.17	0.8
	Hungary	0.17	0.16	0.16	0.16	0.17	0.17	0.18	0.18	1.0

Herfindahl index: $H_j^s = \sum_i (S_{ij}^s)^2$ where S_{ij}^s is the share of employment in industry i in region j in the total employment of region j.

Source: Author's calculations on the basis of the REGSTAT database.

per cent in 1999) towards the production of *machinery and equipment* (from 20 per cent in 1992 to 32 per cent in 1999). A very similar process has occurred in the other EU border region, that is, Győr-Moson-Sopron.[15]

The Balassa index is a relative measure that compares regional specialization with national patterns of specialization. This approach thus allows us to identify regions that are relatively more specialized than the national average.

Table 6.15 shows a different picture with respect to absolute specialization. First of all, the four regions that had the highest Herfindahl indexes in 1999 are at the bottom of the rank. This indicates that the regions that are more specialized in absolute terms actually have a pattern of specialization very similar to that of the whole of Hungary and, probably, have determined this pattern. In relative terms, the most specialized counties in 1999 were: Veszprém, with a high relative specialization in *non-metallic mineral products* (15 per cent of employment versus 4 per cent at a national level); Zala, which also shows a relatively high specialization in *non-metallic mineral products* (10 per cent of employment) and in *other manufacturing*; Csongrád, which has moved from a relative specialization in *other manufacturing* towards a relative specialization in *non-metallic mineral products*; Nógrád, which is also relatively specialized in *non-metallic mineral products* (15 per cent of employment). Thus, evidence from the relative specialization index allows us to conclude that, in Hungary, regional specialization patterns for the manufacture of mineral products differ significantly from the national pattern.

Finally, I take the Krugmann dissimilarity index into consideration. This index allows us to evaluate how much a regional pattern of specialization differs from a national one, taking all the sectors into account. The results shown in Table 6.16 confirm and expand the conclusions drawn from the two previous indices. In fact, I find that the three most differentiated counties in 1999 were: Fejér, because of the previously-mentioned overspecialization in *machinery and equipment* and the under-specialization in *textiles, leather and related products*; Tolna, which, in contrast is overspecialized in *textiles, leather and related products* and under-specialized in *machinery and equipment*; and Borsod-Abaúj-Zemplén, because of its over-specialization in *chemicals* (a residual of formerly planned industrial-ization) and its under-specialization in *machinery and equipment*.[16]

In conclusion, the growth of machinery and equipment production (which includes electric and non-electric machines and the automotive industry) has strongly determined the pattern of absolute specialization of 'winner' counties. The higher level of the absolute specialization index in terms of output (when compared to the index in terms of employment) confirms the increase in the

Table 6.15 Balassa indices of relative regional specialization (1992–99)

Reg*	Counties	1992	1993	1994	1995	1996	1997	1998	1999	1992/99
CH	Budapest	1.01	1.01	1.00	0.98	0.98	0.97	1.00	1.03	1.02
CH	Pest	0.87	0.91	0.92	0.93	0.93	0.98	0.94	0.95	1.08
CT	Fejér	0.94	0.95	0.95	0.93	0.87	0.82	0.90	0.82	0.88
CT	Komárom-Esztergom	1.01	1.08	1.06	1.01	1.04	1.04	1.02	1.01	0.99
CT	Veszprém	1.25	1.22	1.23	1.23	1.25	1.26	1.28	1.31	1.05
WT	Győr-Moson-Sopron	0.88	0.87	0.86	0.87	0.88	0.89	0.91	0.91	1.04
WT	Vas	0.86	0.83	0.81	0.83	0.83	0.82	0.83	0.85	1.00
WT	Zala	1.26	1.27	1.30	1.29	1.35	1.42	1.38	1.31	1.04
ST	Baranya	1.14	1.11	1.15	1.19	1.15	1.11	1.11	1.06	0.93
ST	Somogy	0.90	0.88	0.90	0.88	0.86	0.87	0.82	0.80	0.89
ST	Tolna	0.90	0.88	0.90	0.89	0.90	0.86	0.83	0.81	0.90
NH	Borsod-Abaúj-Zemplén	1.06	1.04	1.07	1.08	1.12	1.16	1.17	1.13	1.07
NH	Heves	0.95	0.99	0.99	1.02	0.97	0.96	0.97	0.99	1.04
NH	Nógrád	1.31	1.29	1.26	1.18	1.12	1.12	1.12	1.19	0.91
NGP	Hajdú-Bihar	0.89	0.92	0.91	0.93	0.94	0.98	0.88	0.85	0.95
NGP	Jász-Nagykun-Szolnok	0.85	0.88	0.91	0.93	0.94	0.91	0.86	0.88	1.04
NGP	Szabolcs-Szatmár-Bereg	0.96	0.94	0.93	0.95	0.96	0.95	0.94	0.99	1.03
SGP	Bács-Kiskun	0.92	0.94	0.93	0.94	0.95	0.95	0.94	0.95	1.03
SGP	Békés	1.02	1.03	1.04	1.14	1.16	1.16	1.19	1.11	1.09
SGP	Csongrád	1.14	1.11	1.12	1.13	1.14	1.12	1.14	1.21	1.07

Balassa index: $RS_j = \frac{1}{I}\sum_i RS_{ij}$ where $RS_j = \frac{s_{ij}}{s_j}$ and where s_{ij} is the share of employment in industry i in region j in the total employment of region j and s_i is the share of the total employment in industry i in total employment and I is the number of industries.

Source: Author's calculations on the basis of the REGSTAT database.

Table 6.16 Dissimilarity indices of regional specialization (1992–99)

Reg*	Counties	1992	1993	1994	1995	1996	1997	1998	1999	1992/99
CH	Budapest	0.39	0.38	0.40	0.41	0.38	0.37	0.35	0.31	0.79
CH	Pest	0.24	0.21	0.24	0.24	0.24	0.23	0.29	0.24	0.97
CT	Fejér	0.62	0.53	0.58	0.64	0.73	0.75	0.71	0.66	1.06
CT	Komárom-Esztergom	0.52	0.49	0.36	0.34	0.32	0.30	0.45	0.41	0.78
CT	Veszprém	0.42	0.42	0.41	0.44	0.47	0.47	0.44	0.42	0.99
WT	Győr-Moson-Sopron	0.29	0.27	0.28	0.27	0.25	0.22	0.22	0.24	0.83
WT	Vas	0.56	0.51	0.49	0.52	0.54	0.48	0.46	0.41	0.73
WT	Zala	0.48	0.49	0.45	0.42	0.56	0.57	0.58	0.54	1.12
ST	Baranya	0.61	0.55	0.57	0.70	0.70	0.65	0.48	0.36	0.59
ST	Somogy	0.38	0.41	0.36	0.33	0.30	0.33	0.32	0.31	0.81
ST	Tolna	0.55	0.55	0.56	0.55	0.58	0.65	0.68	0.62	1.14
NH	Borsod-Abaúj-Zemplén	0.56	0.54	0.60	0.61	0.60	0.56	0.59	0.54	0.96
NH	Heves	0.45	0.55	0.54	0.59	0.53	0.56	0.49	0.39	0.87
NH	Nógrád	0.57	0.53	0.59	0.53	0.50	0.52	0.53	0.41	0.72
NGP	Hajdú-Bihar	0.29	0.30	0.33	0.34	0.41	0.46	0.48	0.51	1.74
NGP	Jász-Nagykun-Szolnok	0.36	0.36	0.31	0.30	0.28	0.25	0.30	0.28	0.78
NGP	Szabolcs-Szatmár-Bereg	0.35	0.37	0.43	0.41	0.39	0.43	0.41	0.40	1.14
SGP	Bács-Kiskun	0.42	0.44	0.47	0.46	0.47	0.45	0.39	0.34	0.82
SGP	Békés	0.60	0.56	0.56	0.56	0.56	0.58	0.50	0.48	0.79
SGP	Csongrád	0.50	0.44	0.41	0.44	0.47	0.51	0.50	0.44	0.88

Dissimilarity index: where $DSR_j = \sum_i \left| s_{ij}^s - s_i \right|$ where s_{ij}^s is the share of employment in industry i in region j in the total employment of region j and s_i is the share of the total employment in industry i in total employment and I is the number of industries.

Source: Author's calculations on the basis of the REGSTAT database.

productivity of industries led by TNCs. In terms of relative specialization, the index shows higher values in the counties that are relatively well-endowed with natural resources or are linked to pre-liberalization manufacturing sectors. This means that the impact of the new emerging industries has not only determined the absolute specialization level of the winner counties, but also of Hungary as a whole. Finally, results on dissimilarity confirm the two previous conclusions, revealing some winners' relative overspecialization in emerging industries and some losers' overspecialization in traditional sectors.

3.4   Location and Concentration of Industrial Activity

This paragraph is devoted to the analysis of the absolute and relative concentration of manufacturing activities in terms of employment. Table 6.17 shows the Herfindahl index. It shows that the most concentrated sectors are *chemical products* and *wood, paper and printing products*, though both sectors have followed a decreasing pattern of geographical concentration. In 1992 around 68 per cent of the labour force employed in chemicals worked in two counties (54 per cent in Budapest and 14 per cent in Borsod-Abaúj-Zemplén). In 1999 the percentages were similar but significantly reduced (41 per cent and 10 per cent, respectively). In the second case, the effect is smaller and the values of the indices are determined only by the declining labour force concentration in Budapest (from 42 per cent in 1992 to 36 per cent in 1999), while in all the other counties there has never been more than 10 per cent total sectoral employment.[18]

In the case of relative concentration, there is not much evidence of particular patterns. It is interesting to note that the relatively high concentration of the *non-metallic mineral products* industry (even if it is decreasing) is due to the fact that this sector is mostly located in counties with very low manufacturing concentration, namely Veszprém (4.1 per cent of total manufacturing employment and 15 per cent in *non-metallic mineral products* in 1999), Nógrád (2 per cent and 7 per cent in 1999), and Csongrád (3.6 per cent and 9.2 per cent in 1999).

Finally, in terms of the Krugman dissimilarity index, *non-metallic mineral products* (because of its relative over-concentration in Veszprém and its under-concentration in Budapest) and *chemical products* (mainly due to its relative over-concentration in Budapest and Borsod-Abaúj-Zemplén) stand out.

Thus, the in depth analysis of concentration of manufacturing activities allow me to conclude that in absolute terms, the most concentrated industries at the county level are mainly related to the former process of heavy

Table 6.17 Herfindahl indices of absolute geographic concentration (1992–99)

	1992	1993	1994	1995	1996	1997	1998	1999	1999/92
Food, beverages and tobacco products	0.06	0.06	0.06	0.06	0.06	0.07	0.07	0.07	1.04
Textiles, apparel, leather products	0.11	0.10	0.09	0.09	0.08	0.08	0.08	0.08	0.71
Wood, paper and printing products	0.20	0.19	0.18	0.17	0.16	0.17	0.16	0.16	0.78
Chemical products	0.33	0.28	0.28	0.26	0.25	0.24	0.23	0.21	0.63
Non-metallic mineral products	0.13	0.11	0.10	0.09	0.09	0.09	0.08	0.08	0.60
Metallurgy and metal products	0.11	0.11	0.11	0.11	0.10	0.10	0.10	0.09	0.77
Machinery and equipment	0.21	0.19	0.17	0.15	0.14	0.13	0.11	0.10	0.50
Other manufacturing, recycling	0.11	0.11	0.10	0.09	0.09	0.09	0.09	0.11	1.00

Herfindal index: where $H_i^s = \sum_j (s_{ij}^s)^2$ where s_{ij}^s is the share of employment in industry i in region j in the total employment in industry i.

Source: Author's calculation on the basis of the REGSTAT database.

Table 6.18   Balassa indices of relative geographic concentration (1992–99)

	1992	1993	1994	1995	1996	1997	1998	1999	1999/92
Food, beverages and tobacco products	1.25	1.21	1.19	1.18	1.16	1.14	1.11	1.12	1.05
Textiles, apparel, leather products	1.11	1.10	1.09	1.09	1.11	1.13	1.12	1.13	0.68
Wood, paper and printing products	0.93	0.93	0.91	0.91	0.92	0.91	0.89	0.89	0.73
Chemical industry	0.61	0.67	0.67	0.68	0.68	0.69	0.70	0.73	0.70
Non-metallic mineral products	1.12	1.16	1.22	1.25	1.28	1.28	1.25	1.24	0.80
Metallurgy and metal products	1.07	1.07	1.05	1.03	1.00	1.01	1.05	1.04	0.90
Machinery and equipment	0.84	0.85	0.88	0.90	0.92	0.92	0.94	0.94	0.60
Other manufacturing, recycling	1.12	1.08	1.09	1.08	1.07	1.08	1.02	0.99	0.98

Balassa index: $RS_i = \frac{1}{J}\sum_J RS_{ij}$ where $RS_{ij} = \frac{s_{ji}}{s_j}$ and where s_{ij} is the share of employment in industry i of region j in the total employment in industry i and s_j is the share of total employment in region j in the total employment and J is the number of regions.

Source: Author's calculations on the basis of the REGSTAT database.

Table 6.19 Dissimilarity indices of relative geographic concentration (1992–99)

	1992	1993	1994	1995	1996	1997	1998	1999	1999/92
Food, beverages and tobacco products	0.42	0.38	0.40	0.40	0.39	0.39	0.36	0.34	0.80
Textiles, apparel, leather products	0.35	0.34	0.34	0.35	0.39	0.42	0.42	0.41	1.17
Wood, paper and printing products	0.41	0.40	0.42	0.42	0.44	0.45	0.43	0.38	0.93
Chemical products	0.64	0.63	0.64	0.65	0.67	0.64	0.62	0.57	0.90
Non-metallic mineral products	0.56	0.62	0.66	0.75	0.83	0.85	0.78	0.71	1.27
Metallurgy and metal products	0.52	0.51	0.55	0.53	0.45	0.45	0.56	0.38	0.72
Machinery and equipment	0.34	0.33	0.31	0.32	0.34	0.33	0.31	0.30	0.89
Other manufacturing, recycling	0.41	0.42	0.47	0.44	0.40	0.40	0.35	0.32	0.79

Dissimilarity index: where $DSR_i = \sum_j \left| s_{ji}^s - s_{ji} \right|$ where s_{ij}^s i is the share of employment in industry i in region j in the total employment in industry i and is the share of total employment of region j in national manufacturing employment.

Source: Author's calculations on the basis of the REGSTAT database.

industrialization in Hungary. In relative terms, industries related to specific natural resources stand out. Also, in this case, results in terms of dissimilarity confirm that there is an over-concentration of heavy manufacturing activities in traditionally industrialized counties and an overspecialization in natural resource processing activities for resource-abundant counties.

4 The Impact of Economic Integration on the Regional Wage Structure

This section is devoted to analysing whether and to what extent Hungary's process of progressive integration with the EU has been responsible for changes in the production structure of Hungarian counties. In the previous analysis I have already pointed out the outstanding performance of the western part of the country and particularly of the two border counties. These results permit us to suppose that integration with the EU has presented a strong incentive for Hungarian economic activities to move as close as possible to the new market, in order to exploit relatively low transport costs and, subsequently, agglomeration economies (Fujita, Krugman, Venables, 2000). A further effect of this process could be the destruction of economic agglomerations created before trade liberalization.

In order to explore the existence and importance of this manufacturing relocation process and of formerly industrialized counties' relative decline, I consider the following model (Hanson 1996):

$$\ln\left(\frac{W_{jt}}{W_{ct}}\right) = \alpha + \beta_1 \ln(DIST_{jc}) + \beta_2 BORD_j + \beta_3 \ln(DIST_{ic})*BORD_j + \varepsilon_{jt}$$

$$(2)$$

where:
j = *counties* ($n = 20$);
t = 1992–98;
W_{jt} is the wage in region j at time t;
W_{ct} is the wage in the core market in the autarky hypothesis (Budapest);
$DIST_{jt}$ is the distance from region j to Budapest;
$BORD_j$ is a dummy variable for EU border regions (it takes a value of one if the region shares borders with the EU and zero otherwise).

Distance is a proxy for unit transport costs (Fujita, Krugman, Venables, 2000), while the border dummy catches the potential time and distance invariant effects specific to regions bordering the EU. The interaction term accounts for the importance of the distance from industry centers for border regions. If the hypothesis that geographical distance from the capital matters, I expect relative wages to be decreasing with it and thus $\beta_1 < 0$.

According to Hanson (1996), trade liberalization and, in the case of Hungary, integration with the EU, should change the geography of economic activities, making distance from existing manufacturing centers less relevant for economic activities. In order to verify this hypothesis, I insert in equation (2) a dummy variable, YEAR, which takes a value of one if the year is equal or greater than 1994 (when the Association Agreement entered into force) and zero otherwise:

$$\ln\left(\frac{W_{jt}}{W_{ct}}\right) = \alpha + \beta_1 \ln(DIST_{jc}) + \beta_2 BORD_j + \beta_3 \ln(DIST_{ic})*BORD_j$$
$$+ \beta_4 \ln(DIST_{ic})*YEAR + \varepsilon_{jt} \tag{3}$$

If trade liberalization actually does eliminate distance's relevance, I expect to be insignificant or positive. Finally, trade liberalization should make distance not only irrelevant for the relative wage structure, but should also reduce the differentiation between border and interior regions. To test this hypothesis, I insert a new factor in (3), which becomes:

$$\ln\left(\frac{W_{jt}}{W_{ct}}\right) = \alpha + \beta_1 \ln(DIST_{jc}) + \beta_2 BORD_j + \beta_3 \ln(DIST_{ic})*BORD_j$$
$$+ \beta_4 \ln(DIST_{ic})*YEAR + \beta_5 \ln(DIST_{ic})*YEAR*BORD\ \varepsilon_{jt} \tag{4}$$

The case $\beta_1 + \beta_2 = \beta_3 + \beta_4$ indicates that distance effects for border and non-border regions converge to similar levels because of trade liberalization and economic integration with the EU.

Table 6.20 shows the estimated effects for equations (2), (3) and (4). They confirm the hypothesis that relative wages are decreasing as distance from Budapest increases. In fact, the estimated coefficient for distance [$\hat{\beta}_1$] is negative and statistically significant in all regressions. In addition, I have also found that distance matters even more for EU border regions, due to the negative coefficient of the ln*(DIST)**BORD variable [$\hat{\beta}_3$].

In contrast, I have no evidence for the hypothesis that, after the integration agreements with the EU, distance from the capital became less relevant. In fact, the estimations of equations (3) and (4) show a negative and significant coefficient for the variable $\ln(DIST)*YEAR$ $[\hat{\beta}_4]$. I also reject the null hypothesis of test H1, and this means that the overall coefficient of distance after trade liberalization $[\hat{\beta}_1 + \hat{\beta}_4]$ is significantly different from zero. These results indicate that distance from the capital city not only has remained a factor that explains differences among relative wages, but since 1994 its importance has, in fact, increased $(\hat{\beta}_1 + \hat{\beta}_4 > \hat{\beta}_1)$. This result shows that trade liberalization and integration with the EU probably did not have the same negative impact on pre-existing manufacturing activities that the integration between Mexico and the USA, as analysed by Hanson, did.[19]

However, integration has had an important impact on EU border regions, as the positive and statistically significant coefficient of the *BORD* dummy $[\hat{\beta}_2]$ indicates. This means that relative wages are definitely higher in EU border regions and the EU border effect is also large enough to offset the negative impact of distance from Budapest.

Further evidence of this positive effect is the different sign of the coefficient for $\ln(DIST)*BORD$ $[\hat{\beta}_4]$ (negative) and for $\ln(DIST)*YEAR*BORD$ $[\hat{\beta}_5]$ (positive), that is, the importance of distance from Budapest before and after the integration agreement. Therefore, after the integration agreement with the EU, the overall coefficient for distance of the EU border regions $[\hat{\beta}_1 + \hat{\beta}_3 + \hat{\beta}_5]$ is still negative, but has decreased in absolute terms. Overall, this means that distance from Budapest matters less for EU border counties than for the other counties.

This conclusion is confirmed by the results of tests H2 and H3: I reject the null hypothesis of test H2, but I fail to reject the null hypothesis of test H3. This indicates that, while before the integration agreements the overall coefficient for distance in border regions $[\hat{\beta}_1 + \hat{\beta}_3]$ was significantly different from zero, after the integration agreements the overall coefficient for distance of EU border counties $[\hat{\beta}_1 + \hat{\beta}_3 + \hat{\beta}_5]$ was not significantly different from all other counties $[\hat{\beta}_1 + \hat{\beta}_4]$, which confirms the decrease in the importance of distance from Budapest for EU border counties and their convergence with the rest of the country.

Given the leading role played by FDI in Hungarian manufacturing activities' reorientation process, I introduce a simple variation in the model previously presented in order to take into account the impact that a relative high concentration of FDI may have on relative wages. To this purpose, I simply add a variable, $\ln(FDI_{jt})$, to the equations (2), (3) and (4).[20]

The results reported in the (b) columns in Table 6.20 confirm FDI's pivotal role in determining productivity and, thus, higher wages. The introduction of

Table 6.20   Regression with robust standard errors of equations (2), (3) and (4)

	Equation (2)		Equation (3)		Equation (4)	
	(a)	(b)	(a)	(b)	(a)	(b)
Constant	−0.07*	−0.09***	−0.07*	−0.09***	−0.06*	−0.09*
	(0.036)	(0.029)	(0.036)	(0.029)	(0.035)	(0.047)
Distance	−0.07***	−0.06***	−0.05***	−0.04***	−0.05***	−0.04***
	(0.008)	(0.008)	(0.008)	(0.007)	(0.008)	(0.10)
EU-border	0.54***	0.51**	0.54**	0.51**	0.54***	0.51***
	(0.192)	(0.205)	(0.253)	(0.241)	(0.180)	(0.191)
Distance*	−0.09**	−0.10**	−0.09*	−0.10**	−0.10***	−0.11***
border	(0.037)	(0.040)	(0.049)	(0.047)	(0.035)	(0.037)
Distance*			−0.02***	−0.02***	−0.03***	−0.03**
year			(0.032)	(0.003)	(0.035)	(0.012)
Distance*					0.02***	0.02***
year*border					(0.005)	(0.005)
FDIs		0.08***		0.08***		0.08***
		(0.023)		(0.020)		(0.020)
Adj. R-squared	0.44	0.44	0.55	0.58	0.55	0.62
F-stat	41.71	41.76	41.83	48.46	41.49	22.22
Prob. F-stat	0.0000	0.0000	0.0000	0.0000	0.0000	0.0000
Observation	160	160	160	160	160	160
F on H1			115.08***	84.38***	115.75***	61.84***
[d.f]			[1, 155]	[1, 154]	[1, 154]	[1, 146]
F on H2	19.27***	15.25***	8.62***	8.60***	19.92***	15.15***
[d.f]	[1, 156]	[1, 155]	[1, 155]	[1, 154]	[1, 154]	[1, 146]
F on H3					2.58	2.60
[d.f]					[1, 154]	[1, 146]

Heteroscedasticity-consistent standard errors in parentheses.
*** significant at 0.01 level; ** significant at 0.05 level; * significant at 0.1 level.
H1 indicates H_0 : distance = − distance*year;
H2 indicates H_0 : distance = − distance*border;
H3 indicates H_0 : distance*year = distance*border + distance*border*year.

Source: Author's calculations on the basis of the REGSTAT database.

the new variable significantly increases R^2 for all regressions. As expected, the estimated coefficient for $\ln(FDI_{jt})$ $[\hat{\beta}_6]$ is positive and statistically significant in all specifications of the model. With reference to the other coefficients, there are no important differences with respect to the case without $\ln(FDI_{jt})$.

Finally, in order to control for counties' fixed effects, I estimate the three equations in first differences. The results presented in Table 6.21 confirm some of the previous conclusions. In fact, the coefficient for distance is negative and statistically significant in all regressions without ln(FDI_{jt}). Equation (2bis) shows that the negative coefficient for distance increased in absolute terms during the 1990s. Furthermore, the positive and statistically significant coefficient for ln*(DIST)*BORD*, indicates that this growth was less relevant in EU border counties than in the others. However, the positive and not statistically significant coefficient for ln*(DIST)*YEAR* in the equation (3bis) indicates that something occurred in the second half of the decade. Finally, considering the equation (4bis), the positive and statistically significant coefficient for distance after 1994 in the border regions (ln*(DIST)*YEAR*BORD*) confirms the inversion of the path of growth in the coefficient for distance in the EU border regions.

In conclusion, the previous analysis demonstrates that trade liberalization and economic integration with the EU, unlike other processes of integration of an emerging economy and a developed market, have had a positive impact not only on counties bordering the EU directly, but also on counties located in the western part of Hungary. This conclusion can be based upon the fact that distance from capital city has increased its relevance in explaining relative wages in the non-EU border counties even after 1994, while it is less relevant for EU border regions. Proximity to the EU border and FDI have a positive impact on productivity and, thus, on relative wages.

5 Conclusions

The evolution and composition of Hungary's external trade and the reconversion of its industrial system towards more market-orientated production indicate that a structural change has occurred during the 1990s as a consequence of increasing trade liberalization and economic integration with the EU. In this period, new emerging industries and new 'winner' counties have emerged. TNCs have played a pivotal role in both of these related processes. FDI and proximity to the EU border seem to be the two main factors explaining successful patterns of industrial reconversion. However, I must underline the important role of Hungarian industrial programs aimed at attracting FDI through special fiscal exemptions for export-oriented production and through real services provided within the industrial parks. This last instrument can also significantly determine agglomeration of homogeneous activities and, thus, local externalities of specialization.

Table 6.21   Regression of first differences of equations (2 bis), (3 bis) and (4 bis)

	Equation (2 bis)		Equation (3 bis)		Equation (4 bis)	
	(a)	(b)	(a)	(b)	(a)	(b)
Constant	0.003	0.001	0.003	0.001	0.003	0.002
	(0.006)	(0.008)	(0.006)	(0.008)	(0.006)	(0.008)
Distance	−0.004**	−0.003	−0.005***	−0.004	−0.005***	−0.004
	(0.002)	(0.003)	(0.002)	(0.004)	(0.002)	(0.004)
Distance*	0.005*	0.004	0.005*	0.004	−0.002	−0.003
Border	(0.003)	(0.004)	(0.003)	(0.004)	(0.003)	(0.004)
Distance*			0.001	0.001	0.000	0.000
Year			(0.002)	(0.002)	(0.002)	(0.002)
Distance*					0.008*	0.008
Year*border					(0.004)	(0.004)
FDIs		0.006		0.006		0.005
		(0.015)		(0.015)		(0.015)
Adj. R-squared	0.03	0.03	0.03	0.03	0.04	0.04
F-stat	3.80	2.76	3.55	2.78	3.08	2.58
Prob. F-stat	0.0247	0.0447	0.0163	0.0341	0.0246	0.0289
Observation	140	140	140	140	140	140
F on H4	3.80**	0.47	5.22***	0.71	4.71**	2.17
[d.f]	[2, 137]	[2, 136]	[2, 136]	[2, 135]	[2, 135]	[2, 134]

Heteroscedasticity-consistent standard errors in parentheses.
*** significant at 0.01 level; ** significant at 0.05 level; * significant at 0.1 level.
H4 indicates H_0 : distance = 0 and distance*border = 0.

Source: Author's calculations on the basis of the REGSTAT dataset.

Trade and FDI liberalization has also had an important impact on the concentration of manufacturing activities within the country. A slight decline in the capital city, the sharp decline in some formerly heavily industrialized counties and the emergence of newly industrialized counties have all modified the concentration of manufacturing activities, determining the clearly leading role of the western part of the country. Within the manufacturing sector, a new pattern of specialization, mainly determined by TNCs, has emerged for the country as a whole and is clearly evident in the new 'winner' counties. Concentration seems to be related to the idiosyncratic initial advantages of some counties, either caused by the presence of natural resources or by the former heavy industrialization process.

Finally, analysing the evolution of relative wages at county level, I have pointed out that the process of integration with EU markets has not generated the negative effects on pre-existing manufacturing activities observed in other cases of integration between an emerging industrial system and a developed and large market. There is some evidence of the impact of integration on productivity and, thus, on relative wages in the EU border regions, and on relative wages in the western part of the country, but there is no clear evidence of a dramatic shift from vertically integrated production, organized around the economic center of the country, to simple product assembly in the proximity of the EU border.

Notes

*   I owe thanks to several people for having helped and encouraged me in my research and studies. In particular, I would like to acknowledge the sound and constructive support I received from Mària Rèdei, Regional Geography Department, Eötvös Lóránd University, Budapest, and Laura Resmini, Bocconi University, Milan: without their excellent cooperation and guidance, it would not have been possible to analyse and understand the complex economic and social changes that occurred in Hungary during the transition. I hope that my work may provide a positive, if small, contribution to the existing knowledge about such a beautiful country.

1   The CMEA organization definitively collapsed in the second half of 1991.

2   This does not mean that there was a collapse in this activity. As Table 6.3 shows, agricultural products' share in total exports to the EU market was already declining. This decrease was significant until 1994, with the most visible contraction occurring between 1994 and 1995. The 1993 contraction occurred in terms of value above all, and was mainly determined by bad weather conditions.

3   1995 was the peak point in the privatization process, which has mainly determined the inflow of FDI during the first half of the decade (UNCTAD, 2001).

4   The data set REGSTAT was created within the framework of the research project 'European Integration, Regional Specialization and Location of Industrial Activity in Accession Countries' undertaken with financial support from the European Community's PHARE ACE Programme 1998.

5   The third county in this region (Zala) ranks fifth.

6   Note that Fejér, Győr-Moson-Sopron, Vas, Budapest and Pest are the only counties that have constantly improved their relative position during the process of economic and political transition (1975–99).

7   The status of 'Industrial Park' was first introduced in 1997 and granted to already established local firm agglomerations. The granting criterion is that the park must integrate at least ten businesses within five years with a total employment of a minimum of 500 people. According to the Ministry of Economic Affairs, in 2001, industrial parks produced one-fifth of the national industrial output and more than one-third of industrial exports, employing more than 100,000 people (Hungarian Ministry of Economic Affairs, 2001b).

Even more impressive performance has characterized the Customs-free Zones. According to the Ministry of Economic Affairs, at the end of 1999, 101 industrial Customs-free Trade Zones were operating in Hungary (though in Hungary one company usually constitutes one Customs-free Zone). Between 1996 and 1999 these firms increased their share of total Hungarian exports from 18 per cent to 43 per cent. In 1999 the ten top Hungarian exporters were operating in eight Customs-free trade zones, accounting for 35 per cent of total Hungarian exports. Yet, a relevant problem will be posed by EU integration: European law allows individual countries to set up Customs-free zones, but scrutinizes each applying firm before granting it access to the areas (Vadász, 2000).

8   It is worth noting that more than the 60 per cent of the industrial Customs-free Trade Zones are located in the Western part of the country. In particular, the industrial Customs-free Trade Zones located in Győr-Moson-Sopron and Fejér account for the 64 per cent of total exports for this kind of firm (Vadász, 2000).

9   In order to understand the potential impact of these investments on the local economy, it might be useful to recall that, after the restructuring of the Philips Group in 1997, the Székesfehérvár factory became the sole Philips VCR and combined TV/VCR production facility in Europe and that, in 2001, this subsidiary also became a regional distribution centre (ITDH, 2002). In the second part of the 1990s, IBM also undertook a huge investment in a plant producing export-only high capacity hard disks (for this reason IBM preferred the form of a Customs-free Zone company). In the last five years IBM has invested more than €200 million in Hungary, and total production exceeded €2.1 billion in 2000. Finally, in Székesfehérvár, there are several subsidiaries of automotive multinationals like Ford (with 1,150 employees and sales of €180 million in 2000) and Denso (with 600 employees).

10   These include, among others, Dunaferr in the steel industry, Dunapack in the paper industry and Elit in the textile sector.

11   During the 1990s, four industrial parks were officially recognized in the county. The Sóstó Industrial Park was funded in 1992 by Loranger. It is located close to Székesfehérvár, at the site of a former Soviet army base. It hosts firms operating in the plastics industry, component manufacturing and electronics (among the largest firms are Loranger, Denso Hungary, Phillips and Nokia). The VIDEOTON industrial park, also in Székesfehérvár, was founded in 1993 by the largest private Hungarian electronics supplier, VIDEOTON, and specializes in electronics (among the largest firms are VIDEOTON and IBM Storage Products). The Alba Industrial Zone, located at the border of the traditional industrial zone of Székesfehérvár, was established in 1996 by international investors in cooperation with the Local Agency of the Hungarian Foundation for Enterprise Promotion and specializes in electronics, the plastics industry and the clothing industry. Finally, the INNOPARK was founded in Dunaújváros in 1996 by the Local Government and hosts Aikawa's only European investment center (Hungarian Ministry of Economic Affairs, 2000b). Customs-free Trade Zones in Fejér county are mainly located in Székesfehérvár, Mór and Bicske.

12   As I have already pointed out, many of these new firms are the result of greenfield investments undertaken by TNCs or domestic firms already connected to them.

13   During the 1990s, Tolna's leather and textile industries significantly advanced and foreign firms specializing in underwear production have started to invest there (Hungarian Ministry of Economic Affairs, 2000).

14   Given the previous consideration of the deep impact that a restricted amount of FDI could have on the concentration of exports (see note 6) and given that the results in terms of output and employment lead to the same conclusions, in this and the following sections, I analyse only data relative to the specialization and concentration of manufacturing activity

in terms of employment. Nevertheless, it is worth noting that indices computed with output data are higher than those computed with employment figures; furthermore, changes in specialization are more marked then than in the case of employment. This result might be due to differences in productivity between new and already-established manufacturing activities.

15   In terms of output, the two EU border regions have significantly increased their indices of absolute specialization thanks to an impressive growth in the output of the *machinery and equipment* industry. In Győr-Moson-Sopron, this sector's share in total manufacturing output increased from 0.22 in 1992 to 0.75 in 1998. It is worth noting the impact of Audi's investment in the city of Győr. In 1994 the plant started its production of engines, becoming one of Audi's main engine suppliers, with total sales of €2.5 billion. In Vas, the share of the total output of the *machinery and equipment* industry increased from 31 per cent in 1992 to 68 per cent in 1998. In this case, much support was given by the subsidiaries of Opel, Packard Electric and Eybl in the automotive industry and Philips and Flextronics in the electronics industry.

16   In terms of output, the dissimilarity index shows some differences: Borsod-Abaúj-Zemplén has an over-specialization in *chemical products* even in terms of output, while the other two most differentiated counties are Zala and Békes. The first is differentiated mainly because of its under-specialization in *machinery and equipment*. The second is differentiated mainly because of its over-specialization in *food, beverages and tobacco* and its under-specialization in *machinery and equipment*.

17   In terms of output, the dissimilarity index shows some differences: Borsod-Abaúj-Zemplén has an over-specialization in *chemical products* even in terms of output, while the other two most differentiated counties are Zala and Békes. The first is differentiated mainly because of its under-specialization in *machinery and equipment*. The second is differentiated mainly because of its over-specialization in *food, beverages and tobacco* and its under-specialization in *machinery and equipment*.

18   It is also interesting to note the significant and increasing values of the absolute concentration index for the *metallurgy and metal products* industry in terms of output. In 1998, 37 per cent of total output was still concentrated in Fejér, while only 13 per cent of total employment in the sector was concentrated there. This result is more interesting if we take into account that, in the county of Borsod-Abaúj-Zemplén, where 13.5 per cent of the labour force was located in 1998, the share of output was just 10 per cent. This is further evidence of the higher productivity of firms located in Fejér.

19   It must be noted that the time series used for estimations started with 1992, when a significant part of the impact of trade liberalization on the Hungarian industrial system probably had already occurred.

20   FDI is the same variable as in equation (1).

References

Balassa, B. (1967), *Trade Liberalization among Industrial Countries*, McGraw Hill, New York.

Eurostat (2000), *Intra- and Extra EU Trade*, COMEXT CD-ROM, Luxembourg, Eurostat.

Eurostat (2001), *European Union Direct Investment Yearbook*, Luxembourg, Eurostat.

Fujita M., Krugman, P. and Venables, A. J. (2000), *The Spatial Economy*, MIT Press, Cambridge, Mass.

Grubel, H.G and Lloyd, P.J. (1975), *Intra-industry Trade, The Theory and Measurement of International Trade in Differentiated Products*, Macmillan, London.

Hanson, G.H. (1996), 'Localization Economies, Vertical Organization and Trade', *The American Economic Review*, Vol. 86, No. 5, pp. 1266–78.

Hungarian Ministry of Economic Affairs (2000), *The Economy of Counties*, http://www.ikm.iif.hu.

Hungarian Ministry of Economic Affairs (2001), 'Foreigners are most Welcome, too', *The Hungarian Economy*, Vol. 30, No. 1, Budapest.

Hungarian Ministry of Economic Affairs (2001b), *Széchenyi Plan Ten-year Development Program for Industrial Parks*.

Hunya, G. (1997), 'Foreign Direct Investment and its Effects in the Czech Republic, Hungary and Poland', The Vienna Institute for International Economic Studies, Reprint Series no. 168, June.

ICE (2002), *Ungheria, Guida Paese,* Editoria Elettronica ICE, Budapest – Roma.

ITDH (2002), *Welcome to Hungary*, http://www.business2hungary.com.

Kaminski, B. (1999), 'Hungary's Integration into the EU Markets: Production and Trade Restructuring', *Working Paper Series 2135*, World Bank, Washington, DC.

Krugman, P. (1991), *Geography and Trade*, MIT Press, Cambridge, Mass.

Krugman, P. and Venables, A.J. (1997), 'Integration, Specialization and Adjustment', *European Economic Review*, No. 40, pp. 959–68.

National Bank of Hungary (2002), *Annual Report*, Budapest, National Bank of Hungary.

Nemes-Nagy, J. (2000), 'The New Regional Structure in Hungary', in G. Petrakos, G. Maier and G. Gorzelak (eds), *Integration and Transition in Europe*, Routledge, London.

Szanyi, M. (2002), 'Spillover Effects and Business Linkages of Foreign-owned Firms in Hungary', *Institute for World Economics Working Papers*, Hungarian Academy of Science, Budapest

Vadász, Z. N. (2000), 'Customs-free Zones in the Lead' *The Hungarian Economy*, vol. 29, no. 2, Budapest.

UNCTAD (2001), *World Investment Report*, Geneva, United Nations.

Chapter 7

The Emerging Economic Geography in Romania[*]

Iulia Traistaru and Carmen Pauna

1 Introduction

During the last decade, the integration of Central and East European countries (CEECs) with the world economy, and, in particular, with the European Union (EU) has increased via trade and foreign investments (as documented in Chapter 1 of this book) and has most likely caused a reallocation of resources across sectors and regions. Sectoral shifts at the national level in the CEECs have been analyzed frequently (see Landesmann and Stehrer, 2002), but the reallocation of resources across regions requires more in-depth investigation. What impact has increased economic integration with the European Union (EU) had on the location of industrial activity in accession countries? Have patterns of regional specialization and geographical concentration of industries changed over the last decade? Does greater specialization imply greater polarization?

Chapter 2 of this book reviews the predictions of location and international trade theories about the implications of trade liberalization and increasing economic integration for the specialization and location of industries and discusses, on the basis of empirical studies, other countries' experiences, particularly those of the current EU member states and the members of the North Atlantic Free Trade Area (NAFTA). Given the existing evidence, one might expect that the structural change that has accompanied increased economic integration in EU accession countries has led to changes in the patterns of regional specialization and geographical concentration of industries in these countries.

This chapter provides empirical evidence about the impact of economic integration with the EU on patterns and changes in regional specialization and geographical concentration of industrial activity in Romania during the period 1991–99. The remainder of this chapter is organized as follows. Section 2 explains data and measurement issues. Section 3 provides evidence of increasing economic integration between Romania and the European Union via trade and foreign direct investments. Section 5 discusses the location and concentration of industrial activity. Section 6 examines the impact of integration and regional specialization on regional relative wages and regional growth. Section 7 concludes.

2 Data and Measurement

This research uses a specially created database, REGSTAT_RO, including regional indicators at the NUTS 2 and NUTS 3 levels for the period 1991–99. The data has been provided by the National Institute for Statistics.

We use employment data for 13 manufacturing industries, eight NUTS 2 regions and 41 NUTS 3 regions, respectively. Data on Gross Domestic Product (GDP) is available only at the NUTS 2 level for the period 1993–98. Unemployment data refers to registered unemployment.

Regional specialization and geographic concentration of industries are defined in relation to production structures.[1] A region j is 'specialized' in a specific industry i if this industry has a high share in the manufacturing employment of region j. The manufacturing structure of a region j is 'highly specialized' if a small number of industries have a large combined share in total manufacturing.

A specific industry i is 'concentrated' if a large part of production is carried out in a small number of regions.

Specialization and concentration can be assessed using absolute and relative measures. Several indicators are proposed in the existing literature, each with certain advantages as well as shortcomings. For our analysis, we have selected an absolute measure (the Herfindahl index) and a relative measure (the dissimilarity index derived from the index proposed by Krugman, 1991). The content and methodology related to these indicators is presented in Box 7.1.

3 Evidence of Increasing Economic Integration with the European Union

3.1 Trade Liberalization and Trade Performance

During the Cold War, Romania had a special relationship with the European Community due to its independent foreign policy. Romania was the only country within the Soviet Bloc to sign a trade agreement with the European

Box 7.1　　Indicators of regional specialization and geographic concentration of industries[2]

$E =$　employment

$s =$　shares

$i =$　industry (sector, branch)

$j =$　region

$s_{ij}^S =$　the share of employment in industry i in region j in total employment of region j

$s_{ij}^C =$　the share of employment in industry i in region j in total employment of industry i

$s_i =$　the share of total employment in industry i in total employment

$s_j =$　the share of total employment in region j in total employment

$$s_{ij}^S = \frac{Eij}{Ej} = \frac{Eij}{\sum_i Eij} \qquad\qquad s_{ij}^C = \frac{Eij}{Ei} = \frac{Eij}{\sum_j Eij}$$

$$s_i = \frac{Ei}{E} = \frac{\sum_j Eij}{\sum_i \sum_j Eij} \qquad\qquad s_j = \frac{Ej}{E} = \frac{\sum_i Eij}{\sum_i \sum_j Eij}$$

The Herfindahl index　　　　　　　**The dissimilarity (Krugman) index**

Regional specialization measure　　　Specialization measure

$$H_j^S = \sum_i (s_y^S)^2 \qquad\qquad DSR_j = \sum_i \left| s_y^S - s_i \right|$$

Geographical concentration　　　　　Concentration measure
measure

$$H_i^C = \sum_j (s_y^C)^2 \qquad\qquad DCR_i = \sum_j \left| s_y^C - s_j \right|$$

Community that granted Romanian exports preferential access to the its market. As a consequence, unlike the other Central and East European countries the volume of Romania's foreign trade with the European Community was high, with exports exceeding imports until 1992.

After 1990, like the other Central and East European countries, Romania liberalized its foreign trade, which led to a reorientation of trade towards the European Union. The signing of the Europe Agreement in 1993 has facilitated the increasing economic integration of Romania and the European Union. The

European Union's share has increased from 31.4 per cent in 1990 to 65.5 per cent in 1999 for exports, and from 19.7 per cent to 60.4 per cent, respectively, for imports.

The analysis of trade performance[3] between Romania and the European Union (EU-15) in the period 1993–98 shows the following patterns of specialization for Romania (see Tables 7.1 and 7.2):

Comparative advantage, with increasing RCA indices:
- live animals and animal products (until 1997, negative RCA in 1998);
- vegetable products (negative RCA in 1993);
- wood and articles of wood;
- base metals and articles of base metals.

Comparative advantage, with decreasing RCA indices:
- textiles and textile articles;
- footwear;
- articles of stone, plaster, cement;
- furniture and miscellaneous articles.

Comparative disadvantage, with increasing RCA indices:
- machinery and mechanical appliances;
- vehicles, aircraft, vessels (positive RCA in 1996);
- optical, precision and medical instruments.

Comparative disadvantage, with decreasing RCA indices:
- animal or vegetable fats (positive RCA in 1993);
- food, beverages, tobacco;
- chemical products;
- plastics and rubber;
- leather, fur skins and products thereof;
- pulp of wood, paper and paperboard.

The increasing degree of integration between Romania and the European Union is also suggested by increasing intra-industry indices (see Table 7.3), from 54.6 in 1993 to 61.5 in 1998. The product groups with the highest intra-industry indices in 1998 were the following: articles of stone, plaster and cement (96.7); vegetable products (95.4); textiles and articles of textiles (89.5); live animals and animal products (84.0); mineral products (83.9); and vehicles, aircraft and vessels (77.7).

Table 7.1   Trade coverage indices (TCI)* in the trade relations of Romania with the EU on product groups according to the Combined Nomenclature

	1990	1991	1992	1993	1994	1995	1996	1997	1998
Total	130.8	110.3	75.7	68.5	86.5	82.6	76.3	80.5	78.5
I Live animals and animal products	14.5	70.6	131.1	120.6	154.4	123.7	140.3	206.8	72.4
II Vegetable products	21.1	31.1	15.1	13.0	98.1	150.5	96.5	121.1	109.6
III Animal or vegetable fats	0.7	11.1	6.6	107.3	12.3	59.2	6.2	8.2	9.1
IV Food, beverages, tobacco	13.0	18.0	18.5	20.3	22.3	16.8	21.7	25.3	15.4
V Mineral products	711.2	322.4	62.6	164.1	212.3	77.1	73.5	47.6	72.3
VI Chemical products	22.9	45.1	46.1	26.4	43.2	36.7	29.6	26.2	16.4
VII Plastics and rubber and articles thereof	46.2	63.0	73.1	35.3	45.0	56.4	44.7	43.8	38.7
VIII Leather, fur skins and products thereof	33.2	42.9	33.7	32.2	33.9	21.8	16.8	19.1	18.5
IX Wood and articles of wood	203.4	225.5	407.5	219.5	306.6	198.9	273.9	305.8	478.4
X Pulp of wood, paper and paperboard and articles	87.3	30.5	14.6	13.2	17.5	23.6	12.9	12.8	9.4
XI Textiles and textile articles	177.1	169.8	152.3	124.0	141.4	131.5	131.9	127.2	123.6
XII Footwear, headgear	403.0	339.2	277.0	423.2	549.3	449.2	428.9	395.7	375.4
XIII Articles of stone, plaster and cement	165.7	239.5	226.9	170.4	162.7	128.0	103.8	108.0	106.8
XV Base metals and articles of base metal	265.2	183.3	209.2	148.7	193.3	244.7	174.5	228.6	208.4
XVI Machinery and mechanical appliances	78.6	36.7	21.1	16.4	20.4	22.6	22.8	25.2	29.6
XVII Vehicles, aircraft, vessels	65.3	52.1	14.4	29.3	54.5	66.6	87.0	63.9	63.5
XVIII Optical, precision and medical instruments	16.5	11.0	10.1	7.9	8.4	6.7	8.5	11.3	13.9
XX Furniture and miscellaneous articles	1,585.0	1194.3	828.3	526.9	377.4	340.4	279.0	255.0	239.5

* $TCI_i = X_i / M_i$;
X_i and M_i are, respectively, exports from and imports to the EU-15 for the product group i.

Source: Authors' calculations based on the EUROSTAT COMEXT database.

Table 7.2 Revealed comparative advantage (RCA) indices in the trade relations of Romania with the EU, 1993–98, on product groups according to the Combined Nomenclature

		1993	1994	1995	1996	1997	1998
I	Live animals and animal products	0.566	0.579	0.404	0.609	0.943	−0.080
II	Vegetable products	−1.658	0.126	0.600	0.234	0.408	0.334
III	Animal or vegetable fats	0.454	−1.949	−0.333	−2.505	−2.279	−2.154
IV	Food, beverages, tobacco	−1.217	−1.357	−1.591	−1.258	−1.156	−1.631
V	Mineral products	0.874	0.898	−0.069	−0.037	−0.525	−0.082
VI	Chemical products	−0.951	−0.695	−0.811	−0.947	−1.121	−1.563
VII	Plastics and rubber and articles thereof	−0.662	−0.654	−0.381	−0.535	−0.609	−0.708
VIII	Leather, fur skins and products thereof	−0.754	−0.938	−1.333	−1.515	−1.438	−1.447
IX	Wood and articles of wood	1.165	1.265	0.879	1.278	1.335	1.808
X	Pulp of wood, paper and paperboard and articles	−1.643	−1.597	−1.253	−1.777	−1.838	−2.119
XI	Textiles and textile articles	0.594	0.491	0.465	0.547	0.457	0.454
XII	Footwear, headgear	1.822	1.848	1.693	1.726	1.592	1.565
XIII	Articles of stone, plaster and cement	0.912	0.631	0.438	0.308	0.294	0.308
XV	Base metals and articles of base metal	0.775	0.804	1.086	0.827	1.043	0.977
XVI	Machinery and mechanical appliances	−1.427	−1.444	−1.298	−1.209	−1.163	−0.975
XVII	Vehicles, aircraft, vessels	−0.850	−0.461	−0.215	0.131	−0.231	−0.211
XVIII	Optical, precision and medical instruments	−2.156	−2.335	−2.516	−2.190	−1.964	−1.730
XX	Furniture and miscellaneous articles	2.041	1.473	1.416	1.296	1.153	1.116

$$
*\quad RCA = \ln \frac{\dfrac{X_i}{M_i}}{\dfrac{\sum_i X_i}{\sum_i M_i}}
$$

X_i and M_i are, respectively, exports from and imports to the EU-15 for the product group i.

Source: Authors' calculations based on the EUROSTAT COMEXT database.

Table 7.3   Intra-industry indices (GL)* in the trade relations of Romania with the EU on product groups according to the Combined Nomenclature

		1993	1994	1995	1996	1997	1998
Total		54.6	57.7	59.3	60.7	60.0	61.5
I	Live animals and animal products	90.7	78.6	89.4	83.2	65.2	84.0
II	Vegetable products	23.1	99.0	79.8	98.2	90.5	95.4
III	Animal or vegetable fats	96.2	21.9	74.4	11.7	15.2	16.7
IV	Food, beverages, tobacco	33.7	36.4	28.8	35.7	40.4	26.6
V	Mineral products	75.7	64.0	87.1	84.8	64.5	83.9
VI	Chemical products	41.8	60.3	53.7	45.7	41.6	28.2
VII	Plastics and rubber and articles thereof	52.2	62.1	72.2	61.8	60.9	55.8
VIII	Leather, fur skins and products thereof	48.7	50.6	35.8	28.7	32.1	31.2
IX	Wood and articles of wood	62.6	49.2	66.9	53.5	49.3	34.6
X	Pulp of wood, paper and paperboard and articles	23.4	29.8	38.2	22.9	22.7	17.2
XI	Textiles and textile articles	89.3	82.9	86.4	86.2	88.0	89.5
XII	Footwear, headgear	38.2	30.8	36.4	37.8	40.3	42.1
XIII	Articles of stone, plaster and cement	73.9	76.1	87.7	98.1	96.1	96.7
XV	Base metals and articles of base metal	80.4	68.2	58.0	72.9	60.9	64.9
XVI	Machinery and mechanical appliances	28.2	33.9	36.8	37.1	40.2	45.7
XVII	Vehicles, aircraft, vessels	45.3	70.6	79.9	93.1	78.0	77.7
XVIII	Optical, precision and medical instruments	14.7	15.5	12.5	15.7	20.3	24.4
XX	Furniture and miscellaneous articles	31.9	41.9	45.4	52.8	56.3	58.9

* $GI_i = 1 - |X_i - M_i| / (X_i + M_i)$
X_i and M_i are, respectively, exports from and imports to the EU-15 for the product group i.

Source: Authors' calculations based on the EUROSTAT COMEXT database.

The product groups with increasing intra-industry indices in the period 1993–98 include:

- vegetable products;
- mineral products;
- plastics and rubber;
- textiles and textile articles;
- footwear;
- articles of stone, plaster and cement;
- vehicles, aircraft and vessels;
- optical, precision and medical instruments;
- furniture and miscellaneous articles.

3.2   Foreign Direct Investment

The cumulated foreign direct investment (FDI) in Romania in the period December 1990–August 2000 reached US$4.89 billion. Compared with other accession countries, this stock level is much lower: at the end of 1999 the FDI stock was US$30 billion in Poland, US$19 billion in Hungary, and US$16 billion in the Czech Republic.[4] The number of firms with foreign investment increased in the period 1991–94 and decreased afterwards, as is shown in Table 7.4. The highest number of firms with foreign capital was registered in 1992 and the lowest in 1995.

The regional distribution of FDI indicates a high concentration of FDI in the capital, Bucharest: 51.3 per cent of the total foreign invested capital and 53.5 per cent of the number of firms with FDI in the period 1990–May 2000.

The FDI distribution in the same period according to the geographical origin of the invested capital is as follows:

Table 7.6 shows the FDI sectoral composition over the period from 1990 to May 2000. Manufacturing and mining account for nearly half of the FDI volume.

4   Regional Specialization Patterns

4.1   The Regional Structure of Romania

Romania is a medium-sized country with a territory of 238,391 km^2, and a population of 22,455.5 thousand inhabitants (on 1 January 2000). The regional

Table 7.4   The evolution of the number of firms with foreign investments

	Number of firms with FDI	Per cent
Total	76,931	100.0
1991	6,326	8.2
1992	12,194	15.9
1993	10,793	14.0
1994	11,521	15.0
1995	3,795	4.9
1996	4,056	5.3
1997	5,752	7.5
1998	9,114	11.8
1999	7,862	10.2
2000*	5,518	7.2

*   until August.

Source: National Office of Trade Registry, Romania.

Table 7.5   The geographical origin of foreign invested capital, 1990–May 2000

Region	Share in per cent
Europe	79.4
Asia	10.2
North America	8.3
South America	1.1
Africa	0.5
Oceania	0.5

Source: National Office of Trade Registry, Romania.

structure of Romania includes 41 counties (*judet*, corresponding to the NUTS 3 level) and the municipality of Bucharest. Each unit has its own local government, as do cities, towns, and *communes* (rural areas), within each county.

With the law 151/1998 on regional development in Romania, eight Development Regions have been created, corresponding to the NUTS 2 statistical level (see Table 7.A4). These regions have been established through

Table 7.6   The sectoral composition of FDI, 1990–May 2000

Sector of activity	Share in per cent
Mining, oil processing, machinery and electrical equipment	25.9
Professional services	20.6
Wholesale trade	14.5
Food industry	13.0
Light industry	10.8
Retail trade	7.8
Agriculture	2.7
Transport	2.0
Construction	1.9
Tourism	0.8

Source: National Office of Trade Registry, Romania.

voluntary cooperation of the counties, but do not have legal status and are not territorial-administrative units.

The territorial-administrative structure of Romania includes 263 towns (of which 84 are municipalities) and 2,688 communes (over 13,000 villages are grouped in these communes). The towns/communes correspond to the NUTS 4 level.

More than half of Romania's towns (152 of 263) have fewer than 20,000 inhabitants and only 23 towns have a population exceeding 100,000 inhabitants. Bucharest has more than two million inhabitants. Urban population represents 54.8 per cent of the total population. Table 7.7 shows the main geographic and demographic characteristics of the NUTS 2 regions in Romania.

Table 7.8 shows the GDP per capita in the eight NUTS 2 regions compared to the national and the EU-15 average for GDP per capita in 1999. The regions with GDP per capita above the national average are Bucharest (142 per cent), west (115 per cent) and Southeast (104 per cent). The poorest region is northeast, with only 76 per cent of the national GDP per capita. Compared to the EU-15's GDP per capita, the richest region, Bucharest, has 38 per cent while the poorest region, northeast, has only 21 per cent.

In Romania, regional disparities have historical, geographical, cultural and economic roots. These disparities, especially the economic ones, have expanded during the transition because, on the one hand, of substantial economic decline (at the end of 1999, the GDP reached only 75 per cent of its 1989 level), and, on the other hand, because of firms' behaviour in an economic environment with very high and long term inflation.

Table 7.7   **Geographic and demographic characteristics of Development Regions (NUTS 2), Romania, 2000**

Region	NUTS 3 component	Area km²	Population (thousands)
Romania	*42 (including Bucharest)*	*238,391*	*22,456*
1. Northeast	Bacau, Botosani, Iasi, Neamt, Suceava, Vaslui	36,850	3,810
2. Southeast	Braila, Buzau, Constanta, Galati, Tulcea, Vrancea	35,762	2,940
3. South	Arges, Calarasi, Dâmbovita, Giurgiu, Ialomita, Prahova, Teleorman	34,453	3,480
4. Southwest	Dolj, Gorj, Mehedinti, Olt, Vâlcea	29,212	2,410
5. West	Arad, Caras-Severin, Hunedoara, Timis	32,034	2,040
6. Northwest	Bihor, Bistrita-Nasaud, Cluj, Maramures, Satu-Mare, Salaj	34,159	2,850
7. Centre	Alba, Braşov, Covasna, Harghita, Mures, Sibiu	34,100	2,645
8. Bucharest-Ilfov	Bucuresti, Ilfov	1,821	2,281

Source: National Commission for Statistics (1999); National Agency for Regional Development (1999); Institute for Economic Forecasting (2000).

Table 7.8   **Regional GDP per capita disparities, Romania, 1999**

Region	Romanian average GDP per capita = 100%	EU-15 average GDP per capita = 100%
Romania	*100*	*27*
1 Northeast	76	21
2 Southeast	104	28
3 South	93	25
4 Southwest	98	26
5 West	115	31
6 Northwest	90	24
7 Centre	103	28
8 Bucurest-Ilfov	142	38

Source: National Institute for Statistics.

Moreover, the transition reveals the economic weakness of poorly developed areas: the strong dependence on a single industry, poor town planning, low locality attractiveness and insufficient utilities infrastructure development. The regions with dominant rural areas are the poorest. They are strongly dependent on agriculture and lack a young adult population (in past decades they have migrated to urban areas).

Beginning in 1997, the unemployment rate increased due to the acceleration of the restructuring process in the mining, chemical and petrochemical sectors and new legislation on compensatory payments.

Over time, some areas became deprived zones, with high unemployment concentrations. These are monoindustrial localities, with development levels below the national average and a lack of job opportunities. Thus, the unemployment rate is far above the national average rate in counties such as Hunedoara, Gorj and Valcea.

Rural areas are more affected by unemployment than urban areas.

Significant disparities also exist within each Development Region. For example, in the Centre Development Region, Braşov and Sibiu counties are significantly more urbanized and wealthier than the other four counties in the region.

4.2   Specialized and Diversified Regions

Table 7.A1 shows the regional structure of manufacturing in Romania in 1991 and 1999 for the eight NUTS 2 regions. The highest share of manufacturing is concentrated in the centre. In 1990 the combined share of the four regions with the highest shares in manufacturing (centre, south, northeast and Bucharest) was 57.52 per cent. The centre region gained 2.06 percentage points in 1999 compared with 1990, while Bucharest has lost 2.19 percentage points. In 1999, the combined share of the four regions with the highest shares in manufacturing was 60.16 per cent, suggesting a tendency towards concentration.

The specialization of regions at the NUTS 2 level is low, as shown in Tables 7.A2 and 7.A3. The highest absolute regional specialization in 1991 is found in Bucharest, northeast, west and southeast. Compared to 1991, in 1999 the southwest replaces the west region in the group of regions with the highest absolute specialization (Table 7.A4). Relative specialization is highest in southeast and northeast (Table A5). Over the period 1991–99, the absolute specialization decreased in six of the eight regions, while the relative specialization increased in five regions (see Table 7.A6 and Fig. 7.1

and 7.2). Both absolute and relative specialization has increased in southeast and northwest and has decreased in southwest, centre and Bucharest.

Table 7.A7 shows the regional structure of manufacturing in 1991 and 1999 at the NUTS 3 level. The regions with the highest shares in manufacturing include: Bucharest, Braşov, Prahova, Arges, Timis and Cluj. Compared with 1991, in 1999 the regional shares of manufacturing had declined most in Bucharest (2.92 percentage points), Prahova (0.53 percentage points), Dambovita (0.46 percentage points), Neamt (0.30 percentage points) and Suceava (0.24 percentage points). The highest increase of regional share in manufacturing in 1999 compared to 1991 occurred in Arges (1.37 percentage points).

The regions with greatest specialization include: Ialomita, Botosani, Caras-Severin, Salaj, Vaslui, Dambovita, Galati and Alba. The most diversified regions include: Iasi, Bihor, Tulcea, Bistrita-Nasaud, Bucharest, Sibiu, Neamt and Timis (see Tables 7.A8 and 7.A9).

At the NUTS 3 level, the regions have higher values for the absolute and relative indicators of regional specialization. The analysis of absolute and relative specialization shown in Tables 7.A8–7.A10 suggests the following patterns of regional specialization:

- *high and increasing specialization*: Ialomita, Valcea, Gorj, Calarasi;
- *high and decreasing specialization*: Caras-Severin, Salaj, Botosani, Dambovita, Braşov, Covasna;
- *diversified and increasing specialization*: Bihor, Teleorman, Dolj, Buzau, Bistrita-Nasaud, Neamt, Timis;
- *diversified and decreasing specialization*: Iasi, Bucharest, Sibiu.

4.3   How Similar/Different are Regional Industrial Structures?

In the above analysis we compared regional industrial (manufacturing) structures with the national structure and identified specialized and diversified regions. In a similar way, we can compare the industrial structures of pairs of regions and assess how similar/different the regional industrial structures are. The smaller the measure of bilateral difference, the more similar the production structures of the two regions.

The measures of bilateral difference between the industrial structures of pairs of regions at the NUTS 2 level for 1991 and 1999 are shown in Table 7.9. The bold figures indicate the most different regions and the bold italics the most similar ones. Over the period 1991–99, bilateral differences have increased in 42 cases out of a total of 56 pairs of regions.

Table 7.9   Bilateral regional specialization indices at the NUTS 2 level, Romania

Bilateral Krugman specialization indices – NUTS 2, Romania, 1991

	Northeast	Southeast	South	Southwest	West	Northwest	Central	Bucharest-Ilfov
Northeast	0.0000	**0.4776**	0.4680	**0.4818**	0.4641	0.4226	0.4069	**0.6079**
Southeast	0.4776	0.0000	0.1940	0.2757	0.3031	0.3914	0.2582	0.4244
South	0.4680	0.1940	0.0000	0.2927	0.3355	0.4517	0.2294	0.4055
Southwest	0.4818	*0.2757*	0.2927	0.0000	0.4483	**0.4722**	0.2646	**0.5431**
West	0.4641	0.3031	0.3355	0.4483	0.0000	0.3261	0.3351	0.2594
Northwest	0.4226	0.3914	0.4517	0.4722	0.3261	0.0000	0.3351	0.4557
Central	0.4069	*0.2582*	*0.2294*	*0.2646*	0.3351	0.3077	0.0000	0.4206
Bucharest-Ilfov	0.6079	0.4244	0.4055	0.5431	*0.2594*	0.4557	0.4206	0.0000

Bilateral Krugman specialization indices - NUTS 2, Romania, 1999

	Northeast	Southeast	South	Southwest	West	Northwest	Central	Bucharest-Ilfov
Northeast	0.0000	**0.5965**	**0.6144**	0.5363	**0.5564**	0.3552	0.3921	**0.5553**
Southeast	0.5965	0.0000	0.3302	0.4196	0.3295	0.4983	0.3124	0.4698
South	0.6144	0.3302	0.0000	0.2160	0.4138	**0.6566**	0.3528	0.4209
South-West	0.5363	0.4196	*0.2160*	0.0000	0.5029	**0.5519**	0.3975	0.4311
West	**0.5564**	*0.3295*	0.4138	0.5029	0.0000	0.3641	0.2489	0.3247
Northwest	0.3552	0.4983	0.6566	0.5519	0.3641	0.0000	0.3288	0.4574
Central	0.3921	*0.3124*	0.3528	0.3975	*0.2489*	0.3288	0.0000	0.3486
Bucharest-Ilfov	0.5553	0.4698	0.4209	0.4311	*0.3247*	0.4574	0.3486	0.0000

Source: Authors' calculations based on REGSTAT_RO database.

The production (manufacturing) structure in the northeast region appears the most different from those of the other regions and the bilateral differences have increased for four of the seven pairings of the northeast with other regions. The centre region seems to have similar production structures to the south, southeast and southwest regions. The west region is most similar to Bucharest and the centre and has converged with thecentre while diverging from Bucharest.

5   Location and Concentration of Industrial Activity

5.1   The Manufacturing Structure in Romania

Table 7.A11 shows the manufacturing structure in Romania in 1991 and 1999. In 1991, the three industries with the highest shares in manufacturing were: textiles and apparel (19.79 per cent), machinery and equipment (18.27 per cent) and metallurgy and metal products (11.18 per cent). Their combined share in manufacturing was 49.24 per cent. In 1999, the three industries with the highest shares in manufacturing were textiles and apparel (20.56 per cent), metallurgy and metal products (11.18 per cent) and food, beverages and tobacco (11.51 per cent). The combined share in manufacturing of the three industries with the highest shares was lower in 1999: 43.78 per cent. The most significant changes in the manufacturing structure in 1999 compared to 1991 were the increase in the share of food, beverages and tobacco (3.53 percentage points) and the decline in the share of machinery and equipment (7.28 percentage points).

5.2   Patterns of Geographic Concentration of Manufacturing

Tables 7.A12–7.A14 show absolute and relative concentration measures for manufacturing in Romania for the years 1991 and 1999. Our research results suggest an increasing geographical concentration of industries in seven out of the thirteen manufacturing branches. The five most concentrated industries include: motor vehicles and transport equipment; electrical machinery; paper and paper products; fuels, chemicals and chemical products; and rubber and plastic products. The five least concentrated industries are: food, beverages and tobacco; furniture and other manufactured goods, metallurgy and metal products; and wood and wood products.

5.3 Spatial Separation of Manufacturing

The indices of geographical concentration used in the above analysis show to what extent each industry is concentrated in a few regions. To understand the factors driving the location of industrial (manufacturing) activity, one would need to know, in addition, whether these (few) regions are close together or distant from each other. Midelfart-Knarvik et al. (2000) propose an index of spatial separation that takes into account the distances between locations. The spatial separation index of industry j (SP^j) is defined as follows:

$$SP^j = C\sum_{k=1}^{n} \sum_{l=1}^{n} (S_{kj}^C S_{lj}^C {}_{kj}^C)$$

where δ_{kl} is a measure of distance between two regions k and l, and C is a constant. SP^j can be interpreted as the weighted average of all bilateral distances between pairs of locations of an industry j, weighted by production shares S_{kj}^C and S_{lj}^C. The index is zero if industrial production is concentrated in a single location. The higher the value of the index, the more spatially separated the production.

Figure 7.1 shows the evolution of the spatial separation index for manufacturing at the NUTS 3 level over the period 1991–99. The spatial separation index shows a U-shaped evolution between 1991–96 with a minimum in 1994. Spatial separation increased between 1997 and 1998, and has slightly decreased in 1999 compared to 1997.

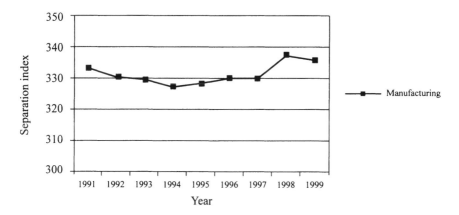

Figure 7.1   Spatial separation index, NUTS 3, Romania, 1991–99

Table 7.10 shows the spatial separation index calculated for the NACE two-digit industries at the NUTS 3 level in Romania. The five most spatially separated industries are: food, beverages and tobacco; textiles and textile products; furniture and other manufactured goods; wood and wood products; and leather and footwear. The five least spatially separated are machinery and equipment; fuels, chemicals and chemical products; paper and paper products; and rubber and rubber products.

The results obtained with the spatial separation index confirm the results we found with the indices of geographical concentration. The most concentrated industries are also the least spatially separated, while the least concentrated industries are the most spatially separated.

6 Does Greater Specialization Imply Greater Polarization?

6.1   *Location and Relocation Patterns of Industrial Activity*

Table 7.11 shows regional manufacturing shares in Romania over the period 1991–99. Manufacturing appears to be evenly distributed across the eight regions. The regions with the biggest shares in 1991 were centre, south, northeast and Bucharest, while the southwest and west regions had the lowest shares.

Table 7.11 indicates the location and relocation patterns of manufacturing in Romania over the period 1991–99. The biggest structural change occurred in the capital region and the centre region. Manufacturing seems to have moved away from Bucharest to the centre region. This change is mainly explained by the increasing shift to the services sector in Bucharest. The northwest region (western border region) increased its share in manufacturing by 1.38 percentage points, while the southwest region (mining and heavy industry) lost 1.45 percentage points, and the west region lost 1.10 percentage points. Population mobility might contribute to the avoidance of polarization.

Values close to one indicate an even spread of manufacturing across the population. The ratio's values increase for south, west, and nortwest, suggesting that these regions are the preferred destinations for relocation while northeast, southeast and Bucharest seem to be the losing regions in terms of population and manufacturing.

Table 7.12 compares the values of coefficients of variation for GDP and GDP per capita for the period 1993–98 at the regional level (NUTS 2). This comparison suggests that regional GDP has a greater concentration than the

Table 7.10 Spatial separation indices for manufacturing branches, Romania

Separation indices, NUTS 3, Romania, 1991 and 1999

Branch		SI		Rank of country (1 = least separation)	
		1991	1999	1991	1999
DA	Food, beverages, tobacco	357.1	355.5	12	11
DB	Textiles, textile products	354.9	357.3	11	12
DC	Leather, leather products	329.0	337.4	8	8
DD	Wood, wood products	343.9	343.2	9	9
DE	Pulp, paper, paper products, publishing & printing	292.6	308.0	4	4
DF+DG	Coke, petroleum products, nuclear fuel, chemicals, chemical products, man-made fibres	286.9	283.3	2	2
DH	Rubber, plastic products	313.6	309.3	5	5
DI	Other non-metallic metal products	324.7	318.3	7	6
DJ+DK	Basic metals, fabricated metal products	322.1	323.6	6	7
DL	Machinery, equipment, n.e.c.	264.6	275.9	1	1
DM	Transport equipment	290.5	286.6	3	3
DN	Manufacturing, n.e.c.	345.5	350.4	10	10

Source: Authors' calculations based on REGSTAT_RO database.

Table 7.11   Ratio of regional manufacturing shares and total population shares

Region	1991	1992	1993	1994	1995	1996	1997	1998	1999
Northeast	0.87	0.85	0.82	0.81	0.82	0.80	0.81	0.79	0.82
Southeast	0.81	0.79	0.80	0.80	0.81	0.78	0.78	0.77	0.80
South	0.95	1.01	1.02	1.02	1.03	1.03	1.03	1.03	1.00
Southwest	0.90	0.85	0.84	0.84	0.87	0.87	0.88	0.88	0.88
West	1.08	1.09	1.09	1.09	1.14	1.11	1.11	1.03	1.03
Northwest	0.96	1.00	1.02	1.02	1.01	1.01	1.01	1.01	1.04
Centre	1.23	1.29	1.33	1.33	1.32	1.37	1.36	1.36	1.46
Bucharest	1.33	1.23	1.18	1.18	1.12	1.11	1.11	1.11	1.02

Source: Authors' calculations based on REGSTAT_RO database.

Table 7.12   Dispersion of GDP and GDP per capita in Romania, 1993–98

Year	GDP	GDP per capita
1993	1.3162	0.1607
1994	1.3151	0.1939
1995	1.3145	0.1571
1996	1.3150	0.1785
1997	1.3142	0.1770
1998	1.3170	0.2413

Source: Authors' calculations based on REGSTAT_RO database.

regional GDP per capita, indicating that a greater concentration of GDP is matched by a greater concentration of population. However, the degree of concentration of the GDP has remained almost the same during the period 1993–98, while the concentration of the GDP per capita has increased, suggesting a tendency towards income polarization.

6.2   The Impact of Economic Integration on the Regional Wage Structure

What is the impact of the increasing economic integration with the EU on the regional wage structure in Romania? What are the determinants of regional relative wages?

Previous theoretical and empirical studies indicate that regional nominal and relative wages are decreasing with transport cost to industry centres (Krugman and Livas, 1996; Hanson, 1996; Hanson, 1997). In a closed economy, the industrial centre is, in most cases, the capital region, which concentrates human and capital resources in the process of industrialization. After trade liberalization, access to foreign markets becomes important and some industrial activity will relocate to border regions. As a consequence of congestion costs, wages in border regions will be driven up. The distance to the capital region will lose its importance in determining regional wage differentials.

Using data on regional manufacturing wages in Mexico before and after trade liberalization, Hanson (1996, 1997) found empirical support for the hypothesis that regional nominal and relative wages decrease with transport cost (proxied with distance) to industry centres (Mexico City and the regions bordering the United States).

Table 7.13 Summary statistics for regional wages relative to the capital region, Romania, 1992–99

	1992	1993	1994	1995	1996	1997	1998	1999
Mean	0.9559	0.9196	0.8346	0.8554	0.83016	0.8572	0.8238	0.6982
Standard deviation	0.0969	0.1006	0.0999	0.1082	0.1117	0.1052	0.0978	0.0772
Min	0.7722	0.7394	0.6995	0.6873	0.6603	0.6826	0.0978	0.0772
Max	1.2639	1.2792	1.1797	1.2490	1.1474	1.1021	1.1290	0.9123
Coefficient of variation	0.1014	0.1094	0.1197	0.1265	0.1346	0.1227	0.1187	0.1105

Source: Authors' calculations based on REGSTAT_RO database.

In the case of Romania, we have shown that manufacturing is concentrated in the capital region of Bucharest. Table 7.A7 indicates that Bucharest's share in total manufacturing employment has decreased, however, from 14.36 per cent in 1991 to 11.46 per cent in 1999. Meanwhile, the share of manufacturing employment has increased in the western border regions (Arad, Bihor, Timis and Satu-Mare). This evidence suggests that trade liberalization has made access to foreign (EU) markets an important factor for the location of industries and, as a consequence, some industrial activity has relocated to western border regions. Regional wages relative to the capital region are shown in Table 7.A15. In 1992, regional wages were lower in most cases. The exceptions were mining regions (Gorj, Valcea and Maramures) and industrial centres specialized in metallurgy (Galati and Hunedoara) or machinery (Braşov and Constanta). In 1999 regional wages in all regions were below the average wages in Bucharest. Summary statistics shown in Table 7.13 indicate decreasing regional wages relative to the capital region. Regional relative wage differentials have increased in the period 1992–96 and decreased in the period 1997–99.

In the following section, we test the hypothesis that regional relative wages decrease with distance from the capital city.

Model Specification Our econometric analysis is based on a model similar to the model in Hanson (1996).

$$\log (WAGE_{jt}/WAGE_{ct}) = \alpha + \beta_t \log(DIST_j) + \varepsilon_{jt} \qquad (1)$$

where:
$WAGE_j$ = annual average wage in region j;
$WAGE_c$ = annual average wage in the capital city;
$DIST_j$ = the distance (road distance in km) between region j (county capital) and capital city;
εjt = the error term with the following structure:
$\varepsilon_{jt} = \omega_j + \nu_t + \eta_{jt}$
ω_j is a fixed effect for region j, ν_t is a fixed effect for year t, η_{jt} is an independently and identically distributed random variable with mean zero and variance σ^2.

We allow distance effects to be different between western border regions and the rest of the regions by including in the regression the interacted variable DIST*BORD.

$$\log (WAGE_{jt}/WAGE_{ct}) = \alpha + \beta_t \log (DIST_j) + \gamma_t \log (DIST*BORD_j)$$
$$+\varepsilon_{jt} \qquad (2)$$

$BORD_j$ = dummy variable for western border regions:
$BORD_j$ =1 if region j is a western border region[5] and zero otherwise

We study the effect of trade liberalization on regional relative wages by interacting the distance variable with a dummy variable (EA) for the years after the signing of the Europe Agreement. EA takes value one for the years after 1993 and zero otherwise.

$$\log (WAGE_{jt}/WAGE_{ct}) = \alpha + \beta_t \log (DIST_j) + \gamma_t \log (DIST*BORD_j) +$$
$$+ \mu_t \log (DIST_j*EA) + \rho_t \log (DIST_j*BORD_j*EA) + \varepsilon_j \qquad (3)$$

We control for fixed time effects including year dummies (1992 is the omitted dummy variable). Since the distance variable varies across regions but not across time, including regional dummies to control for regional fixed effects would introduce perfect multicollinearity. To avoid this problem we instead include dummies for the NUTS 2 regions.

Estimation results The first three columns of Table 7.14 show the estimation results of models 1–3 with fixed year effects while the next three columns report the results obtained by repeating the estimations while controlling, in addition, for regional fixed effects.

We find that, in the period 1992–99, on average and with other factors being constant, regional relative wages are negatively correlated with the distance from the capital city, which confirms the hypothesis that regional relative wages decrease with the distance from the capital city. For example, in model 1 the distance coefficient indicates that a one per cent increase in the distance to the capital city results in a decrease of regional relative wages by three per cent. In the next model we allow distance coefficients to be different for the western border regions and the rest of the regions. The coefficient for distance is still negative and is statistically significant but lower than in model 1. The distance effect for border regions is lower compared to the rest of the regions, which indicates that regional relative wages in western border regions are higher than regional relative wages in the rest of regions.

Estimation results in model 3 confirm the hypothesis that trade liberalization reduces the importance of distance to the capital as a determinant of regional relative wages. The distance coefficients are no longer statistically significant. The distance coefficient for the western border region after trade liberalization is positive, suggesting that regional wages are higher in the regions that are furthest west from the capital city.

In models 4–6, we estimated the same three models with dummies for NUTS 2 regions. The results are similar to those obtained in models 1–3. The explanatory power of regressions is higher compared to the first three models.

In summary, we find empirical support for the hypothesis that regional relative wages decrease with distance to the capital city. This distance effect

Table 7.14 OLS estimation results of distance effects on relative regional wages

Variable	(1)	(2)	(3)	(4)	(5)	(6)
DIST	-0.030**	-0.023**	-0.009	-0.067***	-0.048**	-0.034
	(0.007)	(0.009)	(0.024)	(0.024)	(0.024)	(0.031)
DIST*BORD		-0.006**	-0.007		-0.013**	-0.014**
		(0.003)	(0.006)*		(0.004)	(0.007)
DIST*EA			-0.016			-0.016
			(0.026)			(0.022)
DIST*BORD*EA			0.001			0.001
			(0.007)			(0.007)
Constant	0.118***	0.081	0.004	0.268*	0.158	0.081
	(0.043)*	(0.048)	(0.126)	(0.147)	(0.144)	(0.180)
Year dummies	Yes	Yes	Yes	Yes	Yes	Yes
Region dummies	No	No	No	Yes	Yes	Yes
F statistics	31.59***	28.78***	24.66***	28.79***	27.63***	24.55
Adjusted R^2	0.385	0.383	0.386	0.511	0.529	0.586
N	320	320	320	320	320	320

Heteroscedasticity–consistent errors are given in parentheses.
* statistically significant at a 10 per cent level; ** statistically significant at a 5 per cent level; *** statistically significant at a 1 per cent level.

is lower in the case of western border regions. After trade liberalization the distance to the capital city is no longer significant as a determinant of regional relative wages.

6.3 Regional Specialization and Economic Growth

We investigate the relationship between regional specialization and growth by estimating a classical convergence model[6] to which we add a measure of regional specialization:

$$\log (y_{j,t+1}/y_{jt}) = \alpha + \beta \log y_{jt} + \delta\log(SPEC_{jt}) + \Sigma\gamma_i X_{ijt} + \varepsilon_{jt}$$
y_j = GDP per capita in region j
$SPECjt$ = specialization measure in region j
Xij = structural variables for qualitative regional characteristics
εjt = the error term

Data on regional GDP per capita is available only at the NUTS 2 level. Using data at the NUTS 2 level would, however, limit the degrees of freedom. We therefore first proxied regional GDP per capita at the NUTS 3 level using the regional employment structure, which we applied to the total GDP. We then calculated the regional GDP per capita. We estimate the model for the absolute regional specialization (using the Herfindahl index) and relative specialization (using the Dissimilarity index).

The results of the fixed effect estimation are presented in Table 7.15.

Initial structural variables considered for the model included: the share of employment in the secondary sector, the share of employment in services, the number of firms with foreign capital per 100,000 inhabitants, the number of self-employed per 100,000 inhabitants, the number of students per 100,000 inhabitants, the percentage of population in the age group 15–65 years, the number of telephone lines per 100,000 inhabitants, the density of roads and the public expenditure per capita. After checking for collinearity among the variables, we retained as control variables the self-employed per 100,000 inhabitants (a proxy for private enterprise intensity) and the percentage of working age population in total regional population.

We find that, on average and other factors being constant, regional specialization (absolute and relative measures) is negatively but not significantly related to regional GDP per capita growth. The positive and significant coefficient for the initial level of regional GDP per capita suggests that regions are diverging in terms of regional GDP per capita.

Table 7.15   Estimation results of the effect of specialization on regional growth

	Absolute specialization (Herfindahl index)	Relative specialization (Disimilarity index)
log $SPEC_{jt}$	−0.164	−0.050
	(0.131)	(0.126)
$logy_{jt}$	1.491**	1.482**
	(0.078)	(0.078)
Adjusted R^2	0.946	0.945
N	205	205

7   Conclusions

At the NUTS 2 level a low degree of specialization is found. The highest regional specialization is found for Bucharest, northeast, southwest and south and the lowest in the centre, west and northwest. Over the period 1991–99, regional specialization increased in the southeast and northeast, while Bucharest, the centre and the southwest have become more diversified. At the NUTS 3 level, the regions have higher values for the absolute and relative indicators of regional specialization. We have found the following patterns of regional specialization: regions with the highest and increasing specialization: Ialomita, Valcea, Gorj and Calarasi; regions with the highest specialization and decreasing: Caras-Severin, Salaj, Botosani, Dambovita, Braşov and Covasna; diversified regions with increasing specialization: Bihor, Teleorman, Dolj, Buzau, Bistrita-Nasaud, Neamt and Timis; diversified regions with increasing diversification: Iasi, Bucharest and Sibiu.

Our research results suggest an increasing geographical concentration of industries in seven out of the 13 manufacturing branches. The five most concentrated industries include: motor vehicles and transport equipment; electrical machinery; paper and paper products; fuels, chemicals and chemical products; and rubber and plastic products. The five least concentrated industries are: food, beverages and tobacco; furniture and other manufactured goods, metallurgy and metal products; and wood and wood products.

We find a greater concentration of regional GDP compared to the regional GDP per capita. This result suggests that a greater concentration of GDP is matched by a greater concentration of population. However, the degree of concentration of the GDP has remained almost the same over the period 1993–

98, while the concentration of the GDP per capita has increased, indicating a tendency towards income polarization.

We find empirical support for the hypothesis that regional relative wages decrease with distance from the capital city. This distance effect is lower in the case of western border regions. After trade liberalization the distance to the capital city is no longer significant as a determinant of regional relative wages.

We find that, on average and with other factors being constant, regional specialization (absolute and relative measures) is negatively but not significantly related to regional GDP per capita growth. Our results also suggest that regions are diverging in terms of regional GDP per capita.

Notes

*   This research was undertaken with support from the European Community's PHARE ACE Programme 1998. The content of the publication is the sole responsibility of the authors and it in no way represents the views of the Commission or its services. We thank Anna Iara for her excellent research assistance.

1   See Aiginger et al. (1999) for a survey of theoretical and empirical literature on regional specialization and geographic concentration of industries.

2   Indicators are defined following Aiginger et al. (1999).

3   This analysis is based on the following indicators: revealed comparative advantage (RCA), the trade coverage index (TCI) and the Grubel-Lloyd intra-industry index (GL).

4   Cf. *World Investment Report 2000*, UNCTAD, New York, p. 64.

5   Western border regions in Romania are: Satu-Mare, Bihor, Arad and Timis.

6   See Barro and Sala-i-Martin, 1992; Sala-i-Martin, 1996.

References

Aiginger, K. et al. (1999), 'Specialisation and (Geographic) Concentration of European Manufacturing', Enterprise DG Working Paper No 1, Background Paper for the Competitiveness of European Industry: 1999 Report, Brussels.

Barro, R.X. and Sala-i-Martin (1992), 'Convergence', *Journal of Political Economy*, Vol. 100, No. 2, pp. 223–51.

Hanson, G.H. (1996), 'Localization Economies, Vertical Organization, and Trade', *American Economic Review*, Vol. 87, pp. 1266–78.

Hanson, G.H. (1997), 'Increasing Returns, Trade, and the Regional Structure of wages', *Economic Journal*, Vol. 107, pp. 113–33.

Krugman, P. (1991), *Geography and Trade*, MIT Press, Cambridge.

Krugman, P. and Livas, R. (1996), 'Trade Policy and the Third World Metropolis', *Journal of Development Economics*, Vol. 49, No. 1, pp. 137–50.

Landesman, M. and Stehrer, R. (2002), 'The CEECs in the Enlarged Europe: Convergenge Patterns, Specialization, and Labor Market Implications', The Institute for Comparative Economic Studies (WIIW), Research Report No. 286, Vienna.

Midelfart-Knarvik, K., Overman, H., Redding, S. and Venables, A. (2000), 'The Location of European Industry', *Economic Papers no. 142*, European Commission, Directorate General for Economic and Financial Affairs, Brussels.

Sala-i-Martin, X. (1996), 'The Classical Approach to Convergence Analysis', *The Economic Journal*, Vol. 106, pp. 1019–36.

Appendix

Table 7.A1 Regional structure of manufacturing, Romania, 1991 and 1999, NUTS 2, in per cent

NUTS Regions	s_j1991	s_j 1999	Change
Northeast	14.20	14.63	0.43
Southeast	10.48	11.24	0.77
South	14.72	14.83	0.11
Southwest	9.54	8.09	−1.45
West	10.18	9.08	−1.10
Northwest	12.28	13.66	1.38
Centre	14.97	17.04	2.06
Bucharest	13.63	11.44	−2.19
Total	100.00	100.00	

Source: Authors' calculations based on REGSTAT_RO database.

Table 7.A2   The Herfindahl index for specialization (H_j^S), 1991–99, Romania, NUTS 2

Region NUTS 2	1991	1992	1993	1994	1995	1996	1997	1998	1999
Northeast	0.14261	0.13330	0.13426	0.12810	0.13183	0.13064	0.12773	0.13071	0.14223
Southeast	0.12439	0.13136	0.13418	0.13856	0.14303	0.14808	0.14187	0.15043	0.15044
South	0.12202	0.12995	0.12284	0.12361	0.11774	0.11484	0.11308	0.11291	0.11463
Southwest	0.11908	0.10571	0.10374	0.10151	0.10339	0.10304	0.10453	0.10618	0.10850
West	0.12970	0.12595	0.12638	0.12052	0.11954	0.11881	0.11703	0.11910	0.11590
Northwest	0.11928	0.11509	0.11549	0.10866	0.11109	0.11291	0.10927	0.11658	0.12263
Centre	0.11975	0.11892	0.12195	0.12179	0.11469	0.11752	0.10328	0.10947	0.11162
Bucharest	0.13529	0.11353	0.10533	0.10391	0.10289	0.10057	0.09888	0.10302	0.10197

Source: Authors' calculations based on REGSTAT_RO database.

Table 7.A3   The Krugman specialization index (DSR_j), Romania, 1991–99, NUTS 2

Region NUTS 2	1991	1992	1993	1994	1995	1996	1997	1998	1999
Northeast	0.30532	0.36625	0.38857	0.38179	0.39561	0.38138	0.36716	0.30983	0.33312
Southeast	0.25716	0.33997	0.36602	0.37936	0.41707	0.40656	0.39606	0.43376	0.36892
South	0.14984	0.29806	0.31074	0.34698	0.33813	0.35009	0.36204	0.34266	0.27350
Southwest	0.29415	0.24805	0.20466	0.23105	0.18896	0.22394	0.25891	0.22337	0.28544
West	0.19595	0.22103	0.24757	0.25404	0.24744	0.24660	0.24577	0.25074	0.24370
Northwest	0.24616	0.31944	0.33528	0.32276	0.31758	0.320466	0.32335	0.31960	0.26308
Centre	0.18320	0.19313	0.20986	0.20895	0.20788	0.19418	0.18047	0.16326	0.16669
Bucharest	0.37601	0.35577	0.37138	0.32186	0.32934	0.30171	0.27408	0.30009	0.27545

Source: Authors' calculations based on REGSTAT_RO database.

Table 7.A4   Regional specialization in Romania, 1991, NUTS 2

NUTS 2 Regions	$H_j^S 1991$	Rank	DSR_j	Rank
Northeast	0.142611	1	0.305315	2
Southeast	0.124392	4	0.257164	4
South	0.122017	5	0.149839	8
Southwest	0.119081	8	0.294148	3
West	0.129698	3	0.195953	6
Northwest	0.119283	7	0.246158	5
Centre	0.119746	6	0.183200	7
Bucharest	0.135289	2	0.376010	1

Source: Authors' calculations based on REGSTAT_RO database.

Table 7.A5   Regional specialization in Romania, 1999, NUTS 2

NUTS 2 Regions	$Hj^S 1999$	Rank	DSRj1999	Rank
Northeast	0.1422	2	0.3331	2
Southeast	0.1504	1	0.3689	1
South	0.1146	5	0.2735	5
Southwest	0.1085	7	0.2854	3
West	0.1159	4	0.2437	7
Northwest	0.1226	3	0.2631	6
Centre	0.1116	6	0.1667	8
Bucharest	0.1020	8	0.2754	4

Source: Authors' calculations based on REGSTAT_RO database.

Table 7.A6   Changes in patterns of regional specialization, Romania, 1991–99, NUTS 2

NUTS 2 Regions	$H_j S1991$	$H_j S1999$	Change	$DSR_j 1991$	$DSR_j 1999$	Change
Northeast	0.142611	0.142227	d	0.305315	0.333119	i
Southeast	0.124392	0.150435	i	0.257164	0.368924	i
South	0.122017	0.114630	d	0.149839	0.273497	i
Southwest	0.119081	0.108498	d	0.294148	0.285444	d
West	0.129698	0.115900	d	0.195953	0.243697	i
Northwest	0.119283	0.122627	i	0.246158	0.263081	i
Centre	0.119746	0.111622	d	0.183200	0.166691	d
Bucharest	0.135289	0.101972	d	0.376010	0.275449	d

Source: Authors' calculations based on REGSTAT_RO database.

Table 7.A7 Regional structure of manufacturing in Romania, 1991 and 1999, NUTS 3

NUTS 3 Regions	s_j1991	Rank	s_j1999	Rank	Change
Bacau	3.01	10	3.08	10	0.07
Botosani	1.51	29	1.47	31	-0.04
Iasi	3.20	7	3.42	8	0.22
Neamt	2.79	12	2.49	14	-0.30
Suceava	2.55	15	2.31	16	-0.24
Vaslui	1.79	24	1.85	24	0.07
Braila	1.64	27	1.62	28	-0.02
Buzau	2.22	17	1.95	20	-0.27
Constanta	2.05	19	1.91	23	-0.14
Galati	2.86	11	3.51	7	0.66
Tulcea	0.82	39	0.89	38	0.07
Vrancea	1.25	32	1.37	33	0.12
Arges	3.74	5	5.11	3	1.37
Calarasi	0.94	37	0.84	39	-0.10
Dambovita	2.76	13	2.30	17	-0.46
Giurgiu	0.53	41	0.45	41	-0.08
Ialomita	0.63	40	0.55	40	-0.08
Prahova	4.74	3	4.21	4	-0.53
Teleorman	1.43	30	1.38	32	-0.05
Dolj	2.65	14	2.50	13	-0.15
Gorj	1.10	34	0.96	37	-0.13
Mehedinti	0.92	38	1.13	36	0.21
Olt	1.74	25	1.72	27	-0.02
Valcea	1.52	28	1.78	25	0.26
Arad	1.98	21	2.11	19	0.13
Caras-Severin	1.39	31	1.53	30	0.14
Hunedoara	2.03	20	1.92	22	-0.11
Timis	3.41	6	3.52	6	0.11
Bihor	2.53	16	3.00	12	0.47
Bistrita-Nasaud	1.13	33	1.19	35	0.06
Cluj	4.22	4	4.05	5	-0.17
Maramures	1.80	23	2.11	18	0.31
Satu Mare	1.81	22	1.75	26	-0.07
Salaj	0.98	36	1.56	29	0.58
Alba	2.10	18	2.33	15	0.23
Braşov	4.96	2	5.13	2	0.17
Covasna	1.06	35	1.28	34	0.22
Harghita	1.74	26	1.93	21	0.19

Table 7.A7 cont'd

NUTS 3 Regions	s_j1991		Rank	s_j1999		Rank	Change
Mures		3.02	9	3.02		11	0.00
Sibiu	3.12	8	3.34	9	0.22		
Mun. Bucuresti (including Ilfov)		14.36	1	11.44		1	-2.92
Total	100.00		100.00				

Source: Authors' calculations based on REGSTAT_RO database.

Table 7.A8   Regional specialization in Romania, 1991, NUTS 3

NUTS 3 Regions	H_j^S	Rank	DSR_j	Rank
Bacau	0.132229	35	0.539469	23
Botosani	0.267752	2	0.763619	5
Iasi	0.147921	27	0.348335	41
Neamt	0.125	38	0.431188	34
Suceava	0.178013	15	0.686769	11
Vaslui	0.219384	5	0.565131	21
Braila	0.141406	29	0.500335	27
Buzau	0.070401	41	0.459323	32
Constanta	0.126679	37	0.525896	24
Galati	0.208617	7	0.680286	12
Tulcea	0.161495	22	0.62603	14
Vrancea	0.178225	14	0.500426	26
Arges	0.16717	19	0.569007	20
Calarasi	0.184089	11	0.710226	8
Dambovita	0.214963	6	0.741506	6
Giurgiu	0.181895	12	0.646819	13
Ialomita	0.269558	1	0.921054	1
Prahova	0.169163	18	0.578142	18
Teleorman	0.1482	26	0.394073	39
Dolj	0.135839	32	0.427523	35
Gorj	0.169383	17	0.705513	9
Mehedinti	0.154361	24	0.506587	25
Olt	0.164723	20	0.622628	15
Valcea	0.156371	23	0.778401	4
Arad	0.145621	28	0.467462	31
Caras-Severin	0.266937	3	0.915196	2
Hunedoara	0.187277	10	0.587691	17

Table 7.A8 cont'd

NUTS 3 Regions	H_j^S	Rank	DSR_j	Rank
Timis	0.135292	33	0.445646	33
Bihor	0.129626	36	0.357501	40
Bistrita-Nasaud	0.140752	31	0.394148	38
Cluj	0.111061	40	0.477573	29
Maramures	0.164706	21	0.548487	22
Satu Mare	0.179491	13	0.473293	30
Salaj	0.242224	4	0.907734	3
Alba	0.204334	8	0.574915	19
Braşov	0.188641	9	0.700537	10
Covasna	0.150315	25	0.724977	7
Harghita	0.171626	16	0.618808	16
Mures	0.116052	39	0.484195	28
Sibiu	0.141193	30	0.396728	36
Mun. Bucuresti (including Ilfov)	0.135289	34	0.396461	37

Source: Authors' calculations based on REGSTAT_RO database.

Table 7.A9   Regional specialization in Romania, 1999, NUTS 3

NUTS 3 Regions	H_j^S	Rank	DSR_j	Rank
Bacau	0.172751	19	0.598304	25
Botosani	0.195729	10	0.64544	20
Iasi	0.145526	33	0.327868	40
Neamt	0.124247	38	0.444117	38
Suceava	0.148163	31	0.704313	13
Vaslui	0.264466	5	0.727561	8
Braila	0.174952	17	0.576165	28
Buzau	0.157174	27	0.534074	34
Constanta	0.152045	29	0.660168	19
Galati	0.290045	3	0.844054	4
Tulcea	0.216292	7	0.683399	17
Vrancea	0.30786	1	0.712584	12
Arges	0.187455	12	0.716112	10
Calarasi	0.277752	4	0.764833	6
Dambovita	0.168921	21	0.732718	7
Giurgiu	0.205016	9	0.691644	14

Table 7.A9 cont'd

NUTS 3 Regions	H_j^S	Rank	DSR_j	Rank
Ialomita	0.295854	2	0.950466	1
Prahova	0.160101	25	0.714851	11
Teleorman	0.165778	22	0.549594	32
Dolj	0.145678	32	0.573733	30
Gorj	0.17167	20	0.864616	2
Mehedinti	0.141739	34	0.676243	18
Olt	0.178701	15	0.585977	27
Valcea	0.165171	24	0.857322	3
Arad	0.150017	30	0.448329	37
Caras-Severin	0.212886	8	0.807722	5
Hunedoara	0.174638	18	0.587063	26
Timis	0.132813	36	0.532844	35
Bihor	0.165508	23	0.626324	22
Bistrita-Nasaud	0.121057	39	0.575143	29
Cluj	0.114068	40	0.47028	36
Maramures	0.177732	16	0.550596	31
Satu Mare	0.183585	14	0.625462	23
Salaj	0.192972	11	0.689453	15
Alba	0.153568	28	0.62501	24
Braşov	0.158131	26	0.626656	21
Covasna	0.186293	13	0.687909	16
Harghita	0.231369	6	0.725442	9
Mures	0.127376	37	0.547187	33
Sibiu	0.139152	35	0.353792	39
Mun. Bucuresti (including Ilfov)	0.101972	41	0.282847	41

Source: Authors' calculations based on REGSTAT_RO database.

Table 7.A10 Changes in patterns of regional specialization, Romania, 1991–99, NUTS 3

NUTS 3 Regions	DSRj1991	DSRj1999	Change	H_j^S 1991	H_j^S 1999	Change
Bacau	0.539469	0.598304	i	0.132229	0.172751	i
Botosani	0.763619	0.64544	d	0.267752	0.195729	d
Iasi	0.348335	0.327868	d	0.147921	0.145526	d
Neamt	0.431188	0.444117	i	0.125	0.124247	d
Suceava	0.686769	0.704313	i	0.178013	0.148163	d
Vaslui	0.565131	0.727561	i	0.219384	0.264466	i
Braila	0.500335	0.576165	i	0.141406	0.174952	i
Buzau	0.459323	0.534074	i	0.070401	0.157174	i
Constanta	0.525896	0.660168	i	0.126679	0.152045	i
Galati	0.680286	0.844054	i	0.208617	0.290045	i
Tulcea	0.62603	0.683399	i	0.161495	0.216292	i
Vrancea	0.500426	0.712584	i	0.178225	0.30786	i
Arges	0.569007	0.716112	i	0.16717	0.187455	i
Calarasi	0.710226	0.764833	i	0.184089	0.277752	d
Dambovita	0.741506	0.732718	d	0.214963	0.168921	i
Giurgiu	0.646819	0.691644	i	0.181895	0.205016	i
Ialomita	0.921054	0.950466	i	0.269558	0.295854	i
Prahova	0.578142	0.714851	i	0.169163	0.160101	d
Teleorman	0.394073	0.549594	i	0.1482	0.165778	i
Dolj	0.427523	0.573733	i	0.135839	0.145678	i
Gorj	0.705513	0.864616	i	0.169383	0.17167	i
Mehedinti	0.506587	0.676243	i	0.154361	0.141739	d
Olt	0.622628	0.585977	i	0.164723	0.178701	i
Valcea	0.778401	0.857322	i	0.156371	0.165171	i

Table 7.A10 cont'd

NUTS 3 Regions	DSRj1991	DSRj1999	Change	H_j^S 1991	H_j^S 1999	Change
Arad	0.467462	0.448329	d	0.145621	0.150017	i
Caras-Severin	0.915196	0.807722	d	0.266937	0.212886	d
Hunedoara	0.587691	0.587063	d	0.187277	0.174638	d
Timis	0.445646	0.532844	i	0.135292	0.132813	d
Bihor	0.357501	0.626324	i	0.129626	0.165508	i
Bistrita-Nasaud	0.394148	0.575143	i	0.140752	0.121057	d
Cluj	0.477573	0.47028	d	0.111061	0.114068	i
Maramures	0.548487	0.550596	i	0.164706	0.177732	i
Satu Mare	0.473293	0.625462	i	0.179491	0.183585	i
Salaj	0.907734	0.689453	d	0.242224	0.192972	d
Alba	0.574915	0.62501	i	0.204334	0.153568	d
Braşov	0.700537	0.626656	d	0.188641	0.158131	d
Covasna	0.724977	0.687909	d	0.150315	0.186293	i
Harghita	0.618808	0.725442	i	0.171626	0.231369	i
Mures	0.484195	0.547187	i	0.116052	0.127376	i
Sibiu	0.396728	0.353792	d	0.141193	0.139152	d
Mun. Bucuresti (including Ilfov)	0.396461	0.282847	d	0.135289	0.101972	d

i= increase; d = decrease.

Source: Authors' calculations based on REGSTAT_RO database.

Table 7.A11   The manufacturing structure in Romania, 1991 and 1999, in per cent

Industries	1991	1999	Change
Food, beverages and tobacco	7.98	11.51	3.53
Textiles and apparel	19.79	20.56	0.77
Tanning and dressing of leather, footwear	3.87	4.60	0.73
Wood and wood products	2.74	4.57	1.83
Paper and paper products	1.93	2.30	0.37
Fuels, chemicals and chemical products	6.02	6.44	0.42
Rubber and plastic products	2.37	2.11	-0.26
Other non-metallic products	5.91	5.52	-0.39
Metallurgy and metal products	11.18	11.71	0.54
Machinery and equipment	18.27	10.99	-7.28
Electrical machinery	5.76	4.40	-1.35
Motor vehicles and transport equipment	8.22	8.81	0.59
Furniture and other manufactured goods	5.98	6.47	0.50

Source: Authors' calculations based on REGSTAT_RO database.

Table 7.A12   Concentration of manufacturing, Romania, 1991

Industries	H_i^C 1991	Rank	DCR_i 1991	Rank
Food, beverages and tobacco	0.034408	13	0.313732	13
Textiles and apparel	0.037772	12	0.333558	12
Tanning and dressing of leather, footwear	0.077672	3	0.656876	6
Wood and wood products	0.04509	10	0.692003	5
Paper and paper products	0.109227	2	0.828138	3
Fuels, chemicals and chemical products	0.05849	7	0.732729	4
Rubber and plastic products	0.071259	5	0.619168	8
Other non-metallic products	0.053447	8	0.477695	10
Metallurgy and metal products	0.049521	9	0.50196	9
Machinery and equipment	0.067766	6	0.434446	11
Electrical machinery	0.179281	1	0.870806	2
Motor vehicles and transport equipment	0.077424	4	0.914282	1
Furniture and other manufactured goods	0.041073	11	0.619884	7

Source: Authors' calculations based on REGSTAT_RO database.

Table 7.A13 Concentration of manufacturing in Romania, 1999

Industries	$H_i^C 1999$	Rank	$DCR_i 1999$	Rank
Food, beverages and tobacco	0.038215	12	0.619424	13
Textiles and apparel	0.036513	13	0.670447	12
Tanning and dressing of leather, footwear	0.080031	5	1.178521	3
Wood and wood products	0.049613	10	0.933841	9
Paper and paper products	0.087489	3	1.029469	6
Fuels, chemicals and chemical products	0.065882	6	1.175499	4
Rubber and plastic products	0.080107	4	1.071849	5
Other non-metallic products	0.053619	9	0.945183	8
Metallurgy and metal products	0.057217	8	0.931956	10
Machinery and equipment	0.060611	7	0.962022	7
Electrical machinery	0.122738	1	1.342792	1
Motor vehicles and transport equipment	0.09058	2	1.217353	2
Furniture and other manufactured goods	0.042336	11	0.825127	11

Source: Authors' calculations based on REGSTAT_RO database.

Table 7.A14   Changes in concentration of manufacturing, Romania, 1991–99

Industries	HiC1991	HiC1999	Change	DCRi1991	DCRi1999	Change
Food, beverages and tobacco	0.034408	0.038215	i	0.313732	0.619424	i
Textiles and apparel	0.037772	0.036513	d	0.333558	0.670447	i
Tanning and dressing of leather, footwear	0.077672	0.080031	i	0.656876	1.178521	d
Wood and wood products	0.04509	0.049613	i	0.692003	0.933841	i
Paper and paper products	0.109227	0.087489	d	0.828138	1.029469	i
Fuels, chemicals and chemical products	0.05849	0.065882	i	0.732729	1.175499	d
Rubber and plastic products	0.071259	0.080107	i	0.619168	1.071849	d
Other non-metallic products	0.053447	0.053619	i	0.477695	0.945183	i
Metallurgy and metal products	0.049521	0.057217	i	0.50196	0.931956	i
Machinery and equipment	0.067766	0.060611	d	0.434446	0.962022	i
Electrical machinery	0.179281	0.122738	i	0.870806	1.342792	i
Motor vehicles and transport equipment	0.077424	0.09058	i	0.914282	1.217353	i
Furniture and other manufactured goods	0.041073	0.042336	i	0.619884	0.825127	i

i = increase; d = decrease.

Source: Authors' calculations based on REGSTAT_RO database.

Table 7.A15 Regional relative wages in Romania, 1992–99

NUTS 2 region	1992	1993	1994	1995	1996	1997	1998	1999
Bucuresti-Ilfov	1.0000	1.0000	1.0000	1.0000	1.0000	1.0000	1.0000	1.0000
Bacau	1.0174	0.9442	0.8642	0.8668	0.8444	0.8977	0.8799	0.7538
Botosani	0.7722	0.7394	0.6995	0.6873	0.6603	0.6826	0.7222	0.5748
Iasi	0.9078	0.8687	0.7767	0.7887	0.7550	0.8080	0.7768	0.6565
Neamt	0.9387	0.8962	0.7719	0.8109	0.7824	0.8186	0.7609	0.6250
Suceava	0.8652	0.8456	0.7691	0.8004	0.7731	0.7606	0.7283	0.5978
Vaslui	0.8403	0.8096	0.7245	0.7609	0.7220	0.7433	0.7437	0.6051
Braila	0.9465	0.9228	0.8009	0.8312	0.8085	0.8035	0.7820	0.6305
Buzau	0.8934	0.8550	0.7645	0.8138	0.7569	0.7616	0.8035	0.6633
Constanta	1.0826	1.0705	0.9959	1.0457	1.0461	1.0816	1.0584	0.8828
Galati	1.0686	1.0473	0.9745	1.0381	1.0850	1.1021	0.9878	0.8031
Tulcea	0.9219	0.8689	0.7831	0.7194	0.6885	0.7987	0.7531	0.6452
Vrancea	0.8907	0.8760	0.8169	0.7480	0.7315	0.8175	0.8029	0.6584
Arges	0.9773	0.9381	0.8460	0.9194	0.9332	0.9724	0.8989	0.7305
Calarasi	0.9391	0.8837	0.7844	0.7920	0.7433	0.7666	0.7186	0.6252
Dambovita	0.9532	0.9384	0.8239	0.8808	0.8619	0.8886	0.8742	0.7191
Giurgiu	0.8943	0.8647	0.8496	0.8252	0.7747	0.7868	0.8253	0.6478
Ialomita	0.9303	0.8879	0.7822	0.8220	0.7988	0.7939	0.7814	0.6788
Prahova	1.0468	0.9660	0.8630	0.8761	0.8663	0.9225	0.9250	0.8081
Teleorman	0.9065	0.8909	0.8304	0.8236	0.7901	0.8611	0.8131	0.6814
Dolj	0.9645	0.9162	0.8509	0.9012	0.8685	0.8876	0.8814	0.7201
Gorj	1.2366	1.2792	1.1642	1.1091	1.0513	1.0826	1.1290	0.9123
Mehedinti	0.9810	0.9771	0.9515	0.9033	0.9046	0.9627	0.8570	0.7650
Olt	0.9209	0.8661	0.8254	0.8795	0.8512	0.9159	0.9052	0.7744

Table 7.A15 cont'd

NUTS 2 region	1992	1993	1994	1995	1996	1997	1998	1999
Valcea	1.0454	0.9791	0.8928	0.9019	0.8902	0.8838	0.8478	0.7094
Arad	0.8765	0.8291	0.7600	0.7442	0.7210	0.7181	0.7870	0.6689
Caras-Severin	0.9729	0.9048	0.8169	0.8593	0.8383	0.8122	0.7655	0.7275
Hunedoara	1.2639	1.2281	1.1797	1.2490	1.1474	1.0852	0.9942	0.8409
Timis	0.9677	0.9069	0.8304	0.8204	0.8033	0.8433	0.7164	0.6974
Bihor	0.9456	0.9390	0.8236	0.8882	0.8815	0.9118	0.8348	0.7321
Bistrita-Nasaud	0.8802	0.8604	0.7594	0.7982	0.7569	0.7739	0.7685	0.6347
Cluj	0.9630	0.9235	0.8316	0.8710	0.8314	0.9055	0.8542	0.7359
Maramures	1.0366	0.9285	0.8039	0.8605	0.8105	0.8189	0.7170	0.6589
Satu Mare	0.8545	0.8275	0.7543	0.7495	0.7325	0.7905	0.6730	0.6385
Salaj	0.9567	0.9164	0.8190	0.8214	0.7585	0.8124	0.8167	0.6064
Alba	0.9338	0.9004	0.8017	0.8501	0.8448	0.8421	0.7891	0.6927
Braşov	1.0859	1.0323	0.8898	0.9854	1.0396	0.9919	0.9199	0.7577
Covasna	0.9005	0.8786	0.7894	0.8066	0.7649	0.7500	0.7484	0.6426
Harghita	0.8487	0.8191	0.7490	0.7326	0.7165	0.7199	0.7231	0.6435
Mures	0.8908	0.8564	0.7503	0.7909	0.7758	0.8784	0.8057	0.7231
Sibiu	0.9164	0.9010	0.8191	0.8421	0.7957	0.8351	0.7813	0.6594

Source: Authors' calculations based on REGSTAT_RO database.

Chapter 8

The Emerging Economic Geography in Slovenia*

Jože P. Damijan and Črt Kostevc

1 Introduction

In the beginning of the 1990s, transition countries opened up and reoriented their trade flows towards the European Union (EU). According to the New Economic Geography (EG) theory, extensive trade liberalization with the EU should lead to the inter-regional relocation of manufacturing activities towards border regions close to EU markets. The aim of the present chapter is to analyse the pattern of regional relocation of manufacturing in Slovenia after trade liberalization with the EU. We first provide an in-depth analysis of the regional structure of manufacturing in Slovenia and its relocation pattern after trade with the EU had been opened up. In addition, we test some of the predictions of EG theory for Slovenia. The econometric tests follow the reasoning of the Damijan and Kostevc (2002) model. We assume the following inter-regional adjustment mechanism after trade liberalization has begun to take place. The relocation of factors of production might be, for reasons besides those of market access, driven by extensive FDI inflows from EU countries directed either to existing economic centers or to EU border regions (western and northern regions). Depending on the size of existing economic centers and on inter-regional trade costs, trade liberalization aggravated by FDI inflows may, therefore, either enhance or dampen existing agglomeration effects. Increasing or decreasing differences in relative regional wages may then reverse the agglomeration/deglomeration processes. As a consequence, an (inverted) U-shaped curve of relative regional wages and manufacturing output with respect to the inter-regional trade costs (distance) in the home country might occur in the long run. More specifically, in the first stage of trade liberalization, a divergence (convergence) in relative wages and output is probable, but afterwards this might turn into convergence (divergence). In any case, a non-monotonic relationship between the reduction of foreign trade costs and relative regional wages is expected in the long run. In this chapter, therefore, we analyse the effects of trade liberalization on the regional pattern of FDI inflows, the relocation of manufacturing and the inter-regional adjustment of relative wages in Slovenia.

The structure of this chapter is as follows. In the second section we discuss the evolution of the economic integration of Slovenia with the European Union. The third and fourth sections discuss the patterns of regional specialization and industrial concentration, respectively. Section 5 deals with the regional specialization, polarization and unemployment issue. The effect of economic geography on the relocation pattern of manufacturing output, FDI and wages is tested in section 6. The final section provides some conclusions in the analysis.

2 Evidence on Increasing Integration with the EU

2.1   *Trade Liberalization and Trade Performance*

After the break-up of the former Yugoslavia in 1991 and the subsequent loss of most of the domestic market, Slovenia embarked on an intensive drive to reorient its trade. Slovenia followed a diversified pattern of trade liberalization. In addition to the Cooperation and Europe Agreements with the EU, Slovenia was rapidly entering into free trade agreements (FTAs) with EFTA and CEFTA member states as well as with other European countries. Thus, up to the present day, Slovenia has signed FTAs with 32 European countries, which accounted for 86 per cent of the total Slovenian foreign trade in 2000. Keeping in mind that Slovenia is also a member of the WTO, it is clear that Slovenian foreign trade is almost completely liberalized. This can also be seen in the average import duties, which have been lowered to 2.3 per cent in 1999 for manufacturing goods in general and to 1.8 per cent for imports from the EU.

Table 8.1 demonstrates that the EU is Slovenia's most important trading partner, with a 64 per cent share of exports and a 68 per cent share of imports in 2000. Except for the trade that takes place with Germany, which is Slovenia's largest individual trading partner (about 30 per cent of exports), Slovenia's trade with the EU is characterized by large trade deficits. The trend towards trade deficit is worsening, as the trade coverage ratio has deteriorated to 81 per cent in 2000. Obviously, this might indicate a deterioration of the competitiveness of Slovenia's exports in the EU markets.

In fact, Table 8.2 reveals that in three out of the four largest exporting industries (transport equipment, machinery and electrical and optical

Table 8.1 The role of the EU in Slovenia's foreign trade in 1992–2000 (in €m)

Year	Ex-total	Ex-EU15	EU share	Im-total	Im-EU15	EU share	EU ex/im
1992	5,168	3,145	60.8%	4,751	2,831	59.6%	111.1%
1993	5,208	3,293	63.2%	5,565	3,651	65.6%	90.2%
1994	5,772	3,419	59.2%	6,175	3,523	57.1%	97.0%
1995	6,437	4,315	67.0%	7,347	5,056	68.8%	85.3%
1996	6,636	4,286	64.6%	7,524	5,079	67.5%	84.4%
1997	7,408	4,709	63.6%	8,291	5,588	67.4%	84.3%
1998	8,072	5,288	65.5%	9,018	6,259	69.4%	84.5%
1999	8,023	5,304	66.1%	9,345	6,412	68.6%	82.7%
2000	9,483	6,060	63.9%	10,986	7,446	67.8%	81.4%

Source: Statistical Office of the Republic of Slovenia (SURS).

Table 8.2   The pattern of Slovenia's comparative advantage and intra-industry trade with the EU in 1992–99

Code	Industry	RCA92[a]	RCA99[a]	IIT92[b]	IIT99[b]	%EX92	%EX99
A	AGRIC., HUNTING AND FORESTRY	2.99	0.62	56.19	28.65	1.66	0.51
B	FISHING	1.69	0.11	69.65	16.38	0.02	0.01
CA	MINING AND QUAR. OF EN. PROD.	–	0.00	0.57	0.00	1.37	0.00
CB	MINING AND QUAR. EX. EN. PROD.	0.56	0.08	19.28	10.94	0.11	0.03
DM	Transport equipment	1.63	1.06	69.48	87.85	16.35	18.81
DK	Machinery and equipment	2.65	2.31	47.14	60.47	10.92	14.97
DJ	Basic metals and fab. metal products	1.19	1.28	68.51	71.19	10.38	12.57
DL	Electrical and optical equipment	1.52	1.21	57.60	63.88	8.15	11.04
DB	Textiles and textile products	1.84	1.47	64.74	73.45	18.78	9.78
DN	Furniture and other manuf. products	7.60	5.12	45.45	43.66	5.70	9.04
DG	Chemicals, ch. prod., fibres	0.32	0.62	45.47	42.62	3.63	4.87
DD	Wood and wood products	31.00	7.67	27.70	47.65	5.11	4.15
DH	Rubber and plastic products	4.15	1.42	45.20	60.40	3.66	4.06
DE	Pulp, paper, publishing and printing	2.44	1.18	54.05	80.99	4.55	3.77
DI	Other non-metallic mineral products	10.36	1.79	40.07	48.97	2.63	2.94
DA	Food, beverages and tobacco	2.67	0.84	50.29	37.58	2.62	1.77
DC	Leather and leather products	1.66	1.00	70.77	72.30	3.71	1.66
DF	Coke, ref. petrol. prod., nuclear fuel	0.16	0.01	29.84	2.28	0.65	0.05

[a] Balassa index of revealed comparative advantage, calculated as: $RCA_i = x_{iSLO}/X_{SLO} / x_{iEU}/X_{EU}$, i.e., share of exports of product i in total Slovenia's exports relative to share of exports of product i in total extra EU exports.

[b] Grubel-Lloyd index of intra-industry trade, calculated as: $IIT_i = (1 - (|x_i - m_i| / (x_i + m_i))) * 100$; where x_i and m_i refer to exports and imports of product i.

Note: Both indices were calculated using HS 6-digit trade data (some 5,500 items) and then aggregated to NACE 2-digit sectors.

Source: SURS, authors' calculations.

equipment), which represented 57 per cent of total exports in 1999, the indices of revealed comparative advantage (RCA) have declined over the period from 1992 to 1999. There are only two industries that recorded increased RCA indices in 1999 as compared to 1992. Table 8.2 also indicates a shift away from comparative advantage and towards an increased intra-industry trade pattern in Slovenia's trade with the EU during 1990s. However, a study by Freudenberg (1998) shows that the vast majority of Slovenia's intra-industry trade (as well as that of other advanced CEECs) with the EU is clustered in down-market products characterized by average prices that are more than 15 per cent below the EU average.

The most important issue, therefore, is how to increase the export competitiveness of Slovenia's products in EU markets. One way to do this is to induce the productivity growth of existing exporters, while another way is to create new export products through the attraction of foreign direct investments (FDI).

Table 8.3   Stock of FDI in first-round candidates for EU enlargement in 2000

Country	Inflow (US$m)	Stock (US$m)	Stock of FDI as % GDP (%)	Inflow per capita (US$)	Stock per capita (US$)
Czech Rep.	4,595	21,095	33.0	460	2,110
Estonia	398	2,840	47.9	249	1,775
Hungary	1,957	19,863	39.9	190	1,928
Poland	10,000	36,475	17.2	261	952
Slovenia	181	2,865	13.0	91	1,447
Total	17,131	83,138	21.4	276	1,337

Source: World Investment Report 2001.

2.2   Foreign Direct Investment

Compared to other advanced CEECs, Slovenia has not been very successful in attracting FDI during the 1990s. As shown in Table 8.3, the total stock of FDI at the end of 2000 amounted to some three billion USD, which is, with the exception of Estonia, clearly below the figures presented by the other first round candidates for EU enlargement. With a 13 per cent share of total stock of FDI in GDP, the poor importance of FDI in Slovenia is even more pronounced. A main reason for the low presence of FDI in Slovenia is no doubt the method of privatization chosen by the Slovenian government for formerly socially-owned firms. The law on privatization completely excluded foreign bidders and favoured domestic insiders.

In 2002, the Slovenian government plans to sell the two biggest banks, which are state owned, to strategic foreign partners. Hence, in 2002, the inflow of FDI will be huge. However, one should not forget that none of the above acquisitions is allocated to the manufacturing sector and hence no impact on the export potential of Slovenia's economy can be expected.

A comparative study on the role of FDI in ten transition countries, undertaken by Damijan et al. (2001) using firm-level data, has shown that foreign-owned firms perform better in terms of total factor productivity (TFP) growth in almost all transition countries including Slovenia. This indicates that knowledge, in the form of transfer of technology, transfer of managerial skills, use of intangible assets of the parent firm and more efficient corporate governance, has been successfully transferred to local firms. In addition, for Slovenia, imports of intermediate and capital goods as well as exports of final goods to the EU have been revealed as an important channel for technology transfer in firms without foreign participation.

3 Regional Specialization Patterns

In this chapter we discuss the evolution of regional specialization in the 1990s in Slovenia. We use a database on manufacturing activity at the regional NUTS 3 level, which is indicated as an unofficial database, and which covers the period from 1994 to 2000. As the data for the first period of transition (1990–93) are missing, we are unlikely to be able to uncover the whole process of changes in the regional specialization pattern caused by integration with the EU. Nevertheless, important shifts in the relocation of manufacturing activities between regions can also be discovered for the second part of the transition in Slovenia.

3.1   Pattern of Regional Manufacturing Activity

Distribution of regions by population size Using population data, four large regions with a population share exceeding 10 per cent of the total population of Slovenia can be identified (see Table 8.4). These are: *Osrednjeslovenska,* containing the Slovenian capital Ljubljana, followed by *Podravska,* with the capital of Maribor, *Savinjska,* with the capital of Celje and *Gorenjska* with the capital of Kranj. An additional four regions can be classified as medium-sized, with a population share exceeding 5 per cent: *Dolenjska* with the capital of Novo mesto, *Pomurska* with the capital of Murska sobota, *Goriška* with the capital of Nova Gorica and *Obalno-kraška* with the capital of Koper. Four small regions are: *Koroška* with the capital of Slovenj Gradec, *Spodnjeposavska* with the capital of Krško, *Notranjsko-kraška* with the capital of Postojna and *Zasavska* with the capital of Hrastnik.

Table 8.4   Distribution of NUTS 3 regions by size in Slovenia in 2000

Size rank	Region	Share in geographic area (%)	Population share (%)
1	Osrednjeslovenska	12.6	24.5
2	Podravska	10.7	16.1
3	Savinjska	11.8	12.9
4	Gorenjska	10.5	9.9
5	Dolenjska	13.2	6.9
6	Pomurska	6.6	6.3
7	Goriška	11.5	6.0
8	Obalno-kraška	5.1	5.2
9	Koroška	5.1	3.7
10	Spodnjeposavska	4.4	3.5
11	Notranjsko-kraška	7.2	2.5
12	Zasavska	1.3	2.3

Source: SURS.

Regional GDP per capita disparities Unfortunately, the most recent available official data on GDP per capita at the regional level is for 1997, which makes the analysis somewhat cumbersome. The regional GDP per capita data in Table 8.5 reveal that the central Osrednjeslovenska and the Obalno-Kraška regions exceed the country's average per capita GDP. But only the former substantially surpasses the average, by 33 percent, while the latter's GDP per capita is only marginally higher than the average. The relatively high GDP per capita of the Osrednjeslovenska region, coupled with its share of the population, makes it the largest contributor to the average per capita GDP. GDP per capita levels for another four regions (Goriška, Savinjska, Dolenjska and

Gorenjska) reach 90 percent of the Slovene average. The rest of the regions can be classified into a third group with per capita GDP levels not surpassing 90 percent of the average. The poorest region with respect to per capita GDP is the Pomurska region, where only 77 percent of the average per capita GDP has been reached.

In 1997, the Osrednjeslovenska region reached 62 percent of the European Union average, but none of the other regions' per capita GDPs have surpassed 50 percent of the EU average. The poorest region's GDP is only 36.6 percent of the European Union average. However, note that official data for 2001 indicate that the average Slovene per capita GDP is about 78 per cent of the EU average. According to this figure, the Osrednjeslovenska region in 2001 should have already exceeded the EU average, while the other regions should be in the range of 60–80 per cent of the EU average.

Table 8.5   Distribution of NUTS 3 regions by GDP per capita in Slovenia in 1997

Region	Percent of country average	Percent of EU-15 average
Osrednjeslovenska	132.3	62.3
Obalno-kraška	103.0	48.5
Goriška	98.8	46.5
Savinjska	93.9	44.3
Dolenjska	92.8	43.7
Gorenjska	92.7	43.7
Koroška	86.9	40.9
Spodnjeposavska	86.5	40.8
Notranjsko-kraška	85.6	40.3
Podravska	82.5	38.9
Zasavska	82.4	38.8
Pomurska	77.6	36.6

Source: SURS.

Relocation patterns of manufacturing activity The distribution of economic activity across regions does not completely follow the distribution of regions by population size. Reasons for this are (i) the initial regional specialization pattern and (ii) changes in the regional specialization pattern due to economic integration with the EU. If regional manufacturing employment, output and exports are related to the regional population structure, one can identify regions with over-(under-)proportional manufacturing intensity. Values of corresponding coefficients close to one indicate an even spread of manufacturing across regions. Changes in coefficients over time imply the relocation of manufacturing activity. Shifts of coefficients closer to one imply increased similarity and shifts of coefficients away from one imply increased regional specialization.

Location and relocation patterns of manufacturing activity are presented in Table 8.6 and Figure 8.1. Peaks above one in Figure 8.1 represent the regions Gorenjska, Goriška, Dolenjska, Pomurska and Zasavska, which are characterized by an over-proportional amount of manufacturing activity relative to population share. After 1994, there were evident shifts in manufacturing activities. Some of the regions have benefited from the process of integration with the EU and increased specialization, while the position of other regions has deteriorated somewhat. It is interesting to note that the largest gains in this process have been achieved by the regions with an initial over-proportion of manufacturing activity and that only one of the regions with initially under-proportional manufacturing activity has gained in terms of manufacturing relocation. The results imply that initial production structures may play an important role in determining future patterns of production.

However, not all of the regions with initially over-proportional manufacturing intensity have benefited from integration with the EU. Only three out of the five regions with over-proportional amounts of manufacturing activity have actually benefited through relocation of production. The largest gain from the accession process has been achieved by the Dolenjska region, with a population share of only 6.9 per cent, which witnessed an increase in the share of manufacturing output from 12.9 per cent in 1994 to 17.1 per cent in 2000. This was made possible by the extraordinary export performance of firms in the region, resulting in a 22.5 per cent share in the total of Slovenia's exports (a rise from 17.3 per cent in 1994). Reasons for the region's favourable development can be found primarily in FDI as Renault's subsidiary alone accounts for some 10 per cent of total Slovenian exports (and 40 per cent of the regional performance). This performance has been supported by the achievements of domestic firms in the pharmaceutical industry, electrical equipment and transport equipment, etc. Gains from production relocation have also been observed in the Goriška and Koroška regions, where production shares have risen by 0.9 and three percentage points, respectively. The two regions have also increased their shares in employment and exports, while the Podravska region has managed to increase both production and export shares despite a fall in the share of employment (from 12.2 per cent in 1994 to 11.6 per cent in 2000).

Table 8.6   Distribution of economic activity across regions in Slovenia in 1994–2000 (in per cent)

No.	Region	EMP94	EMP00	PROD94	PROD00	EX94	EX00
6	Dolenjska	8.6	10.6	12.9	17.1	17.3	22.5
9	Podravska	12.2	11.6	11.7	12.4	10.3	12.8
4	Goriška	7.2	8.2	6.3	7.2	6.1	7.5
10	Koroška	3.2	4.8	3.5	6.5	4.3	7.6
1	Osrednjeslovenska	22.2	21.8	25.0	24.2	19.3	19.2
8	Notranjsko-kraška	2.3	2.3	2.0	1.7	2.5	1.7
12	Zasavska	2.7	2.5	1.8	1.7	1.6	1.2
3	Gorenjska	15.2	13.5	13.1	10.7	15.2	10.8
5	Savinjska	12.7	12.8	12.3	10.9	13.1	10.3
7	Pomurska	7.6	6.9	5.4	3.3	4.6	2.8
2	Obalno-kraška	2.9	2.7	4.1	2.9	3.8	2.4
11	Spodnjeposavska	3.1	2.3	1.8	1.4	1.8	1.2

EMP = employment share; PROD = manufacturing output share; EX = exports share.

Source: SURS, authors' calculations.

Regions where major declines in shares have been observed are, on the other hand, the Gorenjska, Savinjska, Pomurska, Obalno-kraška and Spodnjeposavska regions, which have lost their shares in manufacturing employment and output. The Savinjska region, the Gorenjska region and the Pomurska region are still characterized by above average manufacturing intensity in terms of employment. Their position, however, has been rapidly deteriorating over time. Both remaining regions faced below average manufacturing intensities at the beginning of the observed period and their relative manufacturing positions continued to worsen throughout the period. Hence, a trend of deindustrialization in regions that are already less manufacturing-intensive, on one hand, and a trend of increased manufacturing concentration in regions that were initially more manufacturing-intensive, on the other hand, can be observed. Initial production patterns seem to be very important for future regional specialization in Slovenia. The only exception that can be observed is the Koroška region, which experienced an increase in the shares of employment, production and exports despite the low initial shares in manufacturing production. The largest region, the Osrednjeslovenska region, as well as the Notranjsko-kraška and Zasavska regions, has not experienced substantial changes in its share in total manufacturing employment, with the latter two experiencing declines in production and export shares.

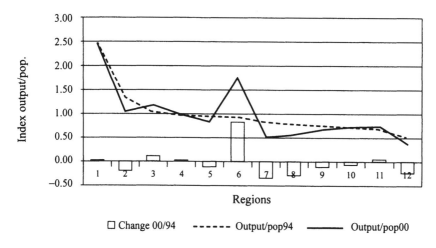

Figure 8.1   **Relocation of manufacturing activity between regions in Slovenia from 1994–2000 (index output/population share)**

An important qualification should be made of the above analysis. One should bear in mind that it is focused solely on manufacturing activity. Some of the regions that have experienced deterioration of manufacturing activity (i.e., the Obalno-Kraška region, etc.) have, in fact, restructured their economic activity towards service industries (tourism, transport, telecommunications, merchandise, etc.). Therefore, despite the relative decline in manufacturing activity some of the regions have experienced substantial catch-up in terms of per capita GDP.

3.2   Changes in Regional Specialization in Manufacturing

In the previous section we discovered important shifts in the relocation of manufacturing activities between regions during the 1990s. In this section it remains to be seen whether these inter-regional shifts were associated with increased regional specialization. In order to save space, in the first subsection regional specialization is analysed, mostly according to manufacturing employment data. In the second subsection we compare evidence of the regional specialization of manufacturing employment with the data for manufacturing output and exports.

Regional concentration of manufacturing employment   First, we show measures of absolute regional specialization as represented by the Herfindahl index. The calculated indices summarized in Table 8.7 show relatively low regional specialization for Slovenia. The average value of Herfindahl indices across the regions is about 0.15,[1] which is relatively low compared to other transition economies. The lowest level of regional specialization is observed in the largest regions (Osrednjeslovenska and Podravska), while the greatest specialization is observed in the smallest regions (Spodnjeposavska and Zasavska) and in the Pomurska region. Manufacturing activity in these small regions is mostly concentrated in three or four sectors, in which firms located in these regions can enjoy a comparative advantage over firms in other regions.

The evolution of Herfindahl indices over the period 1994–2000 reveals that economic integration has stimulated regional specialization most in the small and backward regions like the Koroška, Spodnjeposavska and Pomurska regions. Regional specialization is most evident in the Pomurska region, where the Herfindahl index has risen from 0.22 in 1994 to 0.34 in 2000. Other regions, with the possible exception of Koroška, have not experienced such pronounced changes in their Herfindahl index. Obviously, economic integration with the EU has created even greater specialization in the already specialized smaller regions.

In contrast to the Herfindahl index of absolute specialization, relative measures of regional specialization, such as the Balassa index, compare regional concentrations of manufacturing activity to the national pattern of manufacturing concentration. In other words, relative regional specialization measures search for differences in the patterns of concentration of manufacturing activity between the regional and national levels. Hence, a Balassa index value for region *j* greater than one indicates that the manufacturing concentration in this region is greater than at the national level.

Table 8.8 reveals a different pattern of regional specialization than those based on measures of absolute regional specialization. Regional specialization is now being observed to some extent in larger regions like Osrednjeslovenska, Podravska and Dolenjska, while in some smaller regions like Koroška, Spodnjeposavska, Obalno-kraška and Notranjsko-kraška, a less distinctive pattern of regional specialization can be observed. On the other hand, in only five out of the 12 regions is there evidence of increased regional specialization from 1994 to 2000. Hence, based on this evidence one cannot make any inferences about increasing regional specialization over the period.

Similar conclusions can be drawn based upon the calculated dissimilarity indices presented in Table 8.9. Evidence of increased regional specialization over the period is found in four regions. However, changes observed at the regional level are far from dramatic.

Table 8.7   **Evolution of Herfindahl indices of absolute regional specialization of manufacturing employment in Slovenia from 1994–2000**

Size rank	Region	1994	1995	1996	1997	1998	1999	2000	Ratio 00/94
1	Osrednjeslovenska	0.10	0.10	0.10	0.10	0.10	0.11	0.11	1.1
2	Podravska	0.12	0.11	0.11	0.10	0.10	0.10	0.10	0.8
3	Savinjska	0.14	0.13	0.12	0.11	0.11	0.12	0.12	0.9
4	Gorenjska	0.13	0.13	0.13	0.14	0.14	0.14	0.15	1.2
5	Dolenjska	0.15	0.15	0.14	0.13	0.13	0.12	0.12	0.8
6	Pomurska	0.22	0.23	0.25	0.26	0.31	0.32	0.34	1.5
7	Goriška	0.16	0.16	0.16	0.17	0.17	0.18	0.19	1.2
8	Obalno-kraška	0.14	0.15	0.14	0.14	0.15	0.14	0.16	1.1
9	Koroška	0.03	0.19	0.19	0.19	0.17	0.17	0.18	6.0
10	Spodnjeposavska	0.19	0.22	0.23	0.24	0.23	0.21	0.21	1.1
11	Notranjsko-kraška	0.21	0.20	0.20	0.20	0.20	0.23	0.23	1.1
12	Zasavska	0.23	0.24	0.26	0.24	0.24	0.25	0.24	1.0

Herfindahl index: $H_i^s = \sum_{ij} (s_{ij}^s)^2$ where s_{ij}^s is share of employment in industry i in region j in region j in total employment of region j.

Source: SURS, authors' calculations.

Table 8.8   Evolution of Balassa indices of relative regional specialization of manufacturing employment in Slovenia from 1994–2000

Size rank	Region	1994	1995	1996	1997	1998	1999	2000	Ratio 00/94
1	Osrednjeslovenska	0.95	0.95	0.96	0.97	1.08	1.09	1.09	1.1
2	Podravska	0.92	0.92	0.92	0.94	1.28	1.29	1.31	1.4
3	Savinjska	0.87	0.89	0.89	0.91	0.90	0.90	0.90	1.0
4	Gorenjska	1.01	1.04	1.03	0.98	0.97	0.94	0.92	0.9
5	Dolenjska	1.00	0.98	1.01	1.01	1.01	1.04	1.08	1.1
6	Pomurska	1.63	1.76	1.73	1.64	1.78	1.77	1.77	1.1
7	Goriška	0.87	0.86	0.86	0.86	0.84	0.81	0.79	0.9
8	Obalno-kraška	1.12	1.06	1.09	1.07	1.03	1.09	0.91	0.8
9	Koroška	0.95	0.84	0.83	0.85	0.89	0.91	0.96	1.0
10	Spodnjeposavska	0.83	0.83	0.83	0.81	0.80	0.81	0.85	1.0
11	Notranjsko-kraška	1.08	1.06	1.06	1.04	1.06	1.12	1.10	1.0
12	Zasavska	0.94	0.95	0.91	0.94	0.94	1.15	1.12	1.2

Balassa index: $RS_i = \frac{1}{J}\sum_J RS_{ij}$ where $RS_{ij} = \frac{s_{ji}}{s_j}$, where , where is share of employment in industry i in region j in total employment of region j, and s_i is share of total employment in industry i in total employment and I is the number of industries.

Source: SURS, authors' calculations.

Table 8.9 **Evolution of dissimilarity indices of regional specialization of manufacturing employment in Slovenia from 1994–2000**

Size rank	Region	1994	1995	1996	1997	1998	1999	2000	Ratio 00/94
1	Osrednjeslovenska	0.34	0.37	0.43	0.45	0.46	0.47	0.47	1.4
2	Podravska	1.78	1.82	1.47	1.36	1.42	1.29	1.29	0.7
3	Savinjska	0.80	0.74	0.65	0.54	0.61	0.34	0.34	0.4
4	Gorenjska	1.71	1.75	1.79	1.69	1.59	1.51	1.50	0.9
5	Dolenjska	2.28	2.50	2.57	2.42	2.35	1.90	1.92	0.8
6	Pomurska	1.50	1.55	1.62	1.77	1.78	1.81	1.85	1.2
7	Goriška	2.26	2.64	2.75	2.65	2.67	1.81	1.80	0.8
8	Obalno-kraška	1.00	0.75	0.73	0.72	0.75	0.79	0.75	0.8
9	Koroška	0.78	0.92	0.95	0.97	0.99	0.83	0.93	1.2
10	Spodnjeposavska	2.20	2.16	2.31	2.05	1.92	1.65	1.72	0.8
11	Notranjsko-kraška	1.27	1.21	1.31	1.33	1.30	1.21	1.20	0.9
12	Zasavska	0.98	1.01	1.06	1.03	1.02	1.07	1.00	1.0

Dissimilarity index: $DSR_i = \sum_j \left| s_{ij}^s - s_{ji} \right|$ where s_{ij}^s is share of employment in industry i in region j in total employment of region j, and s_i is share of total employment in industry i in total employment.

Source: SURS, authors' calculations.

Evidence from the dissimilarity index shows some increase in specialization in the Osrednjeslovenska, Pomurska and Koroška regions, while de-specialization can be observed in the Gorenjska, Savinjska, Dolenjska, Podravska and Spodnjeposavska regions. No viable conclusions can therefore be drawn about the effects of economic integration on regional specialization from the above indices.

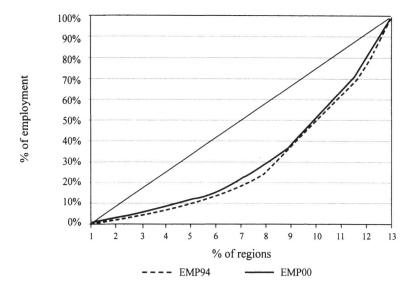

Figure 8.2   The Lorenz curve of regional concentration of manufacturing employment in Slovenia from 1994–2000

The last piece of evidence about the pattern of regional specialization will be given using Lorenz curves. Lorenz curves show, in a very instructive way, whether or not one variable is concentrated. The larger the share of the largest regions in some variable across sectors, the more Lorenz curve will shift away from the diagonal. Figure 8.2 reveals, again, that manufacturing employment in Slovenia is relatively regionally concentrated. However, there is little evidence of increased regional specialization in manufacturing employment over the period from 1994–2000.

Regional concentration of manufacturing output and exports One of the reasons for the lack of evidence of increased regional specialization in manufacturing employment data might be the fact that employment structures are always very rigid and respond only reluctantly to changes in patterns of manufacturing output. Hence, in this subsection we will compare responses to increased integration with the EU in terms of the regional concentration of manufacturing employment, output and exports.

Figures 8.3 and 8.4 do indeed demonstrate much greater responses in manufacturing output and exports to increased integration with the EU during the 1990s than was the case with manufacturing employment. Both manufacturing output and export regional specialization increased in 2000 compared to 1994. Individual regions responded differently to changed demand patterns. The consequence of this is the relocation of manufacturing activities among regions and, hence, greater regional specialization. Evidently, this was not the case with the regional concentration of manufacturing employment. At the beginning of the observed period, in 1994, the concentration of manufacturing employment, output and exports did not differ significantly across regions. At the end of the period, in 2000, however, the concentration of manufacturing output and exports had increased significantly, while the concentration of employment remained virtually unchanged. This means that only output and exports across regions adapted to the changed demand pattern and that employment has yet to follow this pattern.

4 Location and Concentration of Industrial Activity

4.1 Evolution of the Manufacturing Structure (1994–2000)

It is apparent from Table 8.10 that there has been a shift in the structure of manufacturing employment during the period of economic integration. The largest gain in employment share can be observed in electrical and optical equipment (DL), where the share of employment has increased by three percentage points. Other winning industries are basic metals and fabricated metal products (DJ), chemical products (DG) and food, beverages and tobacco (DA). On the other side, textiles and textile products (DB) (with a decrease by about three percentage points), machinery and equipment (DK) and non-metallic mineral products (DI) faced decreased employment shares. The rest of the manufacturing industries were less affected by the accession process in terms of employment.

4.2 Evolution of Industrial Specialization

Trade liberalization does not seem to have influenced labor relocation significantly in the case of Slovenia. No common trend can be found for labor

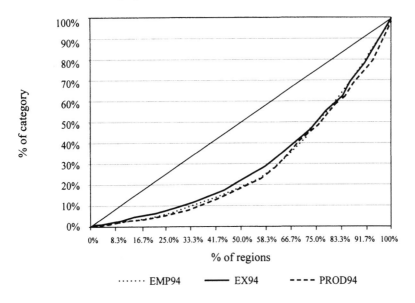

Figure 8.3   Lorenz curve of regional concentration of manufacturing employment, output and exports in Slovenia in 1994

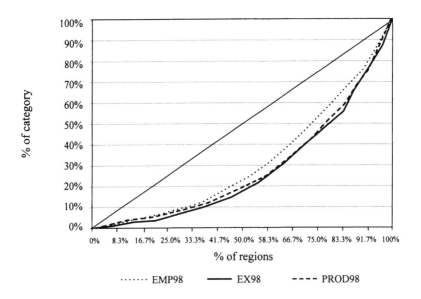

Figure 8.4   Lorenz curve of regional concentration of manufacturing employment, output and exports in Slovenia in 2000

Table 8.10 Distribution of employment in NACE-2 manufacturing industries (as shares of total employment in manufacturing)

Industry	Share of total employment in 1994 (%)	Share of total employment in 2000 (%)
DA	8.65	9.27
DB	16.29	13.16
DC	4.77	3.4
DD	4.8	4.21
DE	6.52	6.52
DF	0.39	0.05
DG	5.88	7.96
DH	4.37	4.14
DI	5.16	4.57
DJ	11.87	13.9
DK	10.05	8.83
DL	11.38	14.39
DM	4.5	3.9
DN	5.38	5.7

Source: SURS; authors' calculations.

flows in Slovenia with respect to the manufacturing specialization indices. The effects of economic integration on the concentration and relocation of industrial activity still have to be explored. Similarly, as in the case of regional specialization, we attempt to analyse the evolution of the location and concentration of industrial activity in this section using concentration and specialization indices.

Table 8.11 represents the evolution of the Herfindahl index measuring the absolute geographic concentration of manufacturing industries. Again, as in the case of regional specialization, most indices show a relatively low absolute geographic concentration, with values of the Herfindahl index averaging about 0.2. There are, however, some notable exceptions, such as the fuel industry (DF), where the index had been, understandably, much higher (it fell from 0.87 in 1994 to 0.51 in 2000), as well as leather products (DC), where the Herfindahl index has decreased substantially from 0.45 in 1994 to 0.27 in 2000. A decrease in geographic concentration can also be observed in the rubber (and plastic products) industry (DH) and in the machinery and equipment industry. On the other hand, there has been an increase in the geographic concentration of the transport equipment industry (DM) and of the furniture industry (DN).

Table 8.11 Evolution of Herfindahl indices of absolute geographic concentration of manufacturing employment by NACE-2 industries in Slovenia from 1994–2000

No.Industry	1994	1995	1996	1997	1998	1999	2000	00/94
1 DA	0.16	0.18	0.18	0.17	0.17	0.18	0.18	1.13
2 DB	0.13	0.13	0.13	0.11	0.13	0.14	0.16	1.23
3 DC	0.45	0.46	0.43	0.36	0.30	0.29	0.27	0.60
4 DD	0.13	0.13	0.13	0.13	0.13	0.13	0.13	1.00
5 DE	0.31	0.30	0.31	0.29	0.30	0.30	0.28	0.90
6 DF	0.87	0.87	0.87	0.64	0.51	0.50	0.51	0.59
7 DG	0.27	0.26	0.27	0.30	0.29	0.28	0.28	1.04
8 DH	0.25	0.25	0.25	0.17	0.18	0.15	0.14	0.56
9 DI	0.14	0.14	0.13	0.17	0.14	0.15	0.14	1.00
10 DJ	0.16	0.15	0.14	0.19	0.14	0.14	0.14	0.88
11 DK	0.16	0.17	0.16	0.18	0.12	0.13	0.12	0.75
12 DL	0.16	0.17	0.17	0.13	0.16	0.16	0.16	1.00
13 DM	0.21	0.22	0.25	0.28	0.28	0.31	0.37	1.76
14 DN	0.03	0.12	0.11	0.10	0.11	0.11	0.11	3.67

Herfindahl index: $H_i^S = \sum_j (s_{ij}^S)^2$ where s_{ij}^S, where is share of employment in industry i in region j in total employment in industry i.

Source: SURS, authors' calculations.

The evolution of the Herfindahl indices of geographic concentration over the period 1994–2000 reveals that the two most concentrated industries, namely fuel and leather products, have become dispersed following economic integration with the EU. On the other hand, the transport equipment and furniture industries have further consolidated as a result of trade liberalization, which has led to an increase in geographical concentration.

The Balassa index, a relative measure of geographical concentration, compares the geographical concentration of employment in an industry to the geographical concentration of total employment. Thus, a value of the Balassa index for industry i greater than one indicates that the geographical concentration of that industry is larger than the geographical concentration of total manufacturing.

Table 8.12 reveals that geographical concentration has increased only for leather and leather products (DC), rubber and plastic products (DH) and basic metals and fabricated metal products (DJ), relative to the concentration of manufacturing in general. In all other industries geographic concentration did not change substantially. Indeed, in the wood industry (DD), fuel products (DF) and transport equipment (DM), it has largely decreased. Based on the above evidence, we cannot make any significant inferences about geographical concentration.

Table 8.12   Evolution of Balassa indices of relative geographic concentration of manufacturing employment by NACE-2 industries in Slovenia from 1994–2000

No.	Industry	1994	1995	1996	1997	1998	1999	2000	00/94
1	DA	0.95	1.43	0.95	1.50	0.93	0.94	0.93	0.98
2	DB	1.01	1.33	1.07	1.04	1.10	1.07	1.06	1.05
3	DC	0.59	0.99	0.60	0.86	0.67	0.68	0.72	1.22
4	DD	1.36	3.11	1.19	1.15	1.14	1.21	1.24	0.91
5	DE	0.73	0.54	0.78	0.69	0.71	0.72	0.74	1.01
6	DF	1.07	2.87	1.21	1.29	0.57	0.57	0.57	0.53
7	DG	0.75	1.53	0.76	1.19	0.71	0.71	0.72	0.96
8	DH	1.08	1.26	0.97	1.91	1.07	1.16	1.21	1.12
9	DI	1.23	1.82	1.28	1.85	1.33	1.36	1.28	1.04
10	DJ	0.98	1.83	1.94	2.23	1.05	1.07	1.10	1.12
11	DK	1.07	1.75	1.02	1.54	1.11	1.00	1.02	0.95
12	DL	0.97	1.15	0.98	1.23	0.95	0.95	0.94	0.97
13	DM	1.12	1.21	1.03	1.12	0.93	0.95	0.72	0.64
14	DN	1.29	1.16	1.26	1.23	1.24	1.28	1.31	1.02

Balassa index: $RS_i = \frac{1}{J}\sum_j RS_{ij}$ where $RS_{ij} = \frac{s_{ji}}{s_j}$ where , where is the share of employment in industry i of region j in total employment in industry i, and s_j is the share of total employment of region j in total employment and J is the number of regions.

Source: SURS, authors' calculations.

The Krugman dissimilarity indices, showing the difference between the regional share of employment in an industry and the average (national) share of that industry's employment, are represented in Table 8.13. Higher dissimilarity indices imply that an industry is either regionally dispersed or concentrated relative to the national average. Krugman indices close to zero therefore imply that the regional structure of manufacturing employment in all industries closely resembles that of the average national structure, while higher values of the indices mean that the regional employment structures differ greatly from the national levels.

Results in Table 8.13 indicate that the dissimilarity index has increased in seven of the 14 industries, most notably in the textile industry (DB), the transport equipment industry (DM) and the furniture industry (DN), while it has decreased in six industries. The largest decreases in dissimilarity of geographic concentration can be observed in fuel products (DF), rubber and plastic products (DH), leather (DC) and paper products (DE) as well as in the machinery sector (DK). The changes in the remaining industries are not significant.

Table 8.13 Evolution of dissimilarity indices of the geographic concentration of manufacturing employment by NACE-2 industries in Slovenia from 1994–2000

No.	Industry	1994	1995	1996	1997	1998	1999	2000	00/94
1	DA	0.53	0.60	0.57	0.52	0.51	0.53	0.55	1.04
2	DB	0.43	0.74	0.47	0.47	0.52	0.53	0.54	1.26
3	DC	1.03	1.02	1.01	1.31	0.82	0.80	0.80	0.78
4	DD	0.64	0.66	0.56	0.47	0.50	0.52	0.56	0.88
5	DE	0.82	0.81	0.74	0.67	0.67	0.66	0.65	0.79
6	DF	1.70	1.56	1.73	1.37	1.32	1.33	1.33	0.78
7	DG	0.80	0.72	0.77	1.08	0.79	0.78	0.77	0.96
8	DH	0.85	0.87	0.80	0.75	0.76	0.67	0.64	0.75
9	DI	0.64	1.03	0.62	0.89	0.70	0.75	0.71	1.11
10	DJ	0.48	0.63	0.49	0.58	0.47	0.47	0.50	1.04
11	DK	0.55	0.67	0.61	0.46	0.45	0.45	0.43	0.78
12	DL	0.51	0.82	0.50	0.53	0.53	0.54	0.53	1.04
13	DM	0.83	0.72	0.93	0.82	0.98	1.00	0.99	1.19
14	DN	0.37	0.55	0.38	0.41	0.39	0.40	0.45	1.22

Dissimilarity index: $DSR_i = \sum_j \left| s_{ji}^s - s_{ji} \right|$ where s_{ij}^s is share of employment of region j in industry i in total employment of industry i, and s_j is share of total employment of region j in national manufacturing employment.

Source: SURS, authors' calculations.

5   Specialization, Polarization and Unemployment

5.1   Does Greater Specialization Imply Greater Polarization?

In this section we compare the variation between regional GDP and per capita GDP in order to determine whether greater concentration of manufacturing is associated with concentration of population. As we have data on regional GDP only for up to 1997, we instead use data on manufacturing value added. For the variables value added (VA) and value added per capita (VA per capita), coefficients of variation have been calculated. Calculated coefficients presented in Figure 8.5 show that variation in VA is in general matched by variation in VA per capita. This indicates that there was almost no relocation of population between regions. The only exceptions occurred in the Podravska and Pomurska regions, where shifts in population were modest but significant. Hence, one can conclude that inter-regional relocation of manufacturing output implies greater polarization, as population does not follow this pattern.

Another interesting feature that can be observed in Figure 8.5 is that variation in VA per capita is positively correlated with the sign of the variation. Regions that have experienced larger variations in VA per capita have also recorded larger absolute growth of value added per capita over the period 1994–98. In other words, more turbulent regions have grown faster. One possible explanation for this interesting fact is that declining regions lack initiatives that would induce positive turbulence into the regions.

5.2   Do Declining Regions Experience Higher Unemployment?

Previous sections have shown that economic integration has induced relocation of manufacturing activity between regions. These shifts, however, have not caused any significant adaptation processes in manufacturing employment or population between regions. Obviously, the excess labor force that was released in the declining regions did not move to the regions with increasing employment, but remained unemployed. Lower growth, in terms of GDP per capita at the regional level, is therefore associated with higher unemployment rates.

In order to test this hypothesis, we use data on value added per capita and unemployment rates at the regional level that are available for the period 1997–99 only. The results of econometric tests are reported in Table 8.14 and illustrated in Figure 8.6. Both OLS and random effects (RE) model estimations[2] confirm that there is a strong negative association between VA per capita

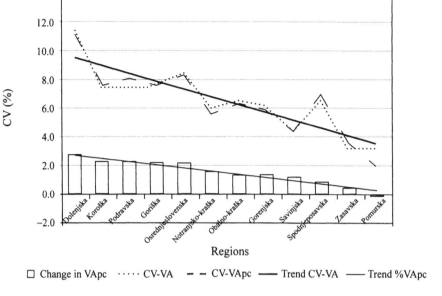

□ Change in VApc   ····· CV-VA   – – CV-VApc   —— Trend CV-VA   —— Trend %VApc

Figure 8.5   Variation of value added per capita across regions in Slovenia from 1994–98 (per cent)

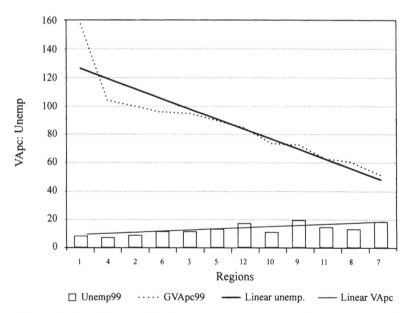

□ Unemp99   ····· GVApc99   —— Linear unemp.   —— Linear VApc

Figure 8.6   Value added per capita and unemployment rates from 1994–98 (per cent)

and unemployment rates (while in fixed effects (FE) model specification the relationship, while negative, is insignificant). Hence, there is clear evidence of greater unemployment in poor regions – poor regions are more likely to suffer from unemployment than rich regions.

Table 8.14   Test: Do poor regions suffer from unemployment?

	OLS	FE	RE
Const.	**21.711	**25.026	**22.222
	(12.23)	(3.52)	(7.57)
VApc	**-0.090	−0.128	**−0.096
	(−4.56)	(−1.56)	(−2.97)
Adj. R^2	0.362	0.095	0.388
No. of obs.	36.0	36	36
Hausman Chi2	–	–	0.18

Dependent variable: rate of unemployment.
t-statistics in parentheses; * and ** denote coefficient estimates significant at 1 and 5 per cent.

6   Is the New Economic Geography Theory Relevant in the Case of Slovenia?

Following the discussion in the introductory section, in the present section we evaluate the implications of the EG models for Slovenia. A basic proposition of the economic geography model by Damijan and Kostevc (2002) for transition countries is that, after trade has been liberalized, the pattern of inter-regional manufacturing relocation will be determined by a trade-off between agglomeration effects, remaining trade (transport) costs and existing differences in relative regional factor costs. With unchanged inter-regional transport costs, regions that are located closer to the EU border (western and northern regions) might benefit from trade liberalization through larger inflows of FDI due to lower trade costs with the EU and lower wages. Some domestic resources might also relocate to border regions. As a result, border regions may converge with the home capital region in terms of relative wages and relative manufacturing output. In non-border regions, however, further divergence might occur. After a certain threshold of trade costs with the EU has been reached, the trend toward convergence of border regions might be reversed. Therefore, regional data for accession countries might exhibit an inverted U-shaped relationship between the relative FDI, relative wages and relative manufacturing output with respect to the foreign trade costs. This process could also be inverted when the agglomeration effects in the existing economic centers prevail over lower factor costs and lower trade costs in the border regions. In this case, a divergence in relative regional wages is expected in the first stage of trade liberalization and a convergence afterwards. In any case, a non-monotonic relationship between reduction of foreign trade costs and relative regional wages is expected to occur in the long run.

A number of papers study these issues for the USA (see Hanson 1996, 1997, 1998; Kim 1995, etc.) and for the EU (refer to Amiti 1998, 1999; Brülhart and Torstensson 1998, etc.). In next subsections we test three basic predictions of EG for Slovenia in order to verify whether regional development in Slovenia after trade liberalization with the EU is occurring according to the predictions of EG theory. The following propositions are tested:

- an initial increase in FDI inflows and a later relative reduction of FDI inflows into western/northern border regions increasing with distance from the capital and decreasing with distance from the W/N border;
- initial dispersion of production (convergence) and later further regional concentration;
- initial convergence and later divergence of relative wages in western/ northern border regions.

6.1   Methodology and Data

The data   It is worth noting that for this chapter 'official' and 'unofficial' regional databases for Slovenia were constructed. The source of the 'official' database is the Slovenian statistical office (SURS). However, the database is only of limited use, since it is incomplete and covers only the years 1995–97. The official data also does not cover many of the indicators needed for the analysis. Slovenia does not have a long tradition of collecting regional statistics, which is due in large part to the fact that it is still not clear what the future statistical regions will be. Slovenian statistical regions have been in use since the mid-nineties. The existing system of regions is being questioned by the EU Economic Commission. Hence, we use the official source of data only when dealing with some special issues, e.g., polarization and unemployment.

For all other purposes we have constructed an 'unofficial' database using firms' balance sheets and income statements (the source of the data is the

Agency for payments). These data cover manufacturing firms only, for the period of 1994–2000. For the purposes of our analyses, these data are aggregated to regional (NUTS 3) and to community (NUTS 5) levels. Here, 170 NUTS 5 regions (communities), classified into 12 broader NUTS 3 regions, are taken as units of observation. NUTS 3 dummy variables have been used in order to control for these broader common regional effects. All the data are recalculated into 1994 constant prices using PPI indices. The data in our database include many aspects of regional performance, but we explore only a small part of it.

The methodology In all of the subsequent analyses and empirical estimations, we use relative regional indicators in order to capture inter-regional relocation patterns in Slovenia. Relative regional indicators are calculated as a ratio of i-th regional performance to the capital (c) region's performance.

When dealing with panel data on regions over a time span of seven years, one should take explicit account of region-specific effects. Without explicit control for this, one might get biased estimates of coefficients since FDI inflows, output growth and changes in relative wages might be correlated over time or subject to random shocks. Using statistical specifications of the models, there are two possible ways to control for this bias. The first and most obvious option is to employ the fixed effects (FE) estimator, which assumes fixed (constant) region-specific effects over time. On the other side, the random effects (RE) estimator assumes that region-specific effects are random and only reflected in the error term; i.e., uncorrelated over time. The FE estimator is usually more robust but quite inaccurate, while the RE estimator is sensitive to assumptions but more accurate. From a substantial point of view, we are interested in observing the pattern of changes in relative regional performance over time induced by external shocks such as trade liberalization. Under different trade regimes, regions respond differently to some inherent forces. Hence, the RE estimator is more suitable for our purposes. Another drawback of the FE estimator in the present case is that some of the crucial variables in our empirical model are time invariant (such as transport costs proxied by road distances in kilometers or the border dummy for regions bordering the EU). These variables are naturally dropped from the estimation procedure when using the FE estimator. Therefore, it makes no sense in the present case to conduct formal Hausman tests for the validity of the model specification. Hausman tests will, of course, be unable to reject the hypothesis of systematic differences among FE and RE estimators. In turn, they would imply the acceptance of the FE specification, which is, as we saw, incorrect. Hence, the RE estimator provides a better specification for our econometric models.

6.2   *The Impact of Economic Geography on Regional Economic Concentration*

In this subsection we explore the pattern of inter-regional manufacturing relocation in Slovenia during the 1994–2000 period. Initially, according to the standard EG hypothesis, regions closer to the capital city would have a greater concentration of manufacturing activity. Along with economic integration with the EU, relocation of manufacturing activity should take place. In line with the Krugman (1991) type of EG models, we should observe further monotonic reduction of manufacturing activity in the more distant regions and further concentration of manufacturing activity in regions closer to the capital region. The Krugman-Venables type of EG model proposes a non-monotonic, U-shaped evolution of relative regional output – convergence of relative output should take place after initial divergence. In addition, the Damijan-Kostevc EG model predicts an even more pronounced U-shaped pattern of relocation of relative regional output as FDI inflows into regions bordering the EU foster quicker adjustment in the EU-bordering regions. Figure 8.7a reveals that in Slovenia there is no pronounced initial geographic pattern of production in line with the standard EG hypothesis. In contrast, the figure suggests an inverse relationship between initial relative regional output and distance to the capital. Figure 8.7b provides little evidence for the convergence of relative production with respect to the initial level. In addition, Figure 8.7c shows that there is no convergence/divergence path for relative regional output with respect to distance from the capital, and Figure 8.7d shows no convergence in western/ northern border (W/N) regions.

The pattern of relative regional production relocation has been estimated using the following linear model:[3]

(1) $ln\ rPROD = \alpha + \beta\ ln\ irPROD + \gamma\ ln\ irWAGE + \delta\ ln\ irVAe + \phi\ ln\ FDI + \xi\ ln\ DIST + \varphi\ BORD + \lambda\ ln\ TC + \upsilon\ ln\ DIST*BORD\ *TC + \rho\Sigma R + \tau\Sigma T + \varepsilon_{it}$, where:
$rPROD$ $rPROD = P_{it}/P_{ct}$ is the relative production of the region i with respect to the central region c in time period t;
$irPROD$ is the initial relative production;
$irWAGE$ is the initial relative wage;
$irVAe$ is the initial relative value added per employee;

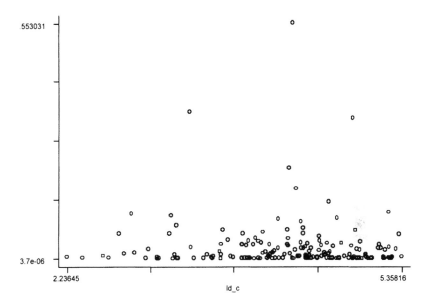

a) Initial relative output and distance

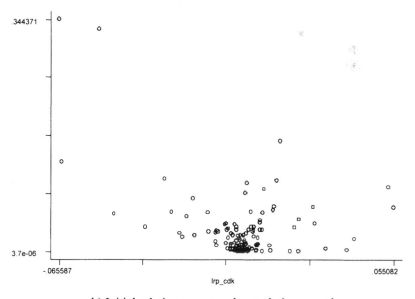

b) Initial relative output and cumulative growth

Figure 8.7   **Relative regional manufacturing output in Slovenia**

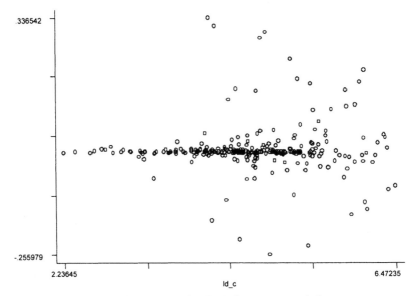

c) Cumulative growth of relative output and distance

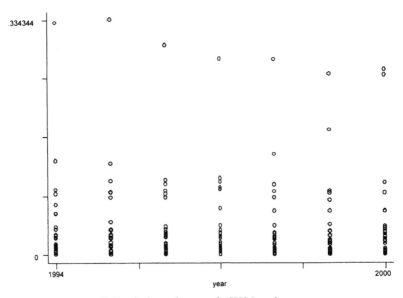

d) Evolution of output in W/N regions

FDI is the share of aggregate FDI in the employment of the region i;

TC is the region's average tariff rate[4] as a proxy for foreign trade costs;

DIST is a proxy for trade/transport costs between home regions; it is measured as the distance (in km) between the i-th region and the capital region;

BORD is a dummy variable for western/northern (W/N) regions, i.e., regions bordering EU member countries;

DIST*BORD*TC is an interaction term, which proxies for distance effects in border regions;

ΣR and ΣT are matrices of regional[5] and time dummies;

ε_{it} is the error term.

The model (1) is estimated using different model specifications as well as different specifications of the data. The model is estimated using NUTS 5 regional data in levels, in differences and then in cumulative differences. The most conclusive results seem to emerge when cumulative differences are used, mainly due to the fact that both data in levels and data in differences fail to show overall long-term trends and shift the focus to short-term changes. The short-term effects vary substantially and fail to encompass properly the underlying trend, which can best be observed using cumulative differences.

Results in Table 8.15 suggest that both year-to-year changes and a cumulative long-term change in relative production are negatively correlated with initial relative production in the region, implying a convergence with respect to the central region. The trend of convergence, however, is very weak. What are the driving forces behind it? Apparently, the initial regional wage differential and the productivity differential do not seem to be responsible for it. Neither is economic geography, since diminishing trade costs with the EU did not result in a stronger convergence/divergence path for relative output. Similarly, border regions did not account for a significantly different convergence/divergence path for relative output. The only exception is internal, inter-regional trade costs (proxied by time invariant distance), where the estimation with cumulative differences reveals a negative correlation between long-term changes in relative production and distance from the capital. This fact, in contrast to the autoregressive convergence path, speaks in favour of divergence in relative regional output.

On the other hand, in all model specifications some conclusive evidence is found that manufacturing concentration in Slovenia is positively correlated with the presence of FDI. Regions with larger relative FDI stocks converge more quickly with the capital region in terms of relative output.

Table 8.15 Does economic geography affect regional concentration of manufacturing?

	Data in levels		First differences		Cumul. differences[1]	
Estimator	*RE*	*RE*	*RE*	*RE*	*OLS*	*OLS*
Const.	***−5.072	***−4.694	0.109	0.317	**1.585	2.239
	(−9.29)	(−7.53)	(0.80)	(1.22)	(2.04)	(1.37)
irPROD	–	–	***−0.030	***−0.033	***−0.245	***−0.266
	–	–	(−2.69)	(−2.83)	(−3.71)	(−3.85)
irWAGE	***3.989	***4.224	−0.013	0.045	0.055	0.415
	(8.92)	(8.98)	(−0.16)	(0.52)	(0.11)	(0.81)
irVA/emp	***−0.980	***−0.990	−0.014	−0.039	−0.155	−0.319
	(−2.79)	(−2.73)	(−0.27)	(−0.71)	(−0.50)	(−1.00)
FDI	***0.065	***0.060	***0.021	**0.018	***0.209	***0.209
	(4.41)	(4.11)	(2.91)	(2.37)	(4.95)	(4.69)
DIST	0.214	0.076	−0.244	−0.338	−1.732	***−2.838
	(0.39)	(0.13)	(−1.41)	(−1.48)	(−1.39)	(−1.67)
BORD	0.764	*1.098	0.198	0.256	−0.973	0.961
	(1.29)	(1.68)	(0.60)	(0.70)	(−0.70)	(0.52)
TC	0.531	0.236	−0.083	−0.179	−1.458	−2.040
	(1.03)	(0.45)	(−0.51)	(−0.89)	(−1.25)	(−1.25)
DIST*TC	−0.469	−0.399	0.164	0.215	1.210	1.995
	(−0.91)	(−0.77)	(1.02)	(1.15)	(1.04)	(1.40)
DIST*BORD*TC	0.008	0.014	−0.028	−0.029	0.208	−0.100
	(0.10)	(0.17)	(−0.50)	(−0.48)	(0.78)	(−0.30)
Year dummies	No	Yes	No	Yes	No	No
Region dummies	No	Yes	No	Yes	No	Yes
Hausman Chi^2	60.5	45.32	19.79	12.51	–	–
Prob chi^2	0.000	0.000	0.001	0.253	–	–
Number of obs.	1134	1134	972	972	162	162
Adj R^2	0.379	0.431	0.029	0.043	0.168	0.157

Dependent variable: relative regional manufacturing output.
t statistics in parentheses; *, ** and *** denote coefficient estimates significant at 10, 5 and 1 per cent.

1   Cumulative difference is a difference in the variable between the last and first period.

In summary, pure trade liberalization does not seem to foster regional manufacturing relocation. In addition, W/N border regions do not seem to benefit or lose in trade integration with the EU. Hence, none of the EG hypotheses found confirmation in the Slovenian regional data for the period of transition.

6.4 The Impact of Economic Geography on the Regional Distribution of FDI

According to the above findings, FDI appears to be quite important for manufacturing relocation. As proposed by the Damijan-Kostevc model, FDI inflows into home regions are determined by (i) differences in relative factor costs, (ii) trade costs between the home country and the foreign country as well as trade (transport) costs between home regions, and (iii) agglomeration effects.

Figure 8.8 shows the pattern of relative regional manufacturing output produced by foreign investment firms (FIEs) in western and northern border regions. Figure 8.8a reveals a monotonic and slightly decreasing trend in relative regional performance of FIE in border regions, while Figure 8.8b shows virtually no correlation between relative FDI and distance from the country's capital.

In order to further verify the above relationship, we have estimated the following model of relative regional FDI performance:

(2) $\ln rFDI = \alpha + \beta \ln irFDI + \gamma \ln rPROD + \delta \ln rWAGE + \phi \ln iVAe + \xi \ln DIST + \varphi BORD + \lambda \ln TC + \upsilon \ln DIST{*}BORDi{*}TC + \rho \Sigma R + \tau \Sigma T + \varepsilon_{it}$ where:

$rFDI$ $rFDI = F_{it}/F_{ct}$ is relative FDI of the region i with respect to the central region c in time period t;
$irFDI$ is the initial relative FDI;
$irPROD$ is the initial relative production;
$rWAGE$ is the relative wage, proxying for differences in relative factor costs;
$irVAe$ is the initial relative value added per employee;
TC is the region's average tariff rate as a proxy for foreign trade costs;
$DIST$ is a proxy for trade/transport costs between home regions; it is measured as a distance (in km) between the i-th region and the capital region;
$BORD$ is a dummy variable for western/northern (W/N) regions, i.e., regions bordering EU member countries;
$DIST{*}BORD{*}TC$ is an interaction term, which proxies for distance effects in border regions;
ΣR and ΣT are matrices of regional[6] and time dummies;
ε_{it} is the error term.

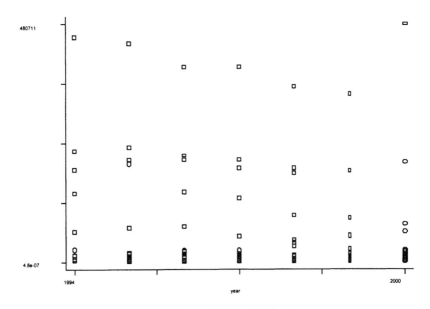

a) Evolution of FDI in W/N regions

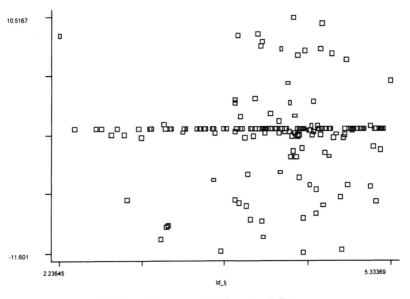

b) Cumulative growth of FDI and distance

Figure 8.8   Regional distribution of FDI in Slovenia

As in the first test, the results in Table 8.16 provide little evidence in favour of the economic geography hypothesis. FDI in Slovenia does not tend to concentrate in regions closer to the capital region (distance does not matter) or to EU borders (W/N border regions do not attract more FDI). Relative FDI is shown to be strongly negatively correlated with relative output and initial FDI stock, implying that existing economic centers tend to attract relatively less FDI.[7] However, differences in relative factor costs (wages) also do not seem to attract more FDI.

In summary, FDI in Slovenia does not seem to work in line with the economic geography models. Hence, the impact of FDI on relocation of manufacturing and relative wages is smaller than has been predicted by Damijan and Kostevc (2002). It is apparent that regional FDI in Slovenia is more likely to be randomly distributed than distributed according to EG models.

6.5   The Impact of Economic Geography on Relative Wages

In this last section, we attempt to clarify the impact of economic geography models on relative regional wages in Slovenia. The orthodox proposition of the EG models assumes that no international trade was present prior to trade liberalization. This is a clear contradiction of reality, especially in Slovenia's case, since Slovenia was relatively open to international trade before 1990. Hence, one could argue that the initial allocation of manufacturing has already been, to some extent, dependent on foreign trade considerations, giving western border regions higher manufacturing concentration than would otherwise be expected. Table 8.17 shows the changes in relative wages (relative to the central region) in Slovenian NUTS 5 regions from 1994 to 2000. It is apparent that relative wages in western and northern border regions have increased more than in other non-border regions. According to this, one might hypothesize divergence between western/northern border regions and non-border regions as predicted by the Damijan-Kostevc model.

Figure 8.9a shows that initial relative regional wages in western/northern regions were slightly decreasing with distance from the capital region. This trend, however, is not statistically significant. Figure 8.9b reveals that the cumulative change in relative wages in W/N regions from 1994 to 2000 shows a decreasing trend with respect to the initial levels, which implies convergence. Figure 8.9c depicts the relationship between cumulative growth in relative regional wages and distance from the capital city. The figure reveals increased dispersion of cumulative growth of relative regional wages with distance from

Table 8.16   Does economic geography affect regional distribution of FDI?

Estimator	Data in levels		First differences		Cumul. differences	
	RE	RE	RE	RE	OLS	OLS
Const.	***−4.854	−2.390	−0.782	−0.104	**−4.111	−1.788
	(−3.86)	(−1.37)	(−1.34)	(−0.09)	(−2.20)	(−0.44)
irFDI	−	−	***−0.117	***−0.120	***−0.715	***−0.725
	−	−	(−4.95)	(−4.82)	(−9.04)	(−8.56)
rPROD	***−0.526	***−0.490	*−0.087	−0.077	***−0.452	***−0.452
	(−6.29)	(−5.67)	(−1.83)	(−1.56)	(−3.08)	(−3.02)
rWAGE	−0.398	−0.589	−0.001	−0.173	−1.337	−1.293
	(−1.07)	(−1.27)	(0.00)	(−0.45)	(−1.03)	(−0.97)
irVA/emp	*0.934	0.855	−0.031	−0.017	0.071	−0.009
	(1.71)	(1.58)	(−0.15)	(−0.08)	(0.10)	(−0.01)
DIST	0.005	−0.757	0.262	−0.211	1.688	−0.810
	(0.00)	(−0.44)	(0.34)	(−0.21)	(0.53)	(−0.19)
BORD	−1.536	−1.448	0.655	1.173	−1.496	1.105
	(−0.84)	(−0.75)	(0.43)	(0.71)	(−0.41)	(0.23)
TC	0.694	−0.298	0.280	−0.162	1.911	0.055
	(0.49)	(−0.20)	(0.39)	(−0.18)	(0.65)	(0.01)
DIST*BORD*TC	−0.008	0.050	−0.137	−0.215	0.128	−0.308
	(−0.03)	(0.17)	(−0.54)	(−0.78)	(0.19)	(−0.34)
Year dummies	No	Yes	No	Yes	No	No
Region dummies	No	Yes	No	Yes	No	Yes
Hausman Chi^2	10.88	11.75	2.3	1.38	−	−
Prob chi^2	0.092	0.466	0.890	1.000	−	−
Number of obs.	1134	1134	972	972	162	162
Adj R^2	0.191	0.245	0.027	0.039	0.328	0.328

Dependent variable: relative FDI output growth.
t-statistics in parentheses; *, ** and *** denote coefficient estimates significant at 10, 5 and 1 per cent.

the capital, but no significant trend can be depicted. Finally, Figure 8.9d reveals a non-monotonic, U-shaped pattern of evolution of relative wages in W/N regions. This fact confirms the predictions of the Krugman-Venables type of EG models, while the predictions of the simple Krugman (1991) type of EG models (monotonically decreasing relative wages) are rejected. It remains to be seen whether these findings can be formally confirmed by the data.

We have verified the above estimations by estimating the following linear model of relative regional wages in Slovenia:

Table 8.17 Changes in relative regional wages in Slovenia from 1994 to 2000

	Non-W/N border regions		W/N border regions		All regions	
	1994	*2000*	*1994*	*2000*	*1994*	*2000*
Mean	0.659	0.680	0.667	0.752	0.660	0.692
Std. error	0.021	0.017	0.042	0.051	0.019	0.016
Std. deviation	0.251	0.199	0.222	0.270	0.246	0.213
Coef. of variation	38.2%	29.2%	33.3%	35.9%	37.3%	30.8%
N	142	142	28	28	170	170

Source: SURS, authors' calculations.

(3) $\ln rWAGE = \alpha + \delta \ln irVAe + \theta \ln rPROD + \phi \ln rFDI + \beta \ln DIST + \gamma BORD + \lambda \ln TC + \upsilon \ln DIST*BORD*TC + \rho\Sigma R + \tau\Sigma T + \varepsilon_{it}$, where:

$rWAGErWAGE = w_{it}/w_{ct}$ is the relative wage of the region i with respect to the central region c in time period t;
$irVAe$ is the initial regional relative value added per employee – it serves as a proxy for initial differences in relative wages,
$rPROD$ is relative production,
$rFDI$ is the initial relative production,
TC is the region's average tariff as a proxy for foreign trade costs,
$DIST$ is a proxy for trade/transport costs between home regions; it is measured as a distance (in km) between the i-th region and the capital region
$BORD$ is a dummy variable for western/northern (W/N) regions, i.e., regions bordering EU member countries
$DIST*BORD*TC$ is an interaction term, which proxies for distance effects in border regions after trade liberalization has been initiated
ΣR and ΣT are matrices of broader regional and time dummies,
ε_{it} is the error term.

Based upon the model's implications, the following predictions regarding the signs of coefficients can be made:

$\theta > 0$ relative wages are increasing in relative manufacturing shares;
$\phi > 0$ relative wages are increasing in relative FDI shares;

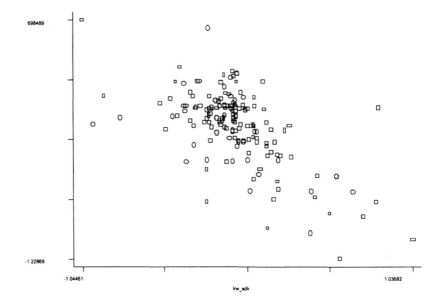

a) Initial wages and distance

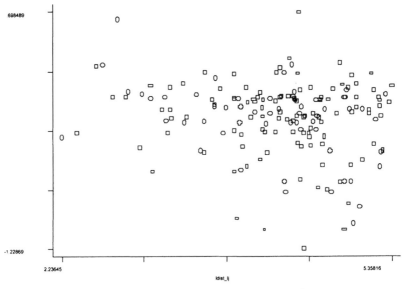

b) Initial wages and cumulative growth of wages

Figure 8.9   Relative regional wages in the western/northern border regions in Slovenia

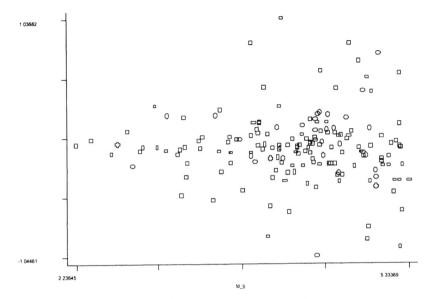

c) Cumulative growth of wages and distance

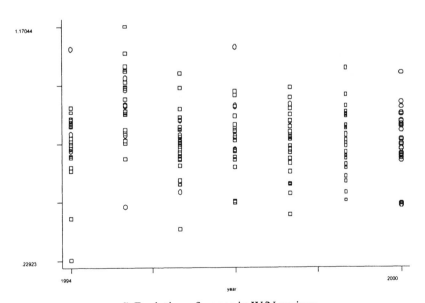

d) Evolution of wages in W/N regions

$\beta < 0$ the distance effects become less important over time,
$\gamma > 0$ W/N border regions converge in terms of relative wages,
$\upsilon > 0$ trade liberalization enhances convergence of relative wages in W/N border regions.

Table 8.18 Does economic geography affect regional relative wages?

	Data in levels		First differences		Cumul. differences	
Estimator	*RE*	*RE*	*RE*	*RE*	*OLS*	*OLS*
Const.	0.601	0.508	1.170		0.069	0.070
	(15.54)	(10.8)	(11.28)		(1.04)	(0.67)
IrVA/emp	**0.010	**0.114	0.003	0.006	0.001	0.070
	(1.96)	(2.19)	(0.27)	(0.59)	(0.16)	(0.84)
rPROD	***0.394	***0.472	−0.548	−0.409	−0.250	−0.007
	(3.44)	(4.22)	(−1.31)	(−0.99)	(−0.72)	(−1.20)
rFDI	−0.027	−0.054	0.310	0.169	0.202	0.322
	(−0.40)	(−0.85)	(1.19)	(0.68)	(0.74)	(1.14)
DIST	**−0.019	0.007	−0.002	−0.011	−0.015	−0.019
	(2.20)	(0.52)	(−0.11)	(−0.24)	(−1.01)	(−0.59)
BORD	0.013	−0.012	0.036	0.054	**0.980	**0.869
	(0.69)	(−0.48)	(0.50)	(0.73)	(2.67)	(2.23)
TC	***−0.013	0.008	***−0.290	***1.069	−0.062	0.004
	(3.11)	(1.13)	(−8.11)	(7.16)	−0.24	0.01
DIST*BORD*TC	0.002	0.001	−0.004	−0.003	***−0.207	**−0.172
	(0.68)	0.68	(−0.22)	(−0.20)	(−2.56)	(−2.03)
Year dummies	No	Yes	No	Yes	No	No
Region dummies	No	Yes	No	Yes	No	Yes
Hausman Chi^2	4.68	1.32	1.75	0.60	–	–
Prob chi^2	0.455	0.997	0.883	0.736	–	–
Number of obs.	1134	1134	972	972	162	162
Adj R^2	0.131	0.211	0.079	0.268	0.033	0.069

Dependent variable: relative regional wages.
Disaggregation: NUTS 5 regions.
t-statistics in parentheses; *, ** and *** denote coefficient estimates significant at 10, 5 and 1 per cent, respectively.

The results in Table 8.18 provide some confirmation of the EG hypothesis for Slovenia, but this remains inconclusive. Distance from the capital does not seem to be very important for determining relative regional wages since the coefficients for distance remain insignificant both in the model with data in levels[8] as well as in the models with data in first and cumulative differences. On the other hand, wages in regions bordering the EU seem to converge with the capital city's wages (the coefficient is positive in all specifications, but significant only in the case of the cumulative differences). At the same time, this is true only for those W/N regions that are located closer to the capital.

Pure trade liberalization (reduction in trade costs) and openness to FDI inflows do not seem to drive the process of convergence of relative wages between regions. It is more likely that the general process of transition to a market economy, coupled with greater openness to trade, has been driving this process. One can therefore conclude that neither the Krugman (1991) model nor the Krugman-Venables type of EG models has found confirmation in the Slovenian regional data for during the transition process. An explanation for the lack of evidence in favour of the Damijan-Kostevc EG model, which stresses the importance of FDI, is the minor role of FDI in the 'Slovenian way of transition'. It remains to be seen, however, whether this model better fits the data for other transition countries that have based their restructuring more heavily on FDI.

7 Conclusions

This chapter studies the relocation pattern of regional manufacturing in Slovenia over the period of 1994–2000. In the first part of the chapter we found a clear pattern of regional relocation of manufacturing activity in Slovenia in terms of manufacturing output and exports. Manufacturing employment, however, did not completely follow this relocation pattern. There is little evidence of the relocation of employment and population between regions, which indicates that declining regions are facing higher unemployment rates. This proposition has also been formally confirmed. Labor markets are clearly not very flexible in Slovenia.

The second part of the present chapter attempts to test whether regional development in Slovenia after integration with the EU is progressing according to the predictions of the EG theory. We test the propositions of three competing EG models. Initially, according to the standard EG hypothesis, with no foreign trade, regions closer to the capital city will have a larger concentration of manufacturing activity. After trade liberalization (i.e., economic integration with the EU), a relocation of manufacturing activity should take place. In line with the Krugman (1991) type of EG models, further monotonic reduction of manufacturing activity in more distant regions and further concentration of manufacturing activity in regions closer to the capital region should be observed. The Krugman-Venables type of EG models propose a non-monotonic,

U-shaped evolution of relative regional output, i.e., a convergence of relative output should take place after initial divergence. In addition, the Damijan-Kostevc EG model predicts an even more pronounced U-shaped pattern of relocation of relative regional output, since FDI inflows into regions bordering the EU foster quicker adjustment in the EU-bordering regions. We evaluate the above propositions using data on relative regional manufacturing output, FDI and wages in Slovenia. Formal tests, however, find little evidence in favour of any of the competing EG hypotheses. The observed relocation pattern after integration with the EU does not correspond to economic geography predictions. Only in the case of relative regional wages in western/northern regions is some evidence found in favour of the U-shaped evolution of wages. But this is not represented by the inter-regional relocation of manufacturing towards the western/northern regions. In fact, the observed evolution of relative regional wages might be purely coincidental.

There are two basic explanations for the lack of evidence in favour of any of the economic geography models for Slovenia. The most obvious one is the small geographic size of Slovenia. It is very likely that in such a small country inter-regional transport costs cannot play as important of a role as in countries that are much larger geographically. Second, Slovenia was already relatively open to international trade before 1990. Hence, one could argue that the initial allocation of manufacturing was already, to some extent, dependent on foreign trade considerations. There would then be less scope for the relocation of manufacturing activity than in other transition countries. It remains to be seen, however, whether the predictions of the economic geography models better fit the data for other transition countries that are larger and have regional production structures that have been affected more heavily by integration with the EU.

Notes

*   This research was undertaken with support from the European Community's PHARE ACE Programme 1998. The content of the publication is the sole responsibility of the authors and it in no way represents the views of the Commission or its services.
1   A maximum value of the Herfindahl index (1) indicates perfect specialization, while the lowest value (0) indicates no specialization.
2   The Hausman specification test indicates that, in this case, the RE estimator provides the least biased estimate.
3   Theoretically, a more proper specification of the regional wages model should also include non-linear terms. Unfortunately, in our model this is not possible because some of the crucial variables (such as border dummies and distance) are non-parametric and time-invariant.

4   A region's average tariff rate is calculated from tariffs at the national level for imports from the EU weighted by the individual region's production structure.
5   NUTS 3 dummy variables have been used in order to control for broader common regional effects.
6   NUTS 3 dummy variables have been used in order to control for broader common regional effects.
7   Note that in this specification, FDI is specified as relative regional FDI stock with respect to the central region, while in the model (1), FDI is specified as the share of employment by foreign investment firms in the total employment of the individual region.
8   The only exception is the estimatation in the model with levels but without time and broad region dummies.

References

Amiti, M. (1998), 'New Trade Theories and Industrial Location in the EU: A Survey of Evidence', *Oxford Review of Economic Policy*, Vol. 14, pp. 45–53.

Amiti, M. (1999), 'Specialisation Patterns in Europe', *Weltwirtschaftliches Archiv*, Vol. 134, pp. 573–93.

Brülhart, Marius and Torstensson, Johan (1996), 'Regional Integration, Scale Economies and Industry Location in the European Union', CEPR Discussion Paper No. 1435.

Damijan, P.J. (1999), 'Interaction of Different Types of Economies of Scale, Monopolistic Competition, and Modern Patterns of Trade', RCEF Working papers series, No. 85.

Damijan, P.J. and Kostevc, Č. (2002),'The Impact of European Integration on the Adjustment Pattern of Regional Wages in Transition Countries: Testing Competitive Economic Geography Models', LICOS Discussion Paper, No. 118/2002.

Davis, Donald R. and Weinstein, David E. (1999), 'Economic Geography and Regional Production Structure: an Empirical Investigation', *European Economic Review*, Vol. 43, pp. 379–407.

De la Fuente, Angel (2000), 'On the Courses of Convergence: a Close Look at the Spanish Regions', Mimeo, Instituto de Analisis Economico.

Forslid, R., Haaland J.I. and Midelfart-Knarvik, K.H. (2001), 'A U-shaped Europe? A simulation Study of Industrial Location', *Journal of International Economics*.

Fujita, M., Krugman, P. and Venables, A.J. (1999), *The Spatial Economy: Cities, Regions, and International Trade*, MIT Press, Cambridge, Mass.

Hallet, Martin (2000), 'Regional Specialization and Concentration in the EU', Economic Papers, European Commission, No. 141, 1-29.

Hanson, Gordon H. (1996), 'Localization Economies, Vertical Organization and Trade', *European Economic Review*, Vol. 86, pp. 1266–78.

Hanson, Gordon H. (1997), 'Increasing Returns, Trade and the Regional Structure of Wages', *The Economic Journal*, Vol. 107, pp. 113–32.

Hanson, Gordon H. (2000a), 'Market Potential, Increasing Returns, and Geographic Concentration', Mimeo, University of Michigan.

Hanson, Gordon H. (2000b), 'Scale Economies and the Geographic Concentration of Industry', NBER Working paper no. 8013.

Hummels, David (1999a), 'Toward a Geography of Trade Costs', Mimeo, Purdue University.

Hummels, David (1999b), 'Have International Trade Costs Declined', Mimeo, Purdue University.

Hummels, David (2001), 'Time as a Trade Barrier', Mimeo, Purdue University.

Kim, S. (1995), 'Expansion of Markets and the Geographic Distribution of Economic Activities: The Trends in US Regional Manufacturing Structure, 1860–87', *Quarterly Journal of Economics*, Vol. 110, pp. 881–908.

Krugman, P. (1991a), '*Geography and Trade*', MIT Press, Cambridge, Mass.

Krugman, P. (1991b), 'Increasing Returns and Economic Geography', *Journal of Political Economy*, Vol. 99, pp. 483–99.

Krugman, P. and Venables, A.J. (1995), 'Globalization and the Inequality of Nations', *Quarterly Journal of Economics*, Vol. 110, pp. 857–80.

Krugman, P. and Venables, A.J. (1997), 'Integration, Specialization, and Adjustment', *European Economic Review*, Vol. 40, pp. 959–67.

Midelfart-Knarvik, K, Overman, H.G. and Venables, A.J. (2000), 'Comparative Advantage and Economic Geography', CEPR Discussion Paper No. 2618.

Overman, H.G., Redding, Stephen and Venables, A.J. (2000), 'The Economic Geography of Trade, Production, and Income: a Survey of Empirics', Mimeo, London School of Economics.

PART III
COMPARATIVE ANALYSIS AND LESSONS

Chapter 9

Specialization of Regions and Concentration of Industries in EU Accession Countries*

Iulia Traistaru, Peter Nijkamp and Simonetta Longhi

1 Introduction

The emerging economies in accession countries will most likely exhibit a high degree of spatial economic dynamics in the years to come, especially if they are increasingly exposed to market forces. The question is whether various regions or industries in these countries have anticipated this transformation, and whether they are already showing the first signs of a shift in their spatial-economic base. Thus, industries may demonstrate a different pattern of regional localization, or alternatively, specific regions may be able to attract new industries. This would mean a drastic change in the location patterns of industries, reflected in changes in the spatial concentration of sectors or firms and in the regional concentration of various industries. The available theoretical frameworks on location of industrial activity and regional growth are not always conclusive, nor are individual country reports from the accession countries. Additional empirical research is therefore needed for a better understanding of the patterns and changes of regional specialization and location of industrial activity in the accession countries.

How specialized/diversified are regions in accession countries? How concentrated/dispersed are industries? Have patterns of regional specialization and geographical concentration of industries changed over the period from 1990–99? What are the determinants of patterns of industrial location?

The aim of this chapter is to identify, explain and compare patterns of regional specialization and location of manufacturing activity in five accession countries, viz. Bulgaria, Estonia, Hungary, Romania and Slovenia.

This chapter is the first comparative analysis of patterns of regional specialization and geographical concentration of manufacturing activity in accession countries. Our research results suggest that, in the five accession countries included in this study, regional relocation of industries has taken place, leading to increasing regional specialization in Bulgaria and Romania and decreasing regional specialization in Estonia. Regional specialization has not changed significantly in Hungary and Slovenia. We find empirical evidence indicating that both factor endowments and geographic proximity to European markets determine the location of manufacturing in accession countries.

The remainder of this chapter is organized as follows. Section 2 discusses the theoretical framework and existing empirical evidence on specialization of countries and regions and geographical concentration of industries. Section 3 gives an overview of the data set and measures used for our analysis. Section 4 analyses patterns of regional specialization in the five accession countries, while Section 5 discusses the geographical concentration of manufacturing in the same countries. Section 6 presents the results of our econometric analysis on determinants of the location of manufacturing activity in the five accession countries included in this study. Section 7 concludes.

2   Analysis Framework

2.1   Theoretical Background

Existing international trade theory about the impact of economic integration on specialization and location of industrial activity could be grouped into three strands of literature.[1] While offering different explanations of patterns of specialization, all three theoretical approaches predict increasing specialization as a result of trade liberalization and economic integration. Neoclassical trade theory explains patterns of specialization on the basis of differences in productivity (technology) or endowments across countries and regions while new trade theory and, more recently, new economic geography models underline increasing returns in production, agglomeration economies and cumulative processes as explanations for the concentration of activities in particular countries and regions.

Neoclassical trade theory explains specialization patterns through differences in relative production costs termed 'comparative advantages' resulting from differences in productivity (technology) (Ricardo, 1817) or endowments (Heckscher, 1919; Ohlin, 1933) between countries and regions. The main features of these models are: perfect competition, homogeneous products and constant returns to scale. The neoclassical theory predicts that trade liberalization and economic integration will result in production relocation and increasing specialization according to comparative advantages. The consequent changes in demands for factors of production will tend to equalize factor prices across countries and regions. A large portion of inter-industry specialization can be explained by neoclassical trade models (see Leamer and Levinsohn, 1995). While relevant, comparative advantage is, however, not sufficient as the sole explanation for specialization. In reality, different production structures are found in regions and countries with similar factor endowments and production technologies. Trade between industrialized countries consists mainly of goods in the same product category, i.e., it is intra-industry trade.

During the 1980s, new trade theory models were developed, mainly for explaining intra-industry trade (Krugman, 1979, 1980, 1981; Helpman and Krugman, 1985; Krugman and Venables, 1990). The main assumptions in these models are increasing returns to scale, product differentiation and imperfect (monopolistic) competition. The new trade theory models focus on the interactions between firms with increasing returns in product markets and explain patterns of specialization and location of industrial activity in terms of the geographical advantage of countries and regions with good market access. When trade barriers fall, activities with increasing returns will locate in countries/regions with good market access ('the center') moving away from remote countries/regions ('the periphery'). Krugman and Venables (1990) suggest that geographical advantage will be greatest at some intermediate trade cost, i.e., the relationship between trade costs and the location of activity has an inverse U-shape. When trade barriers and transport costs are small enough, the geographical advantage of the regions with good market access becomes less important. At this stage, factor production costs will motivate firms to move back to peripheral regions.

The prediction of new trade theory regarding the distribution of economic activity between the core and periphery is relevant in the case of the accession of Central and East European countries to the European Union. The current economic integration situation could be seen as one with 'intermediate trade costs'. Further integration could result in the relocation of manufacturing towards these countries due to factor costs considerations (Hallet, 1998).

The new economic geography models assume that the geographical advantage of large markets is endogenous and suggest that specialization patterns may be the result of the spatial agglomeration of economic activities (Krugman, 1991a, 1991b; Krugman and Venables, 1995, Venables, 1996). The main assumptions of these models are the presence of pecuniary or technological externalities between firms, monopolistic competition and increasing returns to scale. Krugman's analysis focuses on a two sector-two region model similar to that of Krugman and Venables (1990). Unlike in the latter model, the two regions are identical in terms of initial factor endowments and the factor specific to manufacturing (industrial workers) is mobile across regions. Relocation of firms and workers from one region to the other triggers agglomeration via the cumulative effects of demand linkages. With no barriers to the movement of firms or manufacturing workers (like in the Krugman, 1991b model), a bleak scenario could be imagined: the manufacturing sector in the 'donor' region would collapse and manufacturing would concentrate in the 'receiving' region. This scenario could develop gradually following the lowering of trade costs. Initially, when trade costs are high, manufacturing is evenly split between regions (each region produces for its own local market). If trade costs are sufficiently low, demand linkages bring about the agglomeration of activities. Regions with an initial scale advantage in particular sectors would attract more manufacturing activity and thus reinforce their advantage in those sectors. Krugman and Venables (1995) extend these models to include firms with 'supply-side linkages'. Manufacturing firms locate in a region where they benefit from access to suppliers providing specialized inputs.

These new economic geography models imply that, in sectors where supply-side and demand-side linkages are important, European integration would bring massive specialization and concentration. Given the extremely low inter-EU country mobility, this result seems, however, unrealistic (Eichengreen, 1993; Obstfeld and Peri, 1998). Agglomeration effects might still be present if there is sufficient labour mobility within EU countries. In this case, we could observe agglomeration effects emerging around border regions similar to those identified by Hanson (1996, 1997a) for the case of US–Mexican economic integration.

2.2   Empirical Evidence

Empirical literature on the impact of economic integration on production specialization and geographic concentration of industries is still scarce. The most interesting studies have focused on the United States (US) and the European Union (EU) and have established the following stylized facts:

a) regional specialization and industrial concentration are higher in the US than in the EU (Krugman, 1991a; Midelfart-Knarvik et al., 2000);

b) production specialization has increased in EU Member States while trade specialization has decreased (Sapir, 1996; Amiti, 1997; Haaland et al.,

1999; Aiginger et al., 1999; Midelfart-Knarvik et al., 2000; Brülhart, 1996, 2001);

c) slow-growing and unskilled labour-intensive industries have become more concentrated in the EU (Midelfart-Knarvik et al., 2000);

d) medium and high technology industries have become more dispersed in the EU (Brülhart, 1998, 2001);

e) industries with large economies of scale were concentrated close to the European core during the early stages of European integration but have become more dispersed during the 1980s (Brülhart, 1998; Brülhart and Torstensson, 1996).

In a series of papers, Hanson has assessed the locational forces identified by the new economic geography models in the context of US–Mexican integration. Hanson (1996) finds evidence that agglomeration is associated with increasing returns, and shows that integration with the US has led to a relocation of Mexican industry away from Mexico City and towards states with good access to the US market. This is reflected in the falling importance of distance from the capital and the rising importance of distance from the border in explaining interregional wage differentials (Hanson, 1997a, 1997b, 1998). Employment has grown more in regions that have larger agglomerations of industries with buyer/supplier relationships, suggesting that integration has made demand and cost linkages important determinants of industrial location.

Ellison and Glaeser (1997) analyse the geographic concentration of US manufacturing industries. Using a model that controls for industry characteristics, they find that almost all industries seem to be localized. Many industries are, however, only slightly concentrated and some of most concentrated industries are related to natural advantages.

With respect to Europe, Brülhart (1996) and Brülhart and Torstensson (1996) study the evolution of industrial specialization patterns in 11 EU countries (all except Luxembourg and the more recent member states of Austria, Finland, and Sweden) between 1980 and 1990. They find support for the U-shaped relationship between the degree of regional integration and spatial agglomeration predicted by the theoretical models when labour mobility is low: activities with larger scale economies were more concentrated in regions close to the geographical core of the EU during the early stages of European integration, but concentration in the core fell during the 1980s.

Using production data in current prices for 27 manufacturing industries, Amiti (1999) finds that there was a significant increase in specialization between 1968 and 1990 in Belgium, Denmark, Germany, Greece, Italy, and the Netherlands; no significant change occurred in Portugal; and a significant fall in specialization occurred in France, Spain and the UK. There was a significant increase in specialization between 1980 and 1990 in all countries. With more disaggregated data (65 industries) the increase in specialization is more pronounced: the average increase is 2 for all countries except Italy, compared to 1 per cent in the case with 27 manufacturing industries. Other evidence of increasing specialization in EU countries in the 1980s and 1990s based on production data is provided by Hine (1990), Greenway and Hine (1991), Aiginger et al. (1999), and Midelfart-Knarvik et al. (2000). However, analyses based on trade data indicate that EU Member States have a diversified rather than specialized pattern of manufacturing exports (Sapir, 1996; Brülhart, 2001).

In terms of the geographic concentration of industries, Amiti (1999) finds that 17 out of 27 industries experienced an increase in geographical concentration, with an average increase of 3 per cent per year in leather products, transport equipment and textiles. Only six industries experienced a fall in concentration, with paper and paper products and 'other chemicals' showing particularly marked increases in dispersion. Brülhart and Torstensson (1996) compare industry Gini coefficients with the industry centrality indices proposed by Keeble et al. (1986) and find a positive correlation between scale economies and industry bias towards the central EU in both 1980 and 1990. Brülhart (1998) finds that industries such as chemicals and motor vehicles that are highly concentrated and located in central EU countries are subject to significant scale economies. Midelfart-Knarvik et al. (2000) find that many industries have experienced significant changes in their location across EU member states during the period 1970–97. Slow-growing and unskilled labour-intensive industries have become more concentrated, usually in peripheral low wage countries. During the same period, a number of medium and high technology industries have become more dispersed.

With respect to accession countries, existing evidence based on trade statistics suggests that these countries tend to specialize in labour- and resource-intensive sectors, following an inter-industry trade pattern (Landesmann, 1995). Despite the dominance of the inter-industry (Heckscher-Ohlin) type of trade, intra-industry trade has also increased, most evidently in the Czech Republic and Hungary (Landesmann, 1995; Dobrinsky, 1995). This increase, however, may be associated with the intensification of outward processing traffic. It has been claimed that the processes of internationalization and structural change in transition economies tend to favor metropolitan and western regions, as well as regions with a strong industrial base (Petrakos, 1996). In addition, at a macro-geographical level, the process of transition is expected to increase disparities at the European level, by favoring countries near the East-West frontier (Petrakos, 2000). Increasing core periphery differences in Estonia are documented in Raagmaa (1996). Using the 'new economic geography' approach, Altomonte and Resmini (1999) have investigated the role of foreign direct investment in shaping regional specialization in accession countries.

Yet to date, there has been no comprehensive study of the impact of economic integration with the European Union on regional specialization and geographic concentration of industrial activity in accession countries.

3 Data and Measurement

In this chapter we analyse patterns of regional specialization and concentration of manufacturing and their determinants using regional manufacturing employment data and other variables at the NUTS 3 level for Bulgaria, Estonia, Hungary, Romania and Slovenia. The employment data and the other regional variables are part of a specially created data set called REGSTAT.[2] Apart from employment, other variables at the regional level used in our analysis include: geographic and demographic variables, average earnings (wages), Gross Domestic Product (GDP), measures of infrastructure, research and development (R&D), and public expenditures.

The period covered is 1990–99. In most cases, data have been collected from national statistical offices. In the case of Estonia, employment data at the regional level has been estimated using labour force surveys. In Slovenia, employment data at the regional level has been estimated using the information provided in the balance sheets of companies with more than ten employees.

Regional specialization and geographical concentration of industries are defined in relation to production structures.[3] In absolute terms, a region j is 'specialized' in a specific industry i if this industry has a high share in the manufacturing activity of region j. The manufacturing structure of a region j is 'highly specialized', if a small number of industries have a large combined share in the total manufacturing of region j. In relative terms, regional specialization is defined as the distribution of the shares of an industry i in total manufacturing in a specific region j compared to a benchmark.

In absolute terms, a specific industry i is 'concentrated', if a large part of its production is carried out in a small number of regions. In relative terms, geographical concentration of industries is defined as the distribution of the shares of regions in a specific industry i compared to a benchmark.

Several absolute and relative measures of specialization and concentration are proposed in the existing literature, each having certain advantages as well as shortcomings. For our analysis we have selected a relative measure (a dissimilarity index derived from the index proposed by Krugman, 1991a). Notations and definitions are given in Box 9.1.

4 Specialization of Regions

How specialized/diversified are regions in accession countries? Have patterns of regional specialization changed during the 1990s? What is the relationship between regional specialization and economic performance?

Box 9.1   Indicators of regional specialization and geographic concentration of industries[4]

$E =$ employment

$s =$ shares

$i =$ industry (sector, branch)

$j =$ region

$s_{ij}^{s} =$ the share of employment in industry i in region j in total employment of region j

$s_{ij}^{c} =$ the share of employment in industry i in region j in total employment of industry i

$s_i =$ the share of total employment in industry i in total employment

$s_j =$ the share of total employment in region j in total employment

$$s_{ij}^{s} = \frac{Eij}{Ej} = \frac{Eij}{\sum_i Eij} \qquad\qquad s_{ij}^{c} = \frac{Eij}{Ei} = \frac{Eij}{\sum_j Eij}$$

$$s_i = \frac{Ei}{E} = \frac{\sum_j Eij}{\sum_i \sum_j Eij} \qquad\qquad s_j = \frac{Ej}{E} = \frac{\sum_i Eij}{\sum_i \sum_j Eij}$$

The dissimilarity (Krugman) index

Specialization measure   Concentration measure

$$SPEC_j = \sum_i \left| s_{ij}^{s} - s_i \right| \qquad\qquad CONC_i = \sum_j \left| s_{ij}^{c} - s_j \right|$$

Increasing economic integration with the EU and the world economy are likely to result in the relocation of industrial activity and changing specialization patterns across regions in accession countries. Figure 9.1 shows that, during the 1990s, the average regional specialization increased in Bulgaria and Romania and decreased in Estonia and Hungary. In the case of Slovenia, given the short period covered by the available data (1995–98), there is a less clear trend.

In order to check whether regional specialization has changed significantly in the countries under analysis, we have estimated the following trend model:

$$SPEC_{jt} = \alpha + \beta*t + \varepsilon_{jt} \tag{1}$$

where the dependent variable $SPEC_{jt}$ is regional specialization in region j at time t measured by means of the dissimilarity index (see Box 9.1) using employment data on manufacturing branches at the NUTS 3 regional level. The independent variable t is the year to which the data refers, α and β are the parameters to be estimated, and ε_{jt} is the remaining error term.

The trend model has been estimated separately for each country. The results of the OLS estimation with regional fixed effects are shown in Table 9.1.

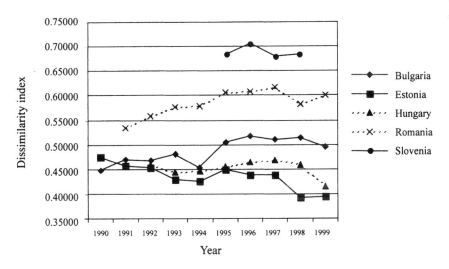

Figure 9.1   **Average regional specialization in accession countries, 1990–99**

Table 9.1   Regional specialization in accession countries, 1990–99

	Bulgaria	Estonia	Hungary	Romania	Slovenia
t	0.0068 ***	−0.0073 **	−0.0019	0.0074 ***	−0.0023
	(0.0011)	(0.0033)	(0.0019)	(0.0012)	(0.0061)
Intercept	0.4488 ***	0.4756 **	0.4638 ***	0.5405 ***	0.7050 ***
	(0.0067)*	(0.0202)	(0.0132)	(0.0077)	(0.0462)
Number of obs.	280	50	160	369	48
R-sq: within	0.1383	0.1029	0.0074	0.1086	0.0039

* significant at 10 per cent; ** significant at 5 per cent; *** significant at 1 per cent.
Standard errors in parentheses.

The above table shows that, on average, regional specialization in the 1990s increased in Bulgaria and Romania, and decreased in Estonia. The estimated coefficient for *t* is not significantly different from zero for Hungary and Slovenia.

The increasing integration of accession countries with the EU may have decreased the importance of internal regions in favor of regions bordering the EU and other accession countries, which were probably less favored in the past. In order to validate this hypothesis, we have classified the regions into the following groups: internal regions (INT), regions bordering the EU (BEU), regions bordering other accession countries (BAC), and regions bordering countries outside the EU enlargement (BEX).[5]

Tables 9.A1–9.A5 show summary statistics for the five accession countries for the specialization of regions as well as economic performance for each of the above groups of regions scaled by the national averages[6] at the beginning and end of the analysed period.

We find that regions bordering the EU are less specialized than the national specialization average in countries closer to the EU accession such as Estonia, Hungary and Slovenia, while they are more specialized in Bulgaria. Regions bordering other accession countries are found more specialized compared to the national averages in Hungary, while in Estonia, Bulgaria and Romania these type of regions are less specialized. Regions bordering countries outside the EU enlargement area are more specialized, with the exception of Romania. Internal (non-border) regions are less specialized in Bulgaria and Hungary and more specialized in Romania and Slovenia.

With a few exceptions, high specialization is associated with inferior economic performance, while regions with low specialization perform better than the national averages. High specialization is associated with superior economic performance in regions bordering countries outside the EU enlargement area in Bulgaria and Slovenia, while lower specialization is associated with inferior economic performance in regions bordering countries outside the EU enlargement area in Romania and in internal regions in Bulgaria.

In summary, our findings suggests that highly specialized regions show an economic performance that is inferior to national averages, while less specialized regions have better economic performance. Over the last decade, regional specialization has increased significantly in Bulgaria and Romania, has decreased in Estonia, and has not changed significantly in Hungary or Slovenia. Proximity to the EU is associated with low specialization in advanced accession countries (Estonia, Hungary and Slovenia) and high specialization in countries lagging behind with the accession (Bulgaria). Proximity to other accession countries is associated with low specialization in Bulgaria, Estonia and Romania, and with high specialization in Hungary. Proximity to countries outside the EU enlargement area is associated with high specialization in Hungary, Bulgaria and Slovenia and with low specialization in Romania. Internal regions have high specialization in Romania and Slovenia and low specialization in Hungary and Bulgaria.

5 Geographical Concentration of Manufacturing

On the basis of the concentration indices calculated for manufacturing branches in Bulgaria, Estonia, Hungary, Romania and Slovenia, we have grouped the industries according to the following characteristics: scale economies, technology levels, and wage levels. The definitions of high-medium-low technology levels, and of high-medium-low wage levels are based on OECD (1994); the definition of high-medium-low levels of scale economies is based on Pratten (1988). The manufacturing classification is according to the Eurostat NACE Rev1 (two-digit classification) for Estonia, Romania, and Slovenia. Employment data have been collected according to national classifications in Hungary and Bulgaria. For these two cases, aggregations have been made to bring these classifications as close as possible to the NACE classification.

Table 9.A6 shows the concentration indices, normalized with the national averages of geographical manufacturing concentration for each of the six industry groups and identified according to the above classifications for Bulgaria, Estonia, Hungary, Romania and Slovenia. We find that industries with low economies of scale had a level of concentration that was stable and very close to the national average in Bulgaria and Romania. In Estonia these sectors were less concentrated than the national average, while in Hungary and Slovenia they were slightly more concentrated than the national average. Slovenian industries belonging to this group were also experiencing a decrease in their level of concentration. The industries with medium economies of scale were below the national average in Bulgaria, Hungary and Slovenia, while they were slightly above the average in Estonia and Romania. In all cases, the level of concentration of these industries seemed to be stable or slightly increasing. Finally, the industries with high economies of scale were much more concentrated than average in all countries. Concentration of these industries seemed to be decreasing slightly, with the exception of Slovenia, in which it seemed to be increasing. In Romania all industries seemed to have the same level of concentration (around the national average); the differences among groups of industries were much more evident for the other countries.

Industries defined as low-tech were usually less concentrated than the national average in all countries, although their level of concentration seems to have increased. The industries defined as medium tech seem to be more concentrated than the average and stable or slightly decreasing in Bulgaria, Estonia and Hungary. In Romania and Slovenia, these industries were as concentrated as the national average, and their level of concentration was stable (in Romania) or increasing (Slovenia). Finally, high-tech industries were less concentrated than the national average in Bulgaria, Hungary and Slovenia. Their level of concentration seemed to be stable or to be increasing (Bulgaria). In Estonia and Romania these industries were more concentrated than the national average. They seemed to become even more concentrated in Estonia, while their level of concentration seemed to be stable or slightly decreasing in Romania.

Industries with the lowest level of wages were the most dispersed ones. Their level of concentration seemed to be stable or slightly increasing. On the other hand, the industries with the highest level of wages were more concentrated than the national average, and their level of concentration seemed to be stable or slightly decreasing. In conclusion, the evidence seems to be in favor of a convergence of concentration levels. The medium-wage industries had a level of concentration that was not far from the national average. Our results suggest that their concentration has increased in Hungary, decreased in Bulgaria and remained stable in the other countries.

Our analysis has been based on the available data for ten years for Bulgaria and Estonia, nine years for Romania, eight for Hungary and only four for

Slovenia. We might not, therefore, be able to capture the impact of regional business cycles on concentration patterns.

At the aggregate level, increasing economic integration with the EU is expected to change patterns of location and concentration of industrial activity in accession countries.

Figure 9.2 shows that, over the period 1990–99, the average concentration of manufacturing has increased in Bulgaria and Romania, decreased in Estonia and Hungary and remained stable in Slovenia.

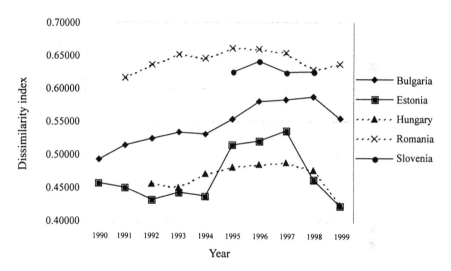

Figure 9.2   **Average concentration of manufacturing in accession countries, 1990–99**

In order to check whether patterns of manufacturing concentration have changed significantly, we have estimated the following model:

$$CONC_{it} = \alpha + \beta * t + \varepsilon_{it} \tag{2}$$

where the dependent variable $CONC_{it}$ is the level of concentration of manufacturing activity in industry i at time t, calculated by means of the dissimilarity index using employment data on manufacturing branches at the NUTS 3 regional level. The independent variable t is the year to which the data refers, α and β are the parameters to be estimated, and ε_{it} is the remaining error term.

The model has been computed separately for each country, using an OLS with industry fixed effects estimation method. The results shown in Table 9.2 indicate that concentration of manufacturing did not change significantly in these countries, with the exception of Bulgaria, in which it increased.

Table 9.2   Geographical concentration of manufacturing in accession countries, 1990–99

	Bulgaria	Estonia	Hungary	Romania	Slovenia
t	0.0092 ***	0.0037	−0.0003	0.0015	−0.0011
	(0.0014)	(0.0037)	(0.0275)	(0.0017)	(0.0061)
Intercept	0.4945 ***	0.4481 ***	0.4690 ***	0.6342 ***	0.6367 ***
	(0.0090)	(0.023)	(0.0189)	(0.0111)	(0.0465)
Num. of obs.	120	130	64	108	48
R-sq: within	0.2773	0.0083	0.0002	0.0077	0.0010

* significant at 10 per cent; ** significant at 5 per cent; *** significant at 1 per cent.
Standard errors in parentheses.

In summary, highly concentrated industries are those with large economies of scale, medium and high technology and high wages. Industries with low technology levels and low wages are dispersed. Geographical concentration of manufacturing has increased significantly in Bulgaria and has not changed significantly in the rest of the accession countries analysed here.

6   Determinants of Manufacturing Location

As pointed out in Midelfart-Knarvik et al. (2000), regional specialization and industrial concentration patterns are determined by the interaction of regional and industry characteristics. Regions differ in size, factor endowments and their geographic position (core or peripheral). Industries differ with respect to economies of scale and factor intensities. The reason for evaluating the interaction between regional and industry characteristics lies in the fact that firms evaluate the same kind of production factors differently (Fujita et al., 1999). Industries will try to locate as close as possible to the place where their most important inputs are available, and will therefore be over-represented in that location. Industries for which the same production factor is less important will instead be under-represented.

To uncover determinants of manufacturing location and explain regional manufacturing production structure differentials in the five accession countries, we estimate a model similar to the model estimated in Midelfart-Knarvik et al. (2000) for EU Member States. We analyse determinants of manufacturing location by regressing the log share of industry i in region j (s_{ij}^S) on regional and industry characteristics, after controlling for the size of regions by means of the log share of the population living in region j (pop_j) and of the log share of total manufacturing employment located in region j (man_j). We use the following specification:

$$ln\ (s_{ij}^S) = c + \alpha\ ln\ (pop_j) + \beta\ ln\ (man_j) + \Sigma_k\ \beta\ [k]\ (y[k]_j - \gamma\ [k])\ (z[k]^i - \kappa\ [k]) \tag{3}$$

where $y[k]_j$ is the level of the kth region characteristic in the jth region and $z[k]^i$ is the level of the kth industry characteristic of industry i. As is clear in (3), the kth region characteristic is matched with the kth industry characteristic. Finally, α, β, $\beta[k]$, $\gamma[k]$, and $\kappa\ [k]$ are the coefficients to be estimated. We have computed the share of industry i in region j (s_{ij}^S) using employment data.

The first two variables appearing on the right hand side (ln (pop_j) and ln (man_j)) capture regional size effects and are therefore needed to correct for the differences in the size of regions. The remaining terms capture the influence of regional and industry characteristics and their interactions. Details of regional and industry characteristics are shown in Table 9.3.

After having defined the regional and the industry characteristics, we interacted them as shown in Table 9.4.

The market potential (MP) characteristic – which has been interacted with the level of scale economies (SE) – may be interpreted as an indicator of proximity to markets. We computed two market potential indicators: the first one (MP1) intends to compare regions inside the same country in the context of a closed economy, while with the second indicator (MP2), we try to get some insights into the consequences of the increasing relationship between each country and the EU. It is plausible that the association agreement with the EU has led to a reduction of the cost of transport into the EU by reducing trade barriers, while transport costs within the country have probably remained unchanged. This had probably favored regions bordering the EU in comparison to central regions, which had had a favorable position before the EU accession agreements. The MP2 variable is used to verify whether increasing integration

Table 9.3   Regional and industry characteristics

Variable name	Description
	Regional characteristics
Market potential (MP1)	Average regional wages (deflated at national level) divided by the distances from country capital (in km)
Market potential (MP2)	Average regional wages (deflated at a national level) divided by a proxy of the distance from EU markets (one if the region borders the EU, two if the region does not border the EU)
R&D (RD)	R&D personnel divided by the number of persons employed for Bulgaria and Hungary; R&D expenditures divided by the value added in manufacturing for Slovenia; no information is available for Estonia and Romania
Labour abundance (LA)	Sum of employment and unemployment, divided by the population of working age (15–65 years)
	Industry characteristics[7]
Scale economies (SE)	1 = low, 2 = medium, 3 = high (definition by Pratten, 1988)
Research oriented (RO)	1 = almost none of the industries of the sector are defined as research oriented; 2 = some industries of the sector are defined as research oriented; 3 = almost all industries of the sector are defined as research oriented (definition by OECD, 1994)
Technology level (TL)	1 = Low technology; 2 = Medium technology; 3 = high technology (definition by OECD, 1994)
Labour intensity (LI)	Labour intensity dummy (definition by OECD, 1994)

Table 9.4   Interaction variables

	Variable name	Regional characteristic	Industry characteristics
J=1	MP1SE	MP1 Market potential (distances from country capital)	SE Scale economies
J=2	MP2SE	MP2 Market potential (distances with EU markets)	SE Scale economies
J=3	RD1RO	RD1 RD2 = RD R&D personnel or expenses	RO Research oriented
J=4	RD2TL		TL Technology level
J=5	LALI	LA labour abundance	LI Labour intensity

with the EU has led to a reallocation of activity (industries) from central regions to those bordering the EU. We introduced the two market potential variables (MP1 and MP2) in two different models in order to keep the two hypotheses (closed versus open economy) separated.

The labour abundance (LA) and the research and development (RD) characteristics are used to identify the relative regional abundance of these different input factors. The RD characteristic is then alternatively interacted with the technology level (TL) and with the importance of R&D inputs in each industry (RO), while the labour abundance (LA) characteristic is interacted with the importance of labour as a production factor (LI).

The two industry characteristics associated with the R&D regional characteristic – research orientation (RO) and technology level (TL) – may in principle seem very similar. However, the industries listed as RO are not the same industries listed as TL. Furthermore, their significance level did not change when we tried to set one of them aside in our estimations.

The interaction variables MP1SE and MP2SE should be interpreted on the basis of the idea that industries with higher economies of scale may tend to concentrate in relatively central locations (Krugman, 1980; Midelfart-Knarvik et al., 2000). Since we expect the central location to be identified as the country capital in the early1990s and as the EU market in the most recent years, we expect the MP1SE and MP2SE variables to capture these changes.

The interaction variables RDRO, RDTL and LALI should be interpreted on the basis of the idea that industries that highly value some production factors (R&D for research-oriented firms and firms with a high technology level; labour abundance for labour-intensive firms) tend to locate in areas in which these production factors are abundant.

After this short illustration of the variables introduced in our estimations, we may now briefly discuss some estimation issues. First of all, since the data collected in the different countries are quite heterogeneous, we estimated equation (3) separately for each country using OLS with White's heteroskedasticity consistent standard errors. The main findings are summarized in Table 9.5.[8] The estimation results for beginning and end years are shown in Tables 9.A7–9.A11.

We estimated our models using yearly data. The first reason for this choice is the limited time period covered by our data set. Secondly, regional differences in business cycles are smaller than differences that may be observed among countries. Finally, this approach may enable us to better identify structural breaks that may occur in our data set (e.g., we may be better able to distinguish between trends before and after certain EU agreements).

**Table 9.5	Summary of the estimations' findings**

		Bulgaria		Estonia		Hungary		Romania		Slovenia	
	lnpop	0	0	0	pos.	pos.	pos.	pos.	0	pos.	pos.
	lnman	pos.	pos.	neg.	pos.+	pos.	0	pos.	pos.	0	0
Regional characteristic	MP1	0	/	neg.	/	neg.	/	neg.	/	neg.	/
	MP2	/	0	/	neg.	/	neg.	/	neg.	/	0
	RD	0	0	pos.	pos.	neg.	neg.	neg.	/	0	0
	LA	0	0	0	neg.	pos.	neg.	neg.	pos.	/	/
Industry characteristic	SE	neg.	0	/	/	/	/	neg.	neg.	neg.	neg.
	RO	pos.	pos.	0	0	0	0	neg.	neg.	neg.	neg.
	LI	0	0	0	0	0	0	0	0	/	/
	TL	pos.	pos.	/	/	/	/	/	/	pos.	pos.
Interaction variables	MP1SE	pos.	/	0	/	pos.	/	pos.	/	pos.	/
	MP2SE	/	pos.	/	pos.	/	pos.	/	pos.	/	0
	RDRO	0	0	pos.	pos.	pos.	pos.	pos.	pos.	pos.	pos.
	LALI	0	0	pos.	pos.	pos.	pos.	pos.	pos.	/	/
	RDTL	0	0	/	/	/	/	/	/	neg.	neg.

The first column represents the results for the model using MP1 for market potential and the second column using MP2 for market potential. (pos.) indicates that the estimated coefficient is significant and positive; (neg.) indicates that the estimated coefficient is significant and negative.

(/) the variable was not available (or was not used) for the model estimation.

(0) the variable was never significant.

(+) the variable was significantly negative in the first period and significantly positive in the last period.

As shown in Table 9.5, the first two independent variables of the model (ln(pop) and ln(man)), which capture the effect of the analysed regions' different sizes, are higher than zero or are not significant (with the exception of the results for Estonia which are significantly negative), confirming that larger regions have larger shares of industrial activity.

The regional characteristics have, in general, the expected signs. We find that the market potential variables – MP1 and MP2 – are positively and significantly related to industry shares (s_{ij}^S). Since MP1 and MP2 are decreasing with distance from the core markets, this result suggests that industry shares (s_{ij}^S) are higher in regions that are located near the core.

The labour abundance (LA) regional characteristic has negative coefficients, confirming that labour abundant regions have larger shares of industries. In Estonia, the LA coefficient is, however, significantly positive, while in Romania, the coefficient of LA is negative when we use MP1 and positive when we use MP2.

Industry characteristics (see Table 9.5) have, in general, the expected signs. With the exception of Hungary, industries with economies of scale are positively and significantly correlated with the shares of industries. Research-oriented industries seem to be concentrated in Slovenia but dispersed in Bulgaria. The technology level (TL) coefficient is either not significant or is positive, although its significance level seems to decrease. Finally, the labour intensity (LI) coefficient is, in general, not significant.

The coefficients of the market potential variables interacted with scale economies are either positive or not significant. While in Hungary and Romania, both MP1SE and MP2SE seem to be significantly higher than zero, in Bulgaria and Slovenia only MP1SE is significantly positive. In Estonia the only coefficient that seems to be positive is MP2SE. Theory predicts that market forces induce industries with high returns to scale to locate near the core, and that these forces are stronger with intermediate levels of transport costs. Although, as mentioned above, more research is needed to better identify the variables influencing the location of manufacturing in EU accession countries, the fact that these forces are not weakening in the countries and over the period of our analysis supports the idea that the transport costs are still at an intermediate level.

The coefficients of the interacted variables RDRO and RDTL have been estimated only for Bulgaria and Slovenia. While for Bulgaria both coefficients do not seem to be significantly different from zero, for Slovenia RDRO becomes significantly positive and RDTL becomes (slightly) significantly negative in the last year (1997). The positive coefficient points out the importance of the supply of researchers in determining the location of research-oriented (RDRO) industries, and is more relevant than for high technology (RDTL) industries. Finally, the coefficient of the interaction variable LALI is either zero (Bulgaria) or positive (Hungary, Romania and, to a lesser extent, Estonia). In Hungary and Romania the coefficient was increasing its significance level during the last periods of observation. We may interpret this finding as support for the idea of country specialization in more labour-intensive industries.

Location shifts take place very slowly and a long time series' worth of data is usually necessary in order to appreciate real changes in industrial relocation and regional specialization. Given the 'young' age of the five accession countries and their data sets, more research is still needed to be able to assess the changes in relocation that their 'transition' is implying.

7 Concluding Remarks

Since 1990, Central and East European economies have experienced increasing economic integration with the world economy, in particular with the EU via trade and foreign direct investments. The spatial implications of this process have not been investigated in-depth so far. In this chapter, we have analysed regional specialization and manufacturing concentration patterns in Bulgaria, Estonia, Hungary, Romania and Slovenia.

Our findings suggest that high specialization of regions is associated with proximity to accession countries and advanced accession, with proximity to EU markets and lagged accession and with proximity to countries outside the EU enlargement area. Low specialization relates to proximity to EU markets and advanced accession and to proximity to accession countries and lagged accession. Regional specialization has increased in Bulgaria and Romania, decreased in Estonia and has not significantly changed in Hungary and Slovenia. Highly specialized regions show inferior economic performance with respect to the national averages, while diversified regions perform better than the national averages.

With respect to manufacturing concentration patterns, we find that highly concentrated industries are those with large economies of scale, high technology and high wages. Industries with low technology and low wages appear to be dispersed. For the majority of industries, there seem to be no significant changes in the level of concentration. During the 1990s, geographic concentration of manufacturing increased in Bulgaria but has not changed significantly in the rest of the accession countries included in our analysis.

Both factor endowments and proximity to industry centers – capital regions and EU markets – explain the emerging economic geography in EU accession countries. Other things being equal, industries are attracted by large markets. Industries with large economies of scale tend to locate close to industry centers. Labour-intensive industries locate in regions endowed with a large labour force while research oriented industries are attracted by regions endowed with researchers.

Notes

*   This research was undertaken with support from the European Community's PHARE ACE Programme 1998. The content of the publication is the sole responsibility of the authors and it in no way represents the views of the Commission or its services. We thank Marius Brülhart, Edgar Morgenroth, Laura Resmini, Christian Volpe Martincus and workshop participants at the Jönköping University and University of Bonn for helpful comments and suggestions.
1   Recent surveys of theoretical literature include: Amiti (1998a), Venables (1998), Brülhart (1998), Aiginger et al. Wolfmayr-Schnitzer (1999), Hallet (2001) and Puga (2002).
2   This data set has been generated in the framework of the PHARE ACE Project P98-1117-R.
3   Overviews of different measurements for specialization and geographic concentration of industries include Ellison and Glaeser (1997), Aiginger et al. (1999), Devereux et al. (1999) and Hallet (2000).
4   The indicators used in this chapter to analyse regional specialization and concentration of industries are defined in a way that is similar to Aiginger et al. (1999). The dissimilarity index is a modified version of the index proposed in Krugman (1991b).
5   This classification is based on Eurostat (1999). Also see chapter 10 of this book.
6   National averages are calculated without the capital regions.
7   Since the available classification of industries is quite aggregated we were sometimes forced to 'average' the qualitative characteristics proposed by Pratten (1988) and by the OECD (1994).
8   Detailed results may be provided by the authors on request.
9   Although some information on R&D was available for Hungary, they are not included in these estimations, since it is available only for a limited number of years.

References

Aiginger, K. et al. (1999), 'Specialisation and (Geographic) Concentration of European Manufacturing', Enterprise DG Working Paper No 1, Background Paper for the 'The competitiveness of European industry: 1999 Report', Brussels.
Altomonte, C. and Resmini, L. (1999), 'The Geography of Transition: Agglomeration versus Dispersion of Firms' Activity in the Countries of Central and Eastern Europe', paper

presented to the European Workshop on 'Regional Development and Policy in Europe', Center for European Integration Studies, Bonn, 10–11 December.

Amiti, M. (1998a), 'New Trade Theories and Industrial Location in the EU: a Survey of Evidence', *The Oxford Review of Economic Policy*, Vol. 14 (2), pp. 45–53.

Amiti, M. (1998b), 'Inter-industry Trade in Manufactures: Does Country Size Matter?', *Journal of International Economics*, Vol. 44, pp. 231–55.

Amiti, M. (1999), 'Specialisation Patterns in Europe', *Weltwirtschaftliches Archiv*, 134(4), pp. 573–93.

Brülhart, M. (1995a), 'Scale Economies, Intra-industry Trade and Industry Location in the New Trade Theory', *Trinity Economic Papers*, No. 95/4.

Brülhart, M. (1995b), 'Industrial Specialisation in the European Union: a Test of the New Trade Theory', *Trinity Economic Papers,* No. 95/5.

Brülhart, M. (1996), 'Commerce et spécialisation géographique dans l'Union Européenne', *Economie Internationale*, Vol. 65, pp. 169–202.

Brülhart, M. and Torstensson, J. (1996), 'Regional Integration, Scale Economies and Industry Location', Discussion Paper No. 1435, Centre for Economic Policy Research.

Brülhart, M. (1998), 'Economic Geography, Industry Location and Trade: The Evidence', *The World Economy*, Vol. 21.

Brülhart, M. (2001), 'Growing Alike or Growing Apart? Industrial Specialisation in EU Countries', in C. Wyplosz (ed.), *The Impact of EMU on Europe and the Developing Countries*, Oxford University Press, Oxford, pp. 169–4.

Constantin, D. (1997), 'Institutions and Regional Development Strategies and Policies in the Transition Period: the Case of Romania', paper presented at the 37th European Congress of the European Regional Science Association, Rome.

Davis D.R., Weinstein, D.E., Bradford, S.C. and Shimpo, K. (1996), 'Using International and Japanese Regional Data to Determine when the Factor Abundance Theory of Trade Works', *American Economic Review*, Vol. 87, pp. 421–46.

Devereux, M., Griffith, R. and Simpson, H. (1999), 'The Geographic Distribution of Production Activity in the UK', *Institute for Fiscal Studies Working Paper*, W99/26.

Dobrinsky, R. (1995), 'Economic Transformation and the Changing Patterns of European East-West Trade', in R. Dobrinsky and M. Landesmann (eds), *Transforming Economies and European Integration*, Edward Elgar, Aldershot, pp. 86–115.

Eichengreen, B. (1993), 'Labour Markets and the European Monetary Unification', in P.R. Masson and M.P. Taylor (eds), *Policy Issues in the Operation of Currency Unions*, Cambridge University Press.

Ellison, G. and Glaeser, E. (1997), 'Geographic Concentration in US Manufacturing Industries: a Dartboard Approach', *Journal of Political Economy*, Vol. 105 (5), pp. 889–927.

Ethier, W.J. (1982), 'National and International Returns to Scale in the Modern Theory of International Trade', *American Economic Review*, Vol. 72, pp. 950–59.

Eurostat (1999), 'Statistical Regions in the EFTA Countries and the Central European Countries (CEC)', European Commission, Eurostat.

Fischer, M. and Nijkamp, P. (1999), *Spatial Dynamics of European Integration*, Springer-Verlag, Berlin/New York.

Fujita, M., Krugman, P. and Venables, A. (1999), *The Spatial Economy*, MIT Press.

Fujita, M. and Thisse, J.F. (1996), 'Economics of Agglomeration', *Journal of the Japanese and International Economics*, Vol. 10, pp. 339–78.

Greenway, D. and Hine, R.C. (1991), 'Intra-industry Specialisation, Trade Expansion and Adjustment in the European Economic Space', *Journal of Common Market Studies*, Vol. 29 (6), pp. 603–22.

Gretschmann, K. and Muscatelli, V.A. (1998), *EMU and the Regions*, Study for Regions of Europe Technology and Industry (RETI).

Haaland, J.I., Kind, H.J., Knarvik, K.H. and Torstensson, J. (1999), 'What Determines the Economic Geography of Europe', CEPR Discussion Paper, No. 2072.

Hanson, G.H. (1996), 'Economic Integration, Intra-industry Trade, and Frontier Regions', *European Economic Review*, Vol. 40, pp. 941–9.

Hanson, G.H. (1997a), 'Localization Economies, Vertical Organization, and Trade', *American Economic Review*, Vol. 87, pp. 1266–78.

Hanson, G.H. (1997b), 'Increasing Returns, Trade, and the Regional Structure of wages', *Economic Journal*, Vol. 107, pp. 113–33.

Hanson, G.H. (1998), 'Market Potential, Increasing Returns, and Geographic Concentration', Working Paper No. 6429, National Bureau of Economic Research.

Hallet, M. (1998), 'The Regional Impact of the Single Currency', paper presented at the 38th Congress of the European Regional Science Association, Vienna, September.

Hallet, M. (2000), 'Regional Specialisation and Concentration in the EU', Economic Papers No 141, European Communities, Brussels.

Hallet, M. (2001), 'Real Convergence and Catching-up in the EU', paper presented to the workshop 'Structural Funds and Convergence', The Hague, 12 June 2001.

Heckscher, E. (1919), 'The Effect of Foreign Trade on Distribution of Income', *Economisk Tidskrift*, pp. 497–512, reprinted in H.S. Ellis and L.A. Metzler (eds) (1949), *Readings in the Theory of International Trade*, Blakiston, Philadelphia, pp. 272–300.

Helpman, E. and Krugman, P. (1985), *Market Structure and Foreign trade: Increasing Returns, Imperfect Competition and the International Economy*, Harvester Wheatsheaf, Brighton.

Hine, R.C. (1990), 'Economic Integration and Inter-Industry Specialisation', CREDIT Research Paper 89/6, University of Nottingham.

Keeble, D., Offord, J. and Walker, S. (1986), *Peripheral Regions in a Community of Twelve Member States*, Commission of European Communities, Luxembourg.

Krugman, P. (1979), 'Increasing Returns, Monopolistic Competition and International Trade', *Journal of International Economics*, Vol. 9, pp. 469–79.

Krugman, P. (1980), 'Scale Economics, Product Differentiation, and the Pattern of Trade', *American Economic Review*, Vol. 70, pp. 950–59.

Krugman, P. (1981), 'Intra-industry Specialization and the Gains from Trade', *Journal of Political Economy*, Vol. 89, pp. 959–73.

Krugman, P. and Venables, A. (1990), 'Integration and the Competitiveness of Peripheral Industry', in C. Bliss and J. Braga de Macedo (eds), *Unity with Diversity in the European Community*, Cambridge University Press, Cambridge.

Krugman, P. and Venables, A. (1995), 'Globalisation and the Inequality of Nations', NBER Working Paper No. 5098.

Krugman, P. (1991a), *Geography and Trade*, MIT Press, Cambridge, Mass.

Krugman, P. (1991b), 'Increasing Returns and Economic Geography', *Journal of Political Economy*, Vol. 99, pp. 484–99.

Landesmann, M. (1995), 'The Patterns of East-West European Integration: Catching up or Falling Behind?', in R. Dobrinsky and M. Landesmann (eds), *Transforming Economies and European Integration*, Edward Elgar, Aldershot, pp. 116–40.

Laursen, K. (1998), 'Do Export and Technological Specialisation co-evolve in Terms of Convergence or Divergence? Evidence from 19 OECD Countries, 1987–1991', DRUID Working Papers, No. 98–18.

Leamer, E.E. and Levinsohn, J. (1995), 'International Trade Theory: the Evidence', in G.M. Grossman and K. Rogoff (eds), *Handbook of International Economics*, Vol. III, Elsevier-North Holland, Amsterdam.

Ludema, R.D. and Wooton, I. (1997), 'Regional Integration, Trade and Migration: are Demand Linkages Relevant in Europe?', CEPR Discussion Paper No. 1656.

Markusen, J.R. and Melvin, J.R. (1981), 'Trade, Factor Prices and Gains from Trade with Increasing Returns to Scale', *Canadian Journal of Economics*, Vol. 14, pp. 450–69.

Martin, P. and Ottaviano, G.I.P. (1996), 'La géographie de l'Europe multi-vitesses', *Economie Internationale*, Vol. 67, pp. 45–65.

Midelfart-Knarvik, K.H., Overman, F.G., Redding, S.J. and Venables, A.J. (2000), 'The Location of European Industry', Economic Papers no. 142 – Report prepared for the Directorate General for Economic and Financial Affairs, European Commission.

Molle, W. (1997), 'The Economics of European Integration: Theory, Practice, Policy', in K. Peschel (ed.), *Regional Growth and Regional Policy within the Framework of European Integration*, Physica Verlag, Heidelberg, pp. 66–86.

Nemes-Nagy, J. (1994), 'Regional Disparities in Hungary during the Period of Transition to a Market Economy', *GeoJournal*, Vol. 32 (4), pp. 363–8.

Nemes-Nagy, J. (1998), 'The Hungarian Spatial Structure and Spatial Processes', *Regional Development in Hungary, 15–26*, Ministry of Agriculture and Regional Development, Budapest.

Nijkamp, P. (2000), 'New Growth, Local Culture and the Labour Market', paper presented to the PHARE ACE workshop 'European Integration, Regional Specialisation and Location of Economic Activity in Accession Countries', University of Bonn.

Obstfeld, M. and Peri, G. (1998), 'Asymmetric Shocks: Regional Non-adjustment and Fiscal Policy', *Economic Policy*, Vol. 28, pp. 206–59.

OECD (1994), *Manufacturing Performance: a Scoreborard of Indicators*, OECD/OCDE, Paris.

Ohlin, B. (1933), *Interregional and International Trade*, Harvard University Press, Cambridge, Mass.

Ottaviano, G.I.P. (1996), 'The Location Effects of Isolation', *Swiss Journal of Economics and Statistics*, Vol. 132, pp. 427–40.

Petrakos, G. (1996), 'The Regional Dimension of Transition in Eastern and Central European Countries: an Assessment', *Eastern European Economics*, Vol. 34 (5), pp. 5–38.

Petrakos, G. (2000), 'The Spatial Impact of East-West Integration', in G. Petrakos, G. Maier and G. Gorzelak (eds), *Integration and Transition in Europe: the Economic Geography of Interaction*, Routledge, London.

Pratten, C. (1988), 'A Survey of the Economies of Scale', in Commission of the European Communities, *Research on the 'Cost of non-Europe', Volume. 2: Studies on the Economics of Integration*, Luxembourg.

Puga, D. (1998), 'Urbanisation Patterns: European versus Less Developed Countries', *Journal of Regional Science*, Vol. 38, pp. 231–52.

Puga, D. (2002), 'European Regional Policies in Light of Recent Location Theories', *Journal of Economic Geography*, Vol. 2, pp. 373–406.

Puga, D. and Venables, A. (1997), 'Preferential Trading Arrangements and industrial location', *Journal of International Economics*, Vol. 43, pp. 347–68.

Puga, D. (1999), 'The Rise and Fall of Regional Inequalities', *European Economic Review*, Vol. 43, pp. 303–34.

Quah, D.T. (1996), 'Regional Convergence Clusters across Europe', *European Economic Review*, Vol. 40, pp. 951–8.

Raagmaa, G. (1996), 'Shifts in regional Development in Estonia during the Transition', *European Planning Studies*, Vol. 4 (6), pp. 683–703.

Ricardo, D. (1817), *On the Principles of Political Economy and Taxation*, London.

Ricci, L.A. (1999), 'Economic Geography and Comparative Advantage: Agglomeration versus Specialisation', *European Economic Review*, Vol. 43, pp. 357–77.

Romer, P.M. (1986), 'Increasing Returns and Long-run Growth', *Journal of Political Economy*, Vol. 94, pp. 1002–37.

Romer, P.M. (1990), 'Endogenous Technological Change', *Journal of Political Economy*, Vol. 98, pp. 71–102.

Sapir, A. (1996), 'The Effects of Europe's Internal Market Programme on Production and Trade: a First Assessment', *Weltwirtschaftliches Archiv*, Vol. 132(3), pp. 457–75.

Spiridonova, J. (1995), *Regional Restructuring and Regional Policy in Bulgaria*, National Center for Regional Development and Housing Policy, Sofia.

Spiridonova, J. (1999), 'Depressed Regions in Bulgaria', in *Proceedings of the 5th Conference of the Network of Spatial Planning Institutes of Central and East European Countries*, Krakow.

Trionfetti, F. (1997), 'Public Expenditure and Economic Geography', *Annales d'Economie et de Statistique*, Vol. 47, pp. 101–25.

Venables, A. (1994), 'Economic Integration and industrial agglomeration', *Economic and Social Review*, Vol. 26, pp. 1–17.

Venables, A. (1996), 'Equilibrium Locations of Vertically Linked Industries', *International Economic Review*, Vol. 37, pp. 341–59.

Venables, A. (1998), 'The Assessment: Trade and Location', *The Oxford Review of Economic Policy*, Vol. 14(2), pp. 1–6.

Walz, U. (1997), 'Growth and Deeper Regional Integration in a Three Country Model', *Review of International Economics*, Vol. 5, pp. 492–507.

Wolfmayr-Schnitzer, Y. (1999), 'Economic Integration, Specialization and the Location of Industries: a Survey of the Theoretical Literature', WIFO Working Paper No. 120. WIFO-Austrian Institute of Economic Research, Vienna.

Appendix

Table 9.A1 Summary statistics for specialization and economic performance of regions in Bulgaria

Type of region: Number of regions:		Overall 28	BEU 3	BAC 6	INT 14	BEX 5
Dissimilarity index	Mean	0.982–0.989	1.204–1.284	0.855–0.786	0.929–0.940	1.150–1.192
over national average	Std. dev.	0.294–0.326	0.042–0.215	0.336–0.315	0.279–0.286	0.285–0.345
GDP per capita	Mean	1.027–1.001	0.949–0.960	0.978–1.003	1.041–0.985	1.093–1.067
over national average	Std. dev.	0.120–0.072	0.078–0.012	0.083–0.043	0.094–0.059	0.208–0.117
GDP per worker	Mean	1.001–0.997	0.961–0.960	0.978–1.033	0.999–0.964	1.058–1.071
over national average	Std. dev.	0.102–0.070	0.096–0.060	0.059–0.047	0.070–0.045	0.198–0.092
Wages	Mean	1.019–1.010	0.995–0.893	0.996–0.991	1.023–1.024	1.049–1.064
over national average	Std. dev.	0.052–0.136	0.016–0.034	0.031–0.165	0.050–0.122	0.080–0.163
Unemployment	Mean	0.939–1.001	1.321–1.023	0.972–1.241	0.847–0.967	0.927–0.838
over national average	Std. dev.	0.233–0.325	0.041–0.315	0.119–0.179	0.194–0.376	0.289–0.197

Notes: the first figure refers to the first year for which the variable is available, while the second figure refers to the last year for which the variable is available.

BEU = regions bordering EU countries; BAC = regions bordering accession countries; BEX = regions bordering countries outside the EU enlargement area; INT =non-border (internal regions).

Source: Authors' calculations based on the REGSTAT data set.

Table 9.A2   Summary statistics for specialization and economic performance of regions in Estonia

Type of region: Number of regions:		Overall 5	BEU 3	BAC 2	INT 0	BEX 0
Dissimilarity index	Mean	0.892–0.980	0.840–0.968	0.969–0.998	–	–
over national average	Std. dev.	0.261–0.163	0.331–0.227	0.182–0.042	–	–
GDP per capita	Mean	1.256–1.318	1.440–1.548	0.978–0.972	–	–
over national average	Std. dev.	0.573–0.716	0.726–0.905	0.074–0.119	–	–
GDP per worker	Mean	1.179–1.238	1.298–1.389	1.000–1.012	–	–
over national average	Std. dev.	0.402–0.539	0.518–0.696	0.074–0.156	–	–
Wages	Mean	1.031–1.107	1.113–1.178	0.907–1.002	–	–
over national average	Std. dev.	0.170–0.240	0.178–0.310	0.042–0.013	–	–
Unemployment	Mean	0.904–0.966	0.852–1.018	0.982–0.888	–	–
over national average	Std. dev.	0.322–0.260	0.442–0.347	0.040–0.094	–	–

Notes: the first figure refers to the first year for which the variable is available, while the second figure refers to the last year for which the variable is available.

BEU = regions bordering EU countries; BAC = regions bordering accession countries; BEX = regions bordering countries outside the EU enlargement; INT =non-border (internal regions).

Source: Authors' calculations based on the REGSTAT data set.

Table 9.A3   Summary statistics for specialization and economic performance of regions in Hungary

Type of region: Number of regions:		Overall 20	BEU 2	BAC 7	INT 8	BEX 3
Dissimilarity index	Mean	0.992–0.986	0.920–0.774	1.023–1.124	0.977–0.991	1.008–0.795
over national average	Std. dev.	0.249–0.279	0.403–0.278	0.240–0.138	0.271–0.368	0.263–0.061
GDP per capita	Mean	1.058–1.065	1.248–1.453	0.968–0.921	1.126–1.159	0.962–0.890
over national average	Std. dev.	0.299–0.374	0.009–0.026	0.149–0.164	0.442–0.519	0.052–0.070
GDP per worker	Mean	1.016–0.996	1.029–1.089	0.973–0.947	1.064–1.040	0.981–0.933
over national average	Std. dev.	0.144–0.150	0.034–0.033	0.071–0.051	0.215–0.221	0.041–0.068
Wages	Mean	1.022–1.028	0.990–1.078	1.000–0.976	1.069–1.093	0.968–0.944
over national average	Std. dev.	0.110–0.147	0.035–0.065	0.050–0.061	0.159–0.207	0.036–0.052
Unemployment	Mean	0.952–0.966	0.395–0.543	1.055–1.155	0.990–0.882	0.980–1.032
over national average	Std. dev.	0.527–0.346	0.134–0.124	0.549–0.342	0.618–0.351	0.192–0.089

Note: the first figure refers to the first year for which the variable is available, while the second figure refers to the last year for which the variable is available.
BEU = regions bordering EU countries; BAC = regions bordering accession countries; BEX = regions bordering countries outside the EU enlargement; INT = non-border (internal regions).

Source: Authors' calculations based on the REGSTAT data set.

Table 9.A4 Summary statistics for specialization and economic performance of regions in Romania

Type of region: Number of regions:		Overall 41	Borders EU 0	Borders AC 11	Internal 23	Border EX 7
Dissimilarity index	Mean	0.993–0.987	—	0.878–0.956	1.015–0.992	1.099–1.018
over national average	Std. dev.	0.263–0.248	—	0.145–0.178	0.259–0.272	0.376–0.283
Wages	Mean	1.001–1.011	—	0.983–1.020	1.018–1.023	0.974–0.956
over national average	Std. dev.	0.100–0.128	—	0.065–0.108	0.112–0.142	0.110–0.110
Unemployment	Mean	0.987–0.986	—	0.861–0.754	0.942–1.056	1.333–1.123
over national average	Std. dev.	0.399–0.292	—	0.296–0.166	0.420–0.287	0.308–0.284

Note: the first figure refers to the first year for which the variable is available, while the second figure refers to the last year for which the variable is available.

BEU = regions bordering EU countries; BAC = regions bordering accession countries; BEX = regions bordering countries outside the EU enlargement;INT =non-border (internal regions).

Source: Authors' calculations based on the REGSTAT data set.

Table 9.A5   Summary statistics for specialization and economic performance of regions in Slovenia

Type of region: Number of regions:		Overall 12	BEU 7	BAC 0	INT 1	BEX 4
Dissimilarity index	Mean	0.954–0.971	0.847–0.864	–	1.380–1.442	1.036–1.040
over national average	Std. dev.	0.354–0.397	0.288–0.380	–	–	0.445–0.425
Wages	Mean	1.015–1.015	0.996–0.993	–	1.048–1.018	1.038–1.053
over national average	Std. dev.	0.078–0.098	0.039–0.065	–	–	0.131–0.155
Unemployment	Mean	0.973–0.968	0.982–0.959	–	1.222–1.321	0.895–0.896
over national average	Std. dev.	0.267–0.294	0.313–0.335	–	–	0.193–0.210

Note: the first figure refers to the first year for which the variable is available, while the second figure refers to the last year for which the variable is available.
BEU = regions bordering EU countries; BAC = regions bordering accession countries; BEX = regions bordering countries outside the EU enlargement; INT =non-border (internal regions).

Source: Authors' calculations based on the REGSTAT data set.

Table 9.A6   Geographical concentration of manufacturing in EU accession countries*

Industry characteristic	Period	Economies of scale			Technology level			Wages		
		Low	Medium	High	Low	Medium	High	Low	Medium	High
Bulgaria	1990	0.999	0.685	1.213	0.965	1.088	0.645	0.801	1.028	1.429
	1999	0.986	0.735	1.210	0.997	1.045	0.742	0.873	0.959	1.421
Estonia	1990	0.822	1.169	1.302	0.834	1.073	1.394	0.695	1.028	1.693
	1999	0.810	1.386	1.186	0.946	0.822	2.342	0.739	1.066	1.487
Hungary	1992	1.018	0.755	1.401	0.861	1.168	0.745	0.867	1.000	1.401
	1999	0.997	0.836	1.344	0.880	1.162	0.711	0.840	1.034	1.344
Romania	1991	0.977	0.972	1.073	0.918	0.999	1.415	0.850	1.015	1.338
	1999	0.973	1.046	1.032	0.928	0.998	1.369	0.893	1.015	1.231
Slovenia	1995	1.070	0.686	1.046	1.067	0.969	0.853	0.946	0.969	1.210
	1998	0.993	0.700	1.217	0.965	1.054	0.849	0.876	0.960	1.410

Note: * Dissimilarity index for geographical concentration normalized with the national averages of manufacturing concentration.

Source: Authors' calculations based on the REGSTAT data set.

Table 9.A7 Estimation results for Bulgaria

	1991	1998	1991	1998
lnpop	-1.35268 (0.87926)	-0.51362 (0.50758)	-1.31498 (0.88700)	-0.36706 (0.49549)
lnman	1.53699 (0.79054)*	0.66672 (0.36593)*	1.51115 (0.77882)*	0.60054 (0.34955)*
Regional characteristics $-\beta_{[k]}\kappa_{[k]}$				
MP1	-0.00048 (0.00032)	-0.00046 (0.00065)		
MP2			0.00114 (0.00129)	0.00242 (0.00265)
RD	8.79373 (17.99799)	28.86050 (51.04526)	-1.85919 (15.26793)	-2.54295 (33.81603)
LA	-0.71955 (4.63843)	1.87620 (8.83097)	-0.68607 (4.89226)	-0.01154 (6.96871)
Industry characteristics $-\beta_{[k]}\gamma_{[k]}$				
SE	-0.89164 (0.48222)*	-0.78143 (0.45625)*	0.00212 (1.20550)	0.96884 (1.31817)
RO	0.79635 (0.43470)*	0.86416 (0.35048)**	0.74754 (0.41307)*	0.80240 (0.33557)**
LI	2.58505 (3.38116)	1.52030 (5.48159)	2.71308 (3.35462)	3.14625 (6.42474)
TL	0.27512 (0.11960)**	-0.02601 (0.15232)	0.26137 (0.11716)**	-0.04843 (0.14532)

Table 9.A7 cont'd

		1991	1998	1991	1998
Interaction variables β[k]	MP1SE	0.00034 (0.00016)**	0.00047 (0.00026)*		
	MP2SE			-0.00075 (0.00102)	-0.00268 (0.00214)
	RDRO	-5.95924 (6.96319)	-7.40478 (16.09887)	0.16887 (4.30971)	9.58402 (12.79504)
	LALI	-1.60266 (4.74261)	0.18726 (7.99241)	-1.79553 (4.67767)	-2.26371 (9.37464)
	RDTL	0.00000 (0.00001)	-0.00001 (0.00003)	0.00001 (0.00001)	0.00002 (0.00003)
	Constant	19.07523 (15.29625)	3.73866 (11.57168)	17.21576 (15.29553)	1.46601 (11.22490)
	Observations	336	324	336	324
	R-squared	0.07899	0.09975	0.08028	0.12315

* significant at 10%; ** significant at 5%; *** significant at 1%; robust standard errors in parentheses

Table 9.A8   Estimation results for Estonia

		1995	1999	1995	1999
	lnpop	15.39278	8.19440	12.95097	−1.79813
		(−8.40e+00)	(−2.51e+00)	(6.03e−08)***	(1.94e−07)***
	lnman	−8.40372	−2.51816	−6.14047	7.55412
		(7.30e−10)***	(1.97e−09)***	(6.70e−08)***	(1.12e−07)***
Regional characteristics −β[k]k[k]	MP1	−0.22220	−0.28560	−0.19256	−0.70371
		(0.10454)	(0.08073)**	(0.19532)	(0.15069)***
	MP2			0.01932	0.01869
				(0.04300)	(0.01177)
Industry characteristics −β[k]γ[k]	LA	−0.00384	0.01569		
		(0.04423)	(0.01026)		
	SE	−0.51690	−1.89560	−0.31428	−3.79792
		(1.03985)	(1.15112)	(1.72897)	(1.62626)*
	LI	−6.25541	1.51038	−5.83142	2.22026
		(6.53039)	(2.24095)	(6.35186)	(2.68356)
Interaction variables β[k]	MP1SE	0.05885	0.08731		
		(0.06272)	(0.04844)		
	MP2SE			0.00131	0.20038
				(0.11719)	(0.09042)*
	LALI	0.13203	0.00676	0.12132	−0.00833
		(0.17693)	(0.04106)	(0.17202)	(0.04710)
	Constant	−103.89694	−51.21848	−86.05118	28.64165
		(2.86119)***	(1.89052)***	(3.38353)***	(2.30799)***

Table 9.A8 cont'd

	1995	1999	1995	1999
Observations	60	60	60	60
R-squared	0.08985	0.19088	0.08669	0.22390

* significant at 10%; ** significant at 5%; *** significant at 1%; robust standard errors in parentheses

Table 9.A9 Estimation results for Hungary[9]

	1992	1999	1992	1999
lnpop	0.73495 (0.68685)	0.32298 (0.15316)**	0.73110 (0.67420)	0.31053 (0.14083)**
lnman	−0.09488 (0.57996)	−0.22712 (0.19176)	−0.20895 (0.65507)	−0.22468 (0.21691)
Regional characteristics −β[k]κ[k]				
MP1	−1.42e−07 (9.78e−08)	−7.23e−08 (2.19e−08)***	−2.91e−07 (2.74e−07)	−1.73e−07 (5.50e−
MP2				
08)***				
LA	0.00531 (0.00667)	−0.00520 (0.00216)**	0.00573 (0.00652)	−0.00517 (0.00214)**
Industry characteristics −β[k]γ[k]				
SE	−0.30466 (0.58040)	0.20570 (0.10023)*	−0.43596 (0.66428)	0.15756 (0.10603)
LI	0.62037 (1.07789)	−0.27511 (0.20881)	0.57081 (1.04342)	−0.29494 (0.20097)
Interaction variables β[k]				
MP1SE	8.30e−08 (4.58e−08)*	4.07e−08 (8.71e−09)***		
MP2SE			2.17e−07 (1.42e−07)	1.02e−07 (2.49e−
08)***				
LALI	0.00326 (0.00632)	0.01080 (0.00328)***	0.00371 (0.00605)	0.01111 (0.00321)***

Table 9.A9 cont'd[9]

	1992	1999	1992	1999
Constant	−7.91446	−5.25921	−8.15732	−5.09790
	(5.95016)	(1.31885)***	(6.44168)	(1.32350)***
Observations	160	160	160	160
R-squared	0.03725	0.10843	0.03855	0.11108

* significant at 10%; ** significant at 5%; *** significant at 1%; robust standard errors in parentheses

Table 9.A10 Estimation results for Romania

	1992	1999	1992	1999
lnpop	2.44492	0.44899	0.28018	−0.35193
	(1.44033)*	(0.51677)	(1.20347)	(0.54741)
lnman	0.22683	1.04845	1.39911	1.36437
	(0.67725)	(0.33505)***	(0.65619)**	(0.35914)***
Regional characteristics −β[k]k[k]				
MP1	−0.00173	−0.00091		
	(0.00050)***	(0.00024)***		
MP2			−0.00082	−0.00098
			(0.00055)	(0.00051)*
LA	−13.19671	−13.65652	−1.79335	−6.97509
	(15.96815)	(10.41864)	(15.46175)	(10.87262)
Industry characteristics −β[k]γ[k]				
SE	−0.97699	−0.21687	−2.91058	−1.43103
	(0.37512)**	(0.26248)	(0.88395)***	(0.58505)**
LI	1.25668	0.19180	0.96440	−0.10677
	(1.10668)	(0.94750)	(1.13672)	(0.97027)
Interaction variables β[k]				
MP1SE	0.00042	0.00018		
	(0.00011)***	(0.00008)**		
MP2SE			0.00107	0.00085
			(0.00031)***	(0.00028)***
LALI	21.74397	13.00172	26.18945	16.80988
	(16.12046)	(12.63318)	(16.95083)	(13.15115)
Constant	−33.68671	−4.50527	−0.04085	8.04545
	(21.06474)	(7.75086)	(17.15426)	(7.91028)

Table 9.A10 cont'd

	1992	1999	1992	1999
Observations	533	533	533	533
R-squared	0.11977	0.06664	0.13720	0.07592

* significant at 10%; ** significant at 5%; *** significant at 1%; robust standard errors in parentheses

Table 9.A11 Estimation results for Slovenia

	1995	1997	1995	1997
lnpop	3.20622 (1.02466)***	2.49112 (1.87178)	2.69214 (0.93231)**	2.01254 (1.72352)
Lnman	−0.69920 (1.08138)	−0.35655 (1.68144)	−0.58153 (0.95082)	−0.14844 (1.57250)
Regional Characteristics −β[k]κ[k]				
MP1	−0.31573 (0.08512)***	−0.30335 (0.07719)***		
MP2			−0.08055 (0.21849)	0.15428 (0.18230)
RD	−24.99003 (76.93151)	−35.65038 (49.29207)	−44.26235 (74.26717)	−51.93411 (48.52253)
Industry Characteristics −β[k]γ[k]				
SE	−0.27865 (0.24045)	−0.38584 (0.19937)*	−0.58693 (1.06643)	0.78827 (0.79933)
RO	−3.78016 (1.28015)**	−5.71808 (1.02469)***	−4.01515 (1.27333)***	−6.12510 (1.02944)***
TL	2.58538 (1.11000)**	3.86484 (1.10604)***	2.66055 (1.11473)**	3.99503 (1.11505)***
Interaction Variables β[k]				
MP1SE	0.10861 (0.03214)***	0.11717 (0.03526)***		
MP2SE			0.04659 (0.11191)	−0.11127 (0.08117)
RDRO	33.46864 (69.51438)	104.94158 (44.14264)**	49.13316 (68.84351)	124.29908 (45.14372)**

Table 9.A11 cont'd

	1995	1997	1995	1997
RDTL	−31.76779	−74.33164	−36.77833	−80.52344
	(51.57103)	(39.38974)*	(51.62353)	(40.11368)*
Constant	−.50988	−14.56367	−16.05070	−13.14851
	(7.86975)**	(12.92564)	(7.40840)*	(11.85207)
Observations	168	168	168	168
R-squared	0.19288	0.22525	0.18655	0.22228

* significant at 10%; ** significant at 5%; *** significant at 1%; robust standard errors in parentheses.

Chapter 10

The Impact of European Integration on the Adjustment Pattern of Regional Wages in Accession Countries*

Jože P. Damijan and Črt Kostevc

1 Introduction

The opening-up of transition countries and their trade integration with the European Union (EU) provides a natural experiment to test new economic geography (EG) models. Before 1990, most transition countries were completely closed in terms of trade with western countries. Trade liberalization after 1990 makes it, therefore, very appealing to study the impact of trade liberalization on the regional relocation of manufacturing activity. Since the beginning of the 1990s, several alternative EG models were established in order to explain the spatial repercussions of trade liberalization in terms of inter-regional manufacturing relocation and the evolution of relative regional wages. Despite the scepticism due to simplifying assumptions and special functional forms that has been expressed by Neary (2001) in an excellent overview of the field, the EG models enable us to analyse the effects of trade liberalization on international as well as intra-national relocation of manufacturing activities. While, in the absence of trade, economic activity is concentrated in locations near home economic centres, trade liberalization may lead to the relocation of manufacturing activities. The exact pattern of relocation of manufacturing activity, however, is ambiguous and dependent on the underlying assumptions. Crucial here is the assumption of inter-regional labour mobility. A first approach, based on Krugman's (1991a, 1991b) model, assumes perfect inter-regional mobility of labour. This approach predicts a monotonic relationship between the reduction of trade costs and the relocation of manufacturing activity. When trade is opened up, larger regions, in terms of industrial activity, will gain from trade liberalization due to existing agglomeration effects. A core-periphery solution, i.e., complete specialization of manufacturing activity in only one region is the likely outcome of this model. This will diminish the initial differences in per capita income levels due to further divergence in relative regional wages.

More recent approaches, starting with Krugman and Venables (1995, 1996) and followed by Puga (1999), Fujita, Krugman and Venables (1999), drop the assumption of perfect labour mobility. For most countries imperfect mobility of labour is characteristic and thus it is necessary to study the spatial repercussions of trade liberalization in a more realistic setup. Depending on transport costs, this approach produces two types of equilibria. For low transport costs, a core-periphery outcome is likely, while for high transport costs a symmetric equilibrium with no agglomeration is possible. On the other hand, wages serve as a spreading force. With increasing agglomeration, wages tend to increase, leading to the dispersion of relative regional manufacturing shares, as firms then tend to relocate to regions with lower labour costs. Typically, the relationship between the regional manufacturing shares and transport costs may take the pattern of a *U-shape*. In the first stage, hence, trade liberalization may increase initial regional differences in income per capita, while further trade liberalization may bring about some convergence in relative regional wages.

Originally, the above approaches were applied to the North-South discussion in order to address the issue of the possible implications of globalization. During the 1970s, many theorists argued that liberalization of world trade generally produced uneven development, i.e., a rise in living standards in western countries at the expense of developing countries. In contrast, since the beginning of the 1990s, many economists as well as political leaders in western nations have claimed that the labour-intensive exports of emerging economies have hurt the competitive position of western countries. In effect, western countries have been hurt in terms of the stagnation of real manufacturing wages and/or higher unemployment.[1]

In the most recent approach, Damijan and Kostevc (2002; henceforth DK) apply the second EG approach to study the within-country regional effects of trade liberalization. In doing so, the authors augment the Fujita-Krugman-Venables (henceforth FKV) type of EG model by breaking the implied regional symmetry and by introducing a second factor of production, capital. Authors analyse a three-region world, with the first region being the large foreign country (EU) and the two home regions being located in a developing country. With the break in regional symmetry, different foreign trade costs for the two home regions are assumed, where one region might benefit from its location closer to the border with the large foreign country. On the other hand, while the restrained mobility of labour does not allow for large inter-regional relocation of manufacturing, the introduction of capital allows either for foreign direct investment (FDI) flows to emerge between the large foreign country and domestic regions or for the domestic relocation of capital. Using a simulation analysis, FDI flows have been shown to accelerate the regional adjustment process in the home country, since they are, due to lower wages and higher returns on capital, attracted to poor regions. In effect, when compared to the FKV approach, the DK model results in a faster convergence of poor regions, demonstrated by a more *upward-and-shifted-to-the-right U-shaped* pattern of regional adjustment of relative regional wages.

The aim of the present chapter is to analyse the effects of trade liberalization with the EU on the inter-regional relocation of manufacturing and the inter-regional adjustment of relative wages in transition (accession) countries. We examine the exact adjustment pattern of relative regional wages, i.e., which of the three competing EG models is the most appropriate approximation of the actual regional adjustment pattern, in selected transition countries. Specifically, we study whether the response of relative regional wages to the reduction of foreign trade costs is monotonic and leads to strong regional polarization, as suggested by the first Krugman approach, or if it is a non-monotonic relation associated with less regional polarization, as suggested by more recent EG approaches. In addition, in the case of a non-monotonic response, we test the propositions of FKV against the DK approach. In doing so, the impacts of FDI, of inter-regional transport costs and of western/northern region dummies (i.e., border effects) on the adjustment pattern of relative regional wages are examined. The implications of the three competing EG approaches are tested using unique regional panel data for five transition countries (Bulgaria, Estonia, Hungary, Romania and Slovenia) for the period 1990–2000. Our results suggest that, in the cases of Estonia and Romania, a Krugman type of divergence of relative regional wages and increased regional polarization might be at work. On the other hand, the adjustment pattern of regional wages in Bulgaria and Hungary seems to be in line with the FKV propositions indicating that market forces by themselves have brought about the upward trend in initially declining relative regional wages. Regional development in Slovenia after trade liberalization, in contrast, does not correspond to any of the examined EG models, as there is clear evidence of a catching-up process in the majority of the regions.

The structure of the chapter is as follows. Section 2 compares basic propositions of the three competing EG models and discusses their implications for transition countries. Section 3 discusses previous empirical studies. Section 4 describes the empirical model, data, and methodology used and discusses the results. The final section summarizes basic findings and provides some policy recommendations.

2 Theoretical Background

A revival of interest in economic geography was launched in 1991 by Krugman's JPE paper. Since then, many attempts have been made to contribute to this eclectic area of economics. According to their assumptions and implications, one can group individual contributions into three approaches.

The first approach, based on Krugman's (1991a, 1991b) basic model, assumes perfect inter-regional mobility of labour. This implies that trade liberalization inevitably leads to core-periphery equilibrium, as agglomeration effects attract all of the manufacturing labour to the largest region. In the real world, however, this approach is rather implausible, as labour is far from being very internationally mobile. Evidence does not confirm the increase in international income inequality in the recent two decades of rapid trade liberalization (see Barro and Sala-i-Martin, 2001) that is the explicit implication of this model.

It is clear that an EG model is needed that is more realistic and less biased in favour of complete agglomeration. Krugman and Venables (1995) and, in a more advanced version, Fujita, Krugman and Venables (1999, FKV) provide such a model by dropping the assumption of perfect labour mobility. Another difference of the FKV model from Krugman's (1991) is that externalities driving the agglomeration now stem from input-output linkages among firms rather than from linkages between firms and consumers (home market effect). Firms benefit from being close to each other by not paying transport costs for intermediate factors of production. Depending on trade (transport) costs, the model produces two types of equilibria. For high trade costs a symmetric regional equilibrium with no agglomeration is likely. When trade costs fall below a certain level, divergence of relative regional wages is likely as a core-periphery pattern spontaneously forms. On the other hand, wages serve as a spreading force. With decreasing trade costs and subsequent increasing agglomeration, wages tend to increase, leading to dispersion of relative regional manufacturing shares, since firms tend to relocate to regions with lower labour costs. Hence, while the first stage of globalization might in fact hurt peripheral countries in terms of decreasing relative regional wages, in the second stage peripheral countries might converge with the core countries.

The most recent approach by Damijan and Kostevc (2002, DK), while built in the tradition of the FKV model, by dropping the assumption of perfect labour mobility revises the latter model in two crucial ways. First, the authors break the implied regional symmetry of the original FKV model by introducing a third region, which enables them to study the effects of trade liberalization on the home country's regional wage inequality. Second, the DK model introduces a second factor of production (capital), which makes it possible to examine the role of foreign direct investment (FDI) as a factor that may accelerate the regional adjustment process in the home country. In the DK approach, a three-region world is being analysed, with the first region being a large foreign country (EU) and the two home regions being located in an accession country (home country). In this approach, different foreign trade costs for the two home regions are assumed, with the smaller region potentially benefiting from its location closer to the border of the large foreign country. Production in manufacturing is characterized by monopolistic competition and internal economies of scale as well as by an interaction between external economies of scale and trade costs. The latter implies that, in the absence of trade, inter-regional trade costs in the home country prevent agglomeration effects from prevailing completely. When trade opens up, there will be a trade-off between agglomeration effects and existing differences in relative factor costs, affecting the pattern of the intra-national as well as inter-regional manufacturing relocation. Immediately after trade liberalization, the small border region will lose manufacturing shares relative to the core region due to agglomeration effects. There are two factors that might turn around the process of complete regional agglomeration in the home country after trade liberalization. First, as in the FKV approach, increasing wages in the larger region, when there is no labour mobility, will prevent complete agglomeration. Second, lower wages and higher returns to capital in the home country will attract FDI flows from the large foreign country. These flows will benefit the small border region due to lower relative wages and lower trade costs with the foreign country. Small border regions might also benefit from the domestic relocation of capital closer to large foreign markets. Hence, a convergence in relative home wages is expected after some threshold of foreign trade costs has been reached. The DK model thus predicts, similarly to the FKV, a U-shaped adjustment pattern of relative regional wages in the home country. The crucial difference between the outcomes of the two competing models, however, lies in the time when the convergence of relative wages starts and in the extent of convergence. In other words, with FDI the convergence of relative regional wages will start at an earlier point in time (i.e., with higher trade costs) and the small region's wages will converge to an absolutely higher level than in the FKV model.

The differences in the implications of both competing models can be easily seen by simulating the adjustment pattern of relative regional wages under different assumptions. The wage equations for both home regions[2] in the DK approach can be written as:

$$(1) \quad w_1 = \alpha \left[\left(\frac{n_1^\sigma \rho \beta^\beta}{i_1^\beta} \right) \left(Y_1 G_1^{\sigma-1} + Y_2 G_2^{\sigma-1} T^{\sigma-1} + Y_3 G_3^{\sigma-1} (TT^*)^{\sigma-1} \right)^{\frac{1}{\sigma-1}} \right]^{\frac{1}{\alpha}}$$

$$(2) \quad w_2 = \alpha \left[\left(\frac{n_2^\sigma \rho \beta^\beta}{i_2^\beta} \right) \left(Y_1 G_1^{\sigma-1} T^{\sigma-1} + Y_2 G_2^{\sigma-1} + Y_3 G_3^{\sigma-1} (T^*)^{\sigma-1} \right)^{\frac{1}{\sigma-1}} \right]^{\frac{1}{\alpha}}$$

where w_1 and w_2 are regional wage rates in the central and peripheral (border) regions, respectively; i_r ($r = 1, 2$) is regional return to capital; α and β are usual labour and capital shares in the Cobb-Douglas (CD) production function, and σ represents the elasticity of substitution between any two varieties of manufactured goods. $Y_r G_r$ is regional real income, representing its market size, and n_r is regional number of firms, representing the scope for external economies of scale.[3] T represents transport costs between both home regions and T^* is transport (trade) costs between the smaller (peripheral, border) region and the foreign country. As the central region is located further from the border of the foreign country, its cost of trade with the foreign country equals TT^*.

According to (1) and (2), relative regional wage w_2 / w_1 (i.e., the wage rate in the peripheral region, relative to the central region) depends on:

- the scope of external economies of scale (number of firms n_r affected through initial factor endowments and factor mobility);
- the aggregate demand for the region's varieties (sum of $Y_r G_r$);
- the return to capital (i) in the region;
- inter-regional (T) as well as international (T^*) trade costs; and
- the elasticity of substitution between varieties of manufactured goods.

Let us assume that the central home region is a bit larger that the peripheral home region in terms of factor endowments region in terms of factor endowments (i.e., $L_1 = 200$, $K_1 = 86$ and $L_2 = 160$, $K_2 = 68$), while the foreign country is much larger ($L_3 = 1000$, $K_3 = 428$), and the crucial elasticities amount to $\alpha = 0.7$, $\beta = 0.3$ and $\sigma = 1.5$. Figure 1 shows some basic simulations.[4] The base simulation reveals a typical U-shaped response of home relative regional wages to the reduction of foreign trade costs, which occurs within reasonable trade costs (in the range $T^* = [1, 2]$).[5] Note that this outcome is not comparable to the FKV outcome, which concentrates on the adjustment of wages between two countries when they liberalize bilateral trade. In contrast, the DK approach produces the same type of adjustment pattern of relative wages between two home regions, after trade barriers with the large foreign country have been removed. At this point it is worth noting that home relative wages cease to decrease (diverge) as foreign trade costs fall below $T^* = 1.45$), i.e. to some 31 per cent.[6] Below this threshold, the smaller (border) region catches up with the larger home region due to its proximity to the large foreign country and due to increasing wages in the larger region.

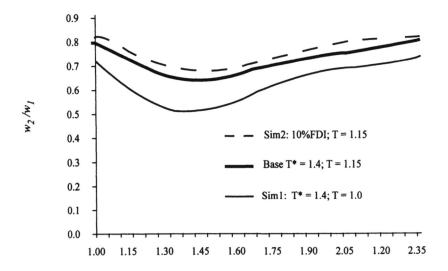

Figure 10.1   Response of home relative regional wages to a reduction in foreign trade costs with different home transport costs

Source: Damijan and Kostevc (2002).

As discussed above, in the absence of foreign trade home transport costs serve to prevent the agglomeration effects from prevailing completely in the home country. In our first exercise, home transport costs were set to a reasonable 13 per cent of the value of shipped goods (T = 1.15). What is the response of home relative wages when there are no home transport costs (T = 1.0)? As revealed in Figure 10.1, the effect of home internal transport costs is substantial. When these transport costs, amounting to 13 per cent of the cost of shipped goods, are removed, agglomeration in the home country will have a larger effect and the small home region will lag behind the larger region in terms of wages by some 20–25 per cent more than in the base scenario. Despite the fact that small home region is only 20 per cent smaller than the large region in terms of factor endowments, in the worst case, its wage rate will only amount to some 50 per cent of large region's wage rate when there are no internal transport costs. This is mentioned only to indicate the strength of the agglomeration effects.

Let us now examine the impact of FDI from the large foreign country. Due to lower wages and higher returns to capital, the FDI will presumably be directed to the smaller home region. FDI increases the capital stock of the small home region by, let us say, 10 per cent and hence expands its production possibilities. Figure 10.1 demonstrates that FDI helps the smaller home region catch up faster and to a larger extent. First, FDI almost immediately increases the wage rate in the small region relative to the central one simply through expansion of its factor endowments. With no labour mobility, manufacturing production becomes more capital-intensive, implying increased productivity and hence, wages. Second, one can also observe that the catch-up in the case of FDI starts earlier – at some 35 per cent of foreign trade costs ($T^* = 1.55$) – than in the base scenario. This second effect, of course, is also caused by the expanded production potential of the small region, which enables it to start catching up earlier, i.e., at higher foreign trade costs.

The above exercises clearly show that FDI is very important for developing countries in order to motivate a more evenly distributed development pattern. Trade liberalization in Mexico and the rise of *maquiladoras* provides an excellent example of the possible positive role of trade liberalization, along with FDI inflows, in inducing a more even geographic pattern of development. Hanson (1997) demonstrates how trade liberalization between Mexico and the US caused manufacturing to relocate towards the Mexico–US border, leading to convergence in relative regional wages. A similar pattern of adjustment might well be expected in transition countries where complete trade liberalization with the EU has been associated with vast inflows of FDI. It remains to be seen, however, in the subsequent sections, whether these expectations are justified.

3   Previous Empirical Studies

Empirical literature gives some, although inconclusive, support to the theoretical EG predictions. There are a number of studies dealing with the implications of EG models. An initial group of papers studies the impact of trade liberalization between countries on international as well as interregional relocation of manufacturing activities. Most of the studies analyse implications for EU specialization patterns. A second group of papers studies the impact of within-country transport costs on the structure of wages, location costs, etc. In this section we provide a short overview of papers from both groups that are relevant to our empirical research purposes in the next section.

Probably the most influential EG empirical study is that of Hanson (1997), who provides evidence of the effects of trade liberalization between Mexico and the USA on regional manufacturing relocation and the evolution of the regional wage pattern in Mexico. In the closed economy, Mexican manufacturing was concentrated near the capital city. Trade liberalization then caused manufacturing to relocate towards the Mexico–US border, leading to a convergence in relative regional wages.

In another study, Hanson (2000a) examines the spatial correlation between wages and consumer purchasing power across US states, analysing whether regional product-market linkages contribute to spatial agglomeration. The structural model of this chapter is heavily based on Krugman's 'home market effect'. The chapter first examines a simple market potential function where proximity to consumer markets is the prime determinant of nominal wages for a given location, as seen in the following estimated model:

$$(3) \quad \Delta \log(w_{jt}) = \alpha_1 \left[\sum_k Y_{kt} e^{\alpha_2 d_{jk}} - \sum_k Y_{kt-1} e^{\alpha_2 d_{jk}} \right] + \varepsilon_{ijt}$$

where w_{jt} represents nominal wages in region j at time t, Y_{kt} is income in region k at time t and d_{jk} is distance between regions j and k.

The data for US states from 1970 to 1990 confirm the predicted negative relationship of nominal wages to transport costs to demand markets and a positive relationship to consumer income in demand markets at a 1 per cent significance level, where this simple market potential function explains around 20 per cent of the nominal wage variation in the observed period. The chapter also examines an augmented market potential function whose parameters reflect the importance of scale economies and transport costs, the stability of spatial agglomeration patterns and their evolution over time. The data conform somewhat better to this function, with the R-square rising in all estimated variants of the model. The inclusion of the two new dependent variables – wages and housing stock – lessens the importance of market potential while enlarging the effect of distance. The impact of personal income and wages in surrounding locations on nominal wages in the given location are expectedly positive, while the variable housing stock in surrounding locations has, in contrast to the theoretical predictions, a positive effect on nominal wages.

Brakman, Garretsen and Schramm (2002) estimate a Helpman-Hanson empirical model (compare Helpman 1998; Hanson 2000a) using data for Germany. An advantage of the Helpman-Hanson model is that it incorporates the fact that the agglomeration of economic activity increases the prices of local (non-tradable) services. The model thus provides a powerful spreading force, which leads to less extreme outcomes than the EG basic model by Krugman (1991). Using specific data for 151 districts for 1994, the authors succeeded in supporting the idea of a spatial nominal wage structure in Germany.

Finally, Overman, Redding and Venables (2001) present a survey of empirics pertaining to the field of EG. The paper focuses on the general effects EG has on the volume of trade, income levels and the structure of production and offers a structural model to provide the basis for the research work in the chapter as well as the research work of other authors. Relevant for present study is the research on the effects of a country's firms' access to world markets and foreign suppliers to the GDP per capita of that country. The effects appear to be very strong (positive and statistically significant at the one per cent significance level) and explain around 35 per cent of the cross-country variation of gross domestic product per capita.

As shown in the above overview of empirical research in the field of EG, only a little attention has been paid to the analysis of the impact of trade liberalization on regional wages. Furthermore, apart from the contributions included in this book we are not aware of any comparative study on economic geography in transition countries or on the spatial repercussions of recent trade liberalization in transition countries. In the subsequent section, we try to fill this gap by verifying the basic implications of alternative EG models for the adjustment pattern of relative regional wages in EU accession countries after trade integration with the EU.

4   Empirical Testing of Integration Effects on the Adjustment Pattern of Relative Regional Wages in EU Accession Countries

In Chapter 9 of this book, Traistaru, Nijkamp and Longhi identify and explain the spatial implications of economic integration in EU accession countries and find evidence for some of the theoretical predictions of the new economic geography about determinants of manufacturing location. In this chapter, we test the predictions of the Krugman, FKV and DK economic geography models about the adjustment pattern of relative regional wages using regional panel data for five accession countries. In the subsequent section we will first discuss the empirical models used and how we expect the different implications of the competing EG models to appear in empirical estimations. Then we proceed to a discussion of the data employed and to an analysis of the time pattern of relative regional wages and FDI inflows in each of the transition countries. Finally, after a short discussion of methodological issues, we provide some results of econometric estimations of our empirical model.

4.1   The Empirical Model

There are clear implications of the above-discussed Krugman, FKV and DK models for transition countries for the adjustment pattern of relative regional wages. The Krugman model predicts a monotonic decline in relative regional wages throughout the period after trade liberalization has been initiated. The FKV model predicts a U-shaped response of relative regional wages to reduced foreign trade barriers in transition countries. In the first years after trade liberalization, relative wages will decline, but after some threshold has been reached, the wages in peripheral regions will start catching up with the central region's wages. The DK model predicts that, after a country has liberalized its trade with the EU, its pattern of inter-regional manufacturing relocation will be determined by a trade-off between agglomeration effects, remaining trade costs and existing differences in relative factor costs. With unchanged inter-regional transport costs, regions that are located closer to the EU border (western and/or northern regions, henceforth W/N regions) will benefit from trade liberalization through larger inflows of FDI due to lower trade costs with the EU and due to lower wages and higher returns to capital relative to the central home region. Some domestic resources might also relocate to border regions. As a result, after an initial downturn, border regions will converge to the home capital region in terms of relative wages (and returns to capital) and relative manufacturing output. In non-border regions this adjustment pattern might be less pronounced. Hence, regional data for transition countries should exhibit a U-shaped curve of relative wages. The crucial differences relative to the FKV model, however, lie in the speed of convergence, in the importance of the FDI factor and in the faster convergence of W/N border regions.

In order to examine the spatial repercussions of trade liberalization in transition countries for relative regional wages and to search for differences between the three competing models, we estimate the following empirical model of relative regional wages:

(4) $\ln rW = \alpha + vt + wt^2 + \delta \ln irVAe + \phi \ln rFDI + \beta \ln DIST + \gamma BORD + \lambda FTA + \mu \ln DIST*FTA + \kappa \ln DIST*BORD + \sigma BORD*FTA + \rho R + \tau \Sigma T + \varepsilon_{it}$

where:

rW relative regional wage (i.e., wage ratio between region r and capital region)

t, t^2 time effects

$irVAe$ initial regional efficiency differential

$rFDI$ relative regional output of foreign firms (as compared to domestic firms)

$DIST$ distance to capital

$BORD$ dummy for western/northern border regions

FTA dummy for enforcement of trade liberalization with EU

ΣR broader regional dummies

ΣT time dummies

ε_{it} error term

According to the model specification (4), one can interpret the predictions of the three competing models in the way summarized in Table 10.1. In general, no firm conclusions can be drawn based upon the initial regional conditions in all of the transition countries (hence, distance and border effects can be of either sign). Initial relative regional wages might depend on the specific regional policies in each of the former socialist countries and on the extent of the initial openness to trade, etc. We can be more conclusive for the period after trade liberalization (FTA) has been initiated. If the Krugman model is to apply to transition countries, one may expect a monotonic negative time pattern of relative wages (negative sign of t) and that negative distance and border effects will become more pronounced over time. In the FKV setting, a U-shaped time pattern of relative wages is expected (hence, a positive sign

Table 10.1   Parameter predictions of EG models

	Krugman	FKV	DK
t	–	ambiguous (+, –)	ambiguous (+, –)
t^2	insignificant	+	+
rFDI	insignificant	insignificant	+
DIST	ambiguous (+, –)	ambiguous (+, –)	ambiguous (+, –)
BORD	ambiguous (+, –)	ambiguous (+, –)	ambiguous (+, –)
FTA	–	+	+
DIST*FTA	–	+	+
DIST*BORD	–	ambiguous (+, –)	+
BORD*FTA	–	ambiguous (+, –)	+

of t^2) and, in general, positive distance and trade liberalization effects on relative regional wages will prevail over time. Similarly, if the DK model is to work in transition countries in the 1990s, one may expect a U-shaped time pattern of relative wages and positive distance and trade liberalization effects on relative regional wages. In addition to this, however, a positive impact from FDI and an upward trend in relative regional wages in the W/N border regions are expected.

4.2   Time Pattern of Relative Wages and FDI Inflows in Transition Countries

The data   We analyse propositions from alternative EG models by using regional data for five transition countries that are eligible for accession to the EU after 2004. These countries are Bulgaria, Estonia, Hungary, Romania and Slovenia. The choice of countries is not arbitrary; it is simply subject to the availability of data. Countries examined in our study are quite heterogeneous both in terms of their level of development and the advancement of the transition process as well as in terms of their distance from the core of the EU. One may thus expect that the distance and border effects in more distant countries like Bulgaria and Romania, which are also less advanced, will be less pronounced than those in the EU-bordering transition countries like Estonia, Hungary and Slovenia.

The data were collected during the PHARE ACE project on the regional pattern of production relocation in transition countries. The data for Bulgaria, Estonia, Hungary and Romania were collected at the NUTS 2 and NUTS 3 levels and cover the period 1990–99 (Bulgaria and Hungary) and the period

1992–99 (Estonia and Romania). For Slovenia, which lacks official regional statistics, the data are aggregated from individual firm-level data to the desired level of regional aggregation (NUTS 3 and NUTS 4 levels) and cover the period 1994–2000. All the data are recalculated into 1994 constant prices using PPI indices. The data in our database include many aspects of regional performance, but we explore only a small part of it. We take account only of data for the manufacturing sector, since other sectors are far less subject to trade liberalization.

As follows from the previous discussion and from the empirical model discussed above, in all of the subsequent analyses and empirical estimations we use relative regional indicators in order to capture inter-regional relocation patterns in a particular transition country. Relative regional indicators for wages and FDI are thus calculated as a ratio of r-th region performance to the capital (c) region performance. In the empirical estimations, regional data at the NUTS 3 (NUTS 5 for Slovenia) level is used for individual observations, while NUTS 2 (NUTS 3 for Slovenia) regional dummies are employed in order to control for broader regional effects.

Table 10.2   Descriptive statistics of regional data by countries

	BG	EST	HU	RO	SLO
Data coverage	1990–99	1992–99	1990–99	1992–99	1994–2000
Enforcement of FTA	1994	1994	1992	1993	1997
No. of NUTS 2 regions*	6	5	7	8	12
No. of NUTS 3 regions**	28	15	20	41	170

* NUTS 3 regions in Slovenia; ** NUTS 5 regions in Slovenia.

While the wage data does not need more discussion, some clarifications should be made with regard to the FDI data. With the exception of Slovenia, we do not have data on the relative importance of FDI in terms of output at the regional level. What we do have is data on the number of foreign-owned and domestic firms by region. The ratio between the two for each region has been taken to proxy for the relative importance of FDI in each region. Of course, this is a very rough approximation, as the study for ten transition countries by Damijan et al. (2001) has demonstrated that foreign-owned firms are much larger in terms of output and employment, more capital intensive, etc. relative to their domestic-owned counterparts in the same sectors. The role of FDI in inter-regional manufacturing relocation in the present study, hence, is, by default, underestimated.

Similarly, with the exception of Slovenia, there is a lack of data on the evolution of foreign trade barriers over the specified period both at the country level as well as at the regional level. Ideally, one should take the time pattern of actual foreign trade barriers (tariffs, NTBs) at the regional level and estimate the impact of their reduction on spatial patterns in each country. Instead, we only have data on the date of the enforcement of the free trade agreement (FTA) with the EU. This, however, imposes several problems. First, in some of the countries an FTA has been enforced at the beginning of the period under examination, which of course eliminates the reference period needed for the comparison of the EG effects before and after trade liberalization. Second, some of the countries examined had unilaterally liberalized their trade even before the enforcement of the FTA. Third, FTAs enforced by the EU were designed asymmetrically in favour of transition countries. Hence, the enforcement date of the FTA does not imply that trade barriers have been reduced linearly from that point on. In all of the countries, trade barriers for the most sensitive goods were eliminated at the end of the examined period. There is little one can do about the limitations of the data. It is, however, necessary to be cautious when discussing the results. On the other hand, we have separately estimated the model with Slovenian data by using either the FTA dummy variable or the data on actual tariffs applied by regions. The results of both estimations, however, do not differ significantly in terms of the signs and significance of the parameters for trade liberalization.

Evolution of relative regional wages   In this section, we examine the evolution of relative regional wages in individual countries. The graphic analysis in Figure 10.1, combined with the descriptive statistics given in Table 10.3, gives us a good insight into the pattern of relative wages during 1990s. Table 10.3 reveals that, with the exception of Slovenia, at the beginning of trade liberalization in the early 1990s there was little dispersion of regional wages in the examined countries. The average relative regional wages in all of the countries exceeded 80 per cent of those of the central region and the coefficient of variation of relative wages was well below ten per cent. The exception here is Slovenia, where the average relative wage amounted to 66 per cent of the central region's wage and the coefficient of variation exceeded 35 per cent. One can think of two possible reasons for the difference between the initial position of Slovenia and those of the other transition countries. The first and most obvious explanation is the

Table 10.3 Changes in relative regional wages in transition countries in the 1990s

	BG (1990)	BG (1999)	EST (1992)	EST (1999)	HU (1990)	HU (1999)	RO (1992)	RO (1999)	SLO (1994)	SLO (2000)
All regions										
Mean	0.92	0.79	0.84	0.67	0.82	0.71	0.96	0.71	0.66	0.69
Std. error	0.01	0.02	0.03	0.03	0.02	0.02	0.02	0.01	0.02	0.02
Std. deviation	0.05	0.11	0.12	0.10	0.08	0.11	0.10	0.09	0.25	0.21
Coef. of variation (%)	5.1	13.5	14.2	14.9	9.8	15.2	10.0	12.7	37.3	30.8
Norm. coef. of variation[7] (%)	1.0	2.5	3.7	3.8	2.2	3.4	1.6	2.0	2.9	2.4
N	28	28	15	15	20	20	41	41	170	170
W/N border regions										
Mean	0.91	0.82	0.85	0.73	0.85	0.76	1.01	0.74	0.67	0.75
Std. error	0.01	0.04	0.04	0.07	0.03	0.03	0.04	0.03	0.04	0.05
Std. deviation	0.04	0.13	0.09	0.16	0.08	0.07	0.14	0.09	0.22	0.27
Coef. of variation (%)	4.4	15.6	10.5	21.3	9.2	9.4	13.6	11.5	33.3	35.9
Norm. coef. of variation (%)	1.5	5.2	4.7	9.5	3.8	3.9	4.3	3.6	6.3	6.8
N	9	9	5	5	6	6	10	10	28	28
Non-W/N border regions										
Mean	0.93	0.77	0.84	0.64	0.81	0.68	0.94	0.70	0.66	0.68
Std. error	0.01	0.02	0.04	0.01	0.02	0.03	0.01	0.02	0.02	0.02
Std. deviation	0.05	0.09	0.14	0.03	0.08	0.11	0.07	0.09	0.25	0.20
Coef. of variation (%)	5.3	12.2	16.3	5.0	9.9	16.7	7.8	13.0	38.2	29.2
Norm. coef. of variation (%)	1.2	2.8	5.2	1.6	2.7	4.5	1.4	2.3	3.2	2.5
N	19	19	10	10	14	14	31	31	142	142

Source: authors' calculations.

fact that, before 1990, Slovenia was relatively more open to international trade than the other transition countries. Exposure to trade and a kind of semi-market economy had affected Slovenian regional development well before 1990 while other transition countries had additionally sheltered their economies by preventing large regional disparities through special regional policies. Another explanation stems from the level of aggregation used in the calculations. For Slovenia, data at the NUTS 5 (community) level is used while for other countries NUTS 3 level data is used, which, of course, levels a lot of the variation. This is confirmed when we apply a coefficient of variation normalized by the square root of the number of observations. In this case, the variation of relative regional wages in Slovenia becomes very similar to that in other transition countries.

More important here, however, is how trade liberalization has affected regional development in individual countries. Again, with the exception of Slovenia, relative wages in all of the countries declined substantially until the end of 1990s. On average, relative regional wages diminished by 15–20 percentage points relative to the central region and the variation has increased to almost 15 per cent, which implies increased regional disparities. In contrast, in Slovenia relative regional wages have increased slightly – on average by some three percentage points – and variation has fallen. It is interesting to note that, in accordance with the propositions of the DK model, the drop in relative regional wages in all countries has been much smaller (in Slovenia, the increase was higher) in W/N border regions, implying that economic geography might be at work here.

Let us now turn to the pattern of changes in relative regional wages throughout the 1990s. Figure 10.2 shows clearly that only in case of Bulgaria can a clearly U-shaped adjustment pattern of regional wages be observed. This pattern is even more pronounced in W/N border regions, where two-tier regional development can be observed. In Hungary, a clear negative trend in relative wages is evident, but the data suggest that the lowest point had been reached by 1998 and that afterwards a rise in regional wages could be expected. For Bulgaria and Hungary, hence, one might expect the FKV and DK models to be at work. On the other hand, in Romania a significant negative trend in relative regional wages is revealed, implying a Krugman type of divergence. The same applies also for Romanian W/N border regions. In Slovenia, no clear adjustment pattern for all regions is visible, but a weak upward trend can be observed for W/N regions after 1997. The latter might speak in favour of the DK explanation of regional adjustment. In contrast to the above four countries, Estonia exhibits a clear picture of an inverted U-

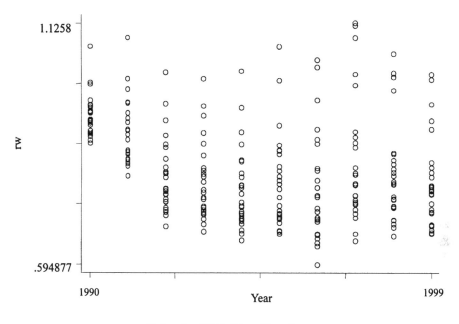

Bulgaria, 1990–99 – all regions

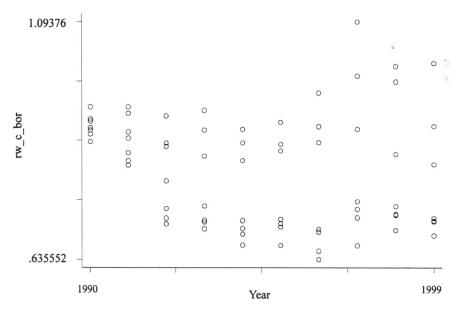

Bulgaria, 1990–99 – W/N regions

Figure 10.2 Evolution of relative regional wages in transition countries

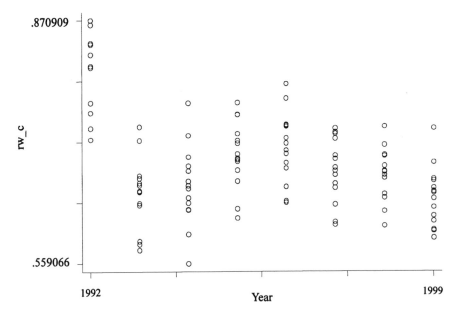

Estonia, 1992–99 – all regions

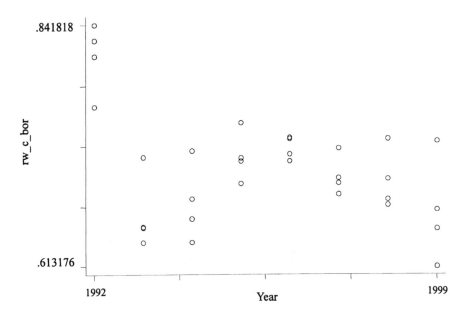

Estonia, 1992–99 – W/N regions

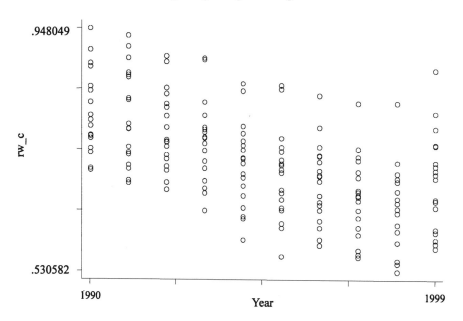

Hungary, 1990–99 – all regions

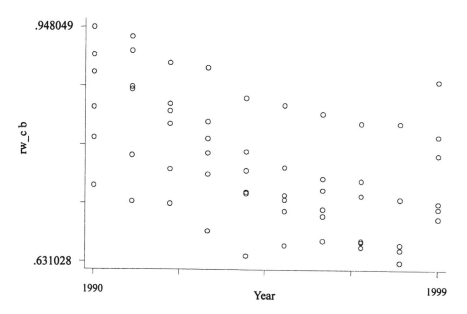

Hungary, 1990–99 – W/N regions

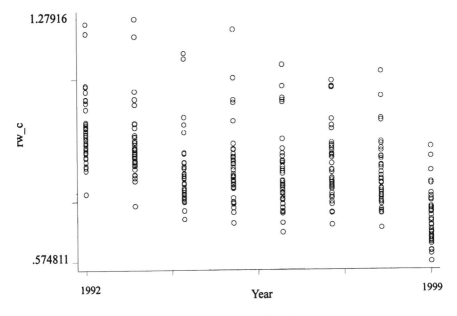

Romania, 1992–99 – all regions

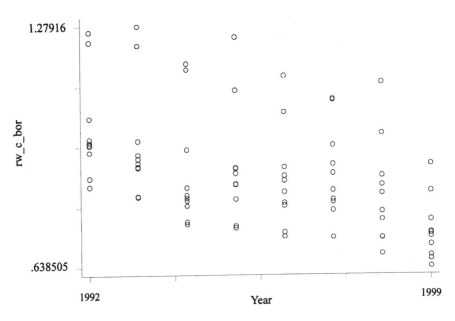

Romania, 1992–99 – W/N regions

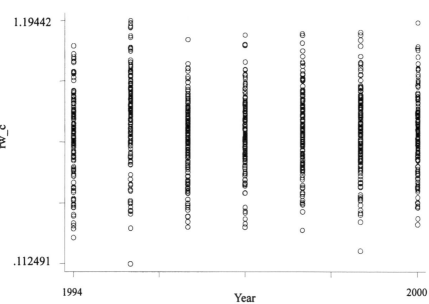

1.19442

rw_c

.112491

1994   Year   2000

Slovenia, 1994–2000 – all regions

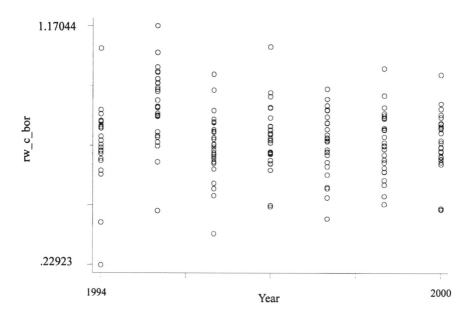

1.17044

rw_c_bor

.22923

1994   Year   2000

Slovenia, 1994–2000 – W/N regions

shaped pattern of regional adjustment. An explanation for this might lie in the fact that the core manufacturing production is based around Tallinn in W/N border regions. Since the early 1990s, these regions benefited enormously from large FDI inflows, especially in non-manufacturing sectors, which triggered a steep rise in wages. Recently, regions that are more distant from the capital are becoming increasingly unable to keep pace with the rapidly expanding capital region of Tallinn. Strong migrations of qualified labour to the central region are apparent, which implies that a typical Krugman type of regional polarization might take place in Estonia.

Evolution of relative regional FDI As revealed in Table 10.4, selected transition countries have been subject to substantial FDI inflows during 1990s. The share of all transition countries in world FDI flows increased from 0.2 per cent in 1990 to 2.3 per cent in 2000. In the countries under examination, the stock of FDI throughout the 1990s added up to some 15-50 per cent of GDP. The major recipient of FDI in absolute terms among the selected countries is Hungary, while in relative terms (as a share of GDP) FDI plays the most important role in Estonia.

Table 10.4   Pattern of FDI inflows to transition countries, 1990–2000 (in US$m)

Year	Bulgaria	Estonia	Hungary	Romania	Slovenia
1990	4	–	311	-18	4
1991	56	–	1,459	37	225
1992	42	58	1,471	73	1,146
1993	40	160	2,339	94	341
1994	105	225	1,146	341	128
1995	90	205	4,453	419	128
1996	109	150	1,983	263	185
1997	498	262	2,085	1,224	321
1998	537	581	2,036	2,031	165
1999	819	305	1,944	1,041	181
2000	1,002	398	1,957	998	181
1999–2000 (as % of GDP)	19.9	47.9	39.9	16.1	13.0

Source: UNCTAD, World Investment Report, 2001.

As proposed by the DK model, the regional pattern of FDI inflows is determined by (i) differences in relative factor costs, (ii) trade costs between the home country and the foreign country as well as trade costs between home regions, and (iii) agglomeration effects. Table 10.5 shows the pattern of the relative regional presence of foreign investment firms (FIEs). Here, in the absence of more appropriate data, the number of FIEs relative to the number of domestic firms serves as an effective measure of the regional importance of FDI.

As discussed earlier, these indicators should be interpreted with a large degree of caution. Since we only deal with data on the number of firms and not with the data on their output, our findings may be biased in an important way. Nonetheless, with some exception for Bulgaria and Estonia, the regional pattern of FDI does not correspond to that suggested by the DK model. In general, the importance of FDI for the regions is quite low. On average, the share of FIEs, by region, is well below 10 per cent and this has not changed much throughout the 1990s. In Bulgaria and Estonia these shares in W/N border regions are substantially higher and amounted to, respectively, 13 and 23 per cent in 1999. In Hungary, Romania and Slovenia, the opposite is true: non-W/N border regions account for substantially higher shares of FDI.

This evidence is in line with the findings of Alessandrini and Contessi (2001), who found that the vast majority of FDI inflows in transition countries had been directed into central regions and traditional economic centres. In summary, with the exception of Bulgaria and Estonia, little evidence, at least using this rough method, is found in favour of the suggestions of the DK model, which proposes that the majority of FDI will flow into W/N border regions. The exact impact of regional FDI on the regional wage structure remains to be seen in the formal tests.

4.3   Results

Estimation issues Before we turn to the estimation results of our empirical model (4), a few words need to be said about the methodology of our estimations. Our data is structured as regional panel data for a time span of seven to ten years, which requires an explicit account of region-specific effects. Without explicit control for this, one might get biased estimates of coefficients since FDI inflows, output growth and changes in relative wages might be correlated over time or subject to random shocks. Using static specification of the model (4), there are two well-known ways of controlling for this bias. The most obvious option is employing the fixed effects (FE) estimator, which assumes fixed (constant) region-specific effects over time. On the other hand,

Table 10.5 Regional pattern of FDI by countries, 1990–2000

	BG (1990)	BG (1999)	EST (1992)	EST (1999)	HU (1990)	HU (1999)	RO (1992)	RO (1999)	SLO (1994)	SLO (2000)
All regions										
Mean	0.07	0.07	0.08	0.09	0.08	0.10	0.04	0.05	0.04	0.03
Std. error	0.04	0.04	0.07	0.07	0.05	0.05	0.02	0.02	0.02	0.01
Std. deviation	0.20	0.19	0.26	0.25	0.22	0.22	0.16	0.15	0.22	0.14
Coef. of variation (%)	271.3	279.1	338.4	279.7	260.6	225.5	366.9	330.4	542.8	466.9
Norm. coef. of variation (%)	51.3	52.7	87.4%	72.2	58.3	50.4	57.3	51.6	41.6	35.8
N	28	28	15	15	20	20	41	41	170	170
W/N border regions										
Mean	0.13	0.13	0.21	0.23	0.04	0.06	0.03	0.03	0.04	0.03
Std. error	0.11	0.11	0.20	0.19	0.01	0.01	0.01	0.01	0.02	0.02
Std. deviation	0.33	0.33	0.44	0.43	0.02	0.02	0.04	0.03	0.10	0.10
Coef. of variation (%)	254.9	253.4	209.7	190.9	56.2	41.4	134.9	74.1	245.9	291.7
Norm. coef. of variation (%)	85.0	84.5	93.8	85.4	22.9	16.9	42.7	23.4	46.5	55.1
N	9	9	5	5	6	6	10	10	28	28
Non-W/N border regions										
Mean	0.05	0.04	0.01	0.02	0.10	0.11	0.05	0.05	0.04	0.03
Std. error	0.02	0.01	0.00	0.01	0.07	0.07	0.03	0.03	0.02	0.01
Std. deviation	0.09	0.06	0.01	0.02	0.26	0.26	0.18	0.18	0.24	0.15
Coef. of variation (%)	199.1	147.0	158.9	106.1	250.9	230.3	380.7	348.8	584.1	500.8
Norm. coef. of variation (%)	45.7	33.7	50.2	33.6	67.1	61.6	68.4	62.7	49.0	42.0
N	19	19	10	10	14	14	31	31	142	142

Source: Authors' calculations.

the random effects (RE) estimator assumes that region-specific effects are random and only reflected in the error term; i.e., uncorrelated over time. The FE estimator is usually more robust but quite inaccurate, while the RE estimator is sensitive to assumptions but more accurate. From a substantial point of view, we are interested in observing the pattern of changes in relative regional performance over time as induced by external shocks such as trade liberalization. Of course, it is straightforward to assume that individual regions will respond homogenously to external shocks throughout the period. Hence, in this case the FE estimator is a natural choice. An important drawback of the FE estimator in the present case, however, is that some of the crucial variables in our empirical model are time invariant (such as border dummies, transport costs proxied by road distances in kilometers or the trade liberalization dummy for countries that had already liberalized their trade with the EU at the beginning of the period covered in our data). When performing regular FE estimations these variables are dropped from the estimation procedure. In order to avoid this, we employed a trivial trick – all of the time invariant variables have been multiplied by the time trend. After differentiating, which is the underlying procedure in FE estimations, this gives normal, time varying values of the parameters under consideration. Of course, one needs to be cautious with the interpretations, since the regression coefficients obtained through this modification are to be interpreted in terms of growth rates. Yet, all that matters in our estimations is the sign and significance of the parameters, which are not altered by the above modifications. Irrespective of the superiority of the FE estimator in the present case, we also conduct formal Hausman tests in order to test for the validity of the model specification.

Results Table 10.6 provides basic estimation results from our empirical model (4) using the FE estimator. F-tests performed confirm that strong individual (regional) effects are present and justify the use of panel data techniques instead of OLS. However, irrespective of the superiority of the FE estimator over the RE estimator, the Hausman specification tests in the cases of Bulgaria, Estonia and Hungary cannot be used to reject the null hypothesis that individual effects are random. In these cases, RE specification of the model is more appropriate. Nevertheless, in Table 10.6 we report only the results obtained by using the FE estimator since we believe that, in our case, it is a more appropriate estimator. On the other hand, close examination of the coefficients obtained using both FE and RE estimators does not reveal any reversals in sign and significance. The interpretation of the results below therefore does not suffer under the misspecification of the model.

Table 10.6 The impact of trade liberalization on the adjustment pattern of relative regional wages in transition countries (results of FE estimations)*

	BG	EST	4.3.1.0.0.0.0.1HU	RO	SLO
t	−0.021	**−0.301**	−0.031	–	0.013
	(−0.81)	(−2.90)	(−0.62)		(0.64)
t^2	−0.001	**−0.004**	**0.001**	0.002	**−0.004**
	(−1.10)	(−7.72)	(7.57)	(0.64)	(−2.70)
irVA/emp	**0.024**	**−0.164**	0.000	**−0.033**	**−0.019**
	(2.76)	(−3.10)	(−0.03)	(−2.56)	(−6.01)
rFDI	−0.089	**0.581**	0.180	**−0.377**	−0.086
	(−0.62)	(1.51)	(0.97)	(−2.18)	(−0.97)
DISTANCE	−0.0029	**0.058**	0.005	−0.003	0.004
	(−0.63)	(3.15)	(0.54)	(−0.74)	(0.84)
BORDER	**0.042**	**0.086**	**0.046**	−0.009	0.003
	(2.02)	(2.18)	(2.64)	(−0.36)	(0.08)
FTA	0.015	**0.503**	0.006	0.0001	**0.071**
	(0.51)	(4.96)	(0.11)	(0.01)	(2.58)
DIST*FTA	0.005	**−0.070**	−0.006	0.0003	**−0.014**
	(0.91)	(−3.60)	(−0.59)	(0.68)	(−2.35)
DIST*BORD	**−0.007**	−0.006	**−0.010**	0.002	−0.001
	(−2.01)	(−0.84)	(−3.51)	(0.39)	(−0.17)
å	−0.002	**−0.070**	0.011	−0.0003	0.009
	(−0.30)	(−3.92)	(1.20)	(−0.43)	(0.80)
Broad region dummies	Yes	Yes	Yes	Yes	Yes
Number of obs.	260	112	180	320	1169
Adj R^2	0.615	0.862	0.858	0.812	0.185
F test for individual effects	22.2	15.66	33.4	14.1	14.0
Hausman chi^2 test	12.2	2.6	11.0	243.5	114.1
Prob>chi^2	0.6676	0.9996	0.6876	0.0000	0.0000

* dependent variable: relative regional wage, i.e., wage in the r-th region relative to the capital region.

The expected U-shaped (i.e., positive sign of t^2) adjustment pattern of relative wages is revealed only in the case of Hungary, while for Bulgaria, contrary to the graphic representation, the existence of such an adjustment pattern is not confirmed. For both countries there is strong evidence of the catching-up of W/N border regions, which has not been affected by trade liberalization. On average, wages in W/N border regions in both countries grow annually at about four per cent faster than in the central region. The distance from the capital does not seem to have affected regional wages significantly either before or after trade liberalization, but it has some impact on the slower growth of wages in more distant W/N border regions. Similarly, FDI inflows in both countries do not seem to affect the manufacturing relocation and adjustment pattern of regional wages. When compared to the predicted signs of parameters according to three competing EG models, one may conclude that the adjustment pattern of regional wages in Bulgaria and Hungary is much in line with the FKV model.

For Estonia, as expected from the preliminary analysis of the data, an inverted U-shaped (i.e., negative sign of t^2) adjustment of relative wages is found. In addition, while initially the distance and border effects were found to be positive, after trade liberalization both parameters became negative. This simply indicates that, after initial favourable regional development, individual regions that are distant from the capital have recently become unable to keep pace with the rapidly expanding capital region of Tallinn. The impact of FDI is found to be positive, but it seems that it is mainly directed towards the central region. In effect, strong migrations of qualified labour to the central region are apparent and a typical Krugman type of regional polarization might be at work. A stronger engagement of the government in designing proper regional policies is needed in order to prevent further regional polarization.

Surprisingly similar results in terms of the coefficient of the long-run adjustment pattern of regional wages (i.e., an inverted U-shape) are found in Slovenia. This, however, is probably a consequence of the misspecification of the model, since one of the included variables is picking up the time effects. A likely candidate for this is the initial efficiency differential, which shows that initially less efficient regions do catch up, in terms of wages, with the central region. Trade liberalization is shown to have a strong positive impact on relative regional wages since, after 1997, non-central regions have been growing annually about seven per cent faster than the central region. In more distant regions, this process of the catch-up of regional wages has slowed down to 1.5 per cent annually. Similarly to Bulgaria and Hungary, FDI inflows in Slovenia do not seem to affect the manufacturing relocation and adjustment pattern of regional wages. Finally, the results for Slovenia do not seem to conform to any of the three competing EG models.

While the preliminary analysis for Romania has shown a clear negative trend in relative regional wages, this was not confirmed in our empirical estimations. After serious economic problems in Romania in the mid-1990s, the initial negative trend in regional wages might have been stopped by the newly elected socialist government. This, however, was temporary, since at the end of the 1990s the trend of relative wages again turned downward. In effect, external policy shocks in Romania, which caused a lot of mismatch in the economic performance of regions, have probably prevented the strong regional polarization effects of the Krugman type initiated by trade liberalization. Similarly to Estonia, a stronger engagement of government in regional policy is needed in the future in order to prevent emerging income inequalities between regions.

5 Conclusions

The present chapter analyses the effects of trade liberalization with the EU on the inter-regional relocation of manufacturing and the inter-regional adjustment of relative wages in transition countries. We start with an overview of the implications of the three alternative EG models. The Krugman model predicts a monotonic response of relative regional wages to the reduction of foreign trade costs, strong migration flows of labour towards the core region, and thus a typical core-periphery regional polarization. Fujita-Krugman-Venables (FKV) and Damijan-Kostevc (DK) models argue that labour is imperfectly mobile between regions, which, due to the increasing wages in the core region, prevents complete agglomeration in the home country. Both models result in a non-monotonic, U-shaped response of relative regional wages to trade liberalization. The major difference between the two approaches is that the DK model introduces a second factor of production, capital, which allows for FDI flows between countries. FDI inflows are shown to accelerate the regional adjustment process in the home country, as they are initially attracted to poor border regions characterized by lower wages and higher returns to capital. The DK model therefore results in a faster convergence of relative regional wages, i.e., compared to the FKV approach, in a more upward and shifted-to-the-right U-shaped response curve of relative wages.

In the second part of the chapter we turn to the examination of the exact adjustment pattern of relative regional wages in five transition countries after they have liberalized their trade with the EU. We study which of the three alternative EG models is a more appropriate approximation of the actual regional adjustment pattern in the selected transition countries. More specifically, we examine whether the response of relative regional wages to trade liberalization is monotonic and leading to strong regional polarization as suggested by the first Krugman approach, or is non-monotonic and associated with less regional polarization as suggested by the more recent EG approaches.

In addition, in the case of a non-monotonic response, we test the propositions of the FKV approach against the DK approach. In this process, the impacts of FDI, of inter-regional transport costs and of western/northern region dummies on the adjustment pattern of relative regional wages are examined. Implications of the three competing EG approaches are tested using unique regional panel data for five transition countries (Bulgaria, Estonia, Hungary, Romania and Slovenia) for the period 1990–2000.

Summing up our empirical findings, one can speculate that a Krugman type of divergence of relative regional wages and increased regional polarization might be at work in Estonia and Romania. A stronger engagement of government in terms of designing proper regional policies is probably needed in these countries in order to prevent serious regional polarization. On the other hand, one may conclude that the adjustment pattern of regional wages in Bulgaria and Hungary is much in line with the FKV model indicating that market forces by themselves have brought about the upward trend after the initial decline in relative regional wages. Here, Slovenia looks like a special case since its regional development pattern does not correspond to any of the examined EG models. In general, a clear catching-up process in most of the regions in Slovenia after trade liberalization is evident. There is some evidence of polarization in a few of the more distant regions, which shows the need for a stronger regional policy. However, in contrast to Estonia and Romania, there is probably no need for a general regional policy aimed at preventing greater income inequalities between regions, but a need for a more specific, tailor-made regional policy. By the latter, we mean special government programs customized to meet the needs of individual regions (such as supporting regions with traditional sectors) or general policy measures that might be applied only in specific regions (i.e., economic zones in regions with high unemployment).

Notes

*   This research was undertaken with support from the European Community's Phare ACE Programme 1998. The content of the publication is the sole responsibility of the authors and it in no way represents the views of the Commission or its services.
1   For an excellent discussion of both issues, see Krugman and Venables (1995) and Krugman (1996).
2   Here, we neglect the impact of trade liberalization on the adjustment pattern of wages between the foreign country and the home country, which was the major concern of FKV and other previous studies, but instead concentrate on the inter-regional adjustment pattern in the home country.

3   By standard assumption, each firm produces one intermediate and one final consumption good. The final consumption good is costlessly assembled from intermediate goods and all intermediate goods are used up in this process. The price of intermediate goods falls with the number of firms due to large-scale production at the firm level. As all of the intermediate goods enter each firm's production function, decreasing prices of intermediate goods induce a downward sloping firm's marginal cost curve (for more details on modeling the interaction between internal and external scale economies through intermediate goods, see Damijan (1999)).

4   Mathematica version 4.1 has been used to perform all of the model simulations.

5   A reader should note that this outcome is not general per se. First, it holds for a very limited range of low transport costs only. When transport costs exceed the value of $T^* = 5$ the shape of the relative wage curve becomes very complex. Second, the above outcome holds only when the size differential between both regions is sufficiently large but not too large. For low size differentials, the relative wage curve is subject to multiple equilibria. Hence, Neary's (2001) critique is extremely relevant.

6   Actual transport cost is calculated as $t = (T-1)/T$, i.e. when $T^* = 1.45$, t is equal to $(1.45-1)/1.45 = 31$ per cent.

7   Normalized coefficient of variation is a coefficient of variation normalized by a square root of number of observations (N). This is to ensure better comparability of variation across countries. Since the variation in data for countries with more disaggregated data is biased upward one should take account of this.

References

Alessandrini, S. and Contessi, S. (2001), 'The Regional Location of FDI in Central Europe: an Empirical Analysis in a Comparative Perspective', Mimeo, Bocconi University.

Barro, R. and Sala-i-Martin, X. (2001), *Economic Growth*, McGraw-Hill, New York.

Brakman, S., Garretsen, H. and Schramm, M. (2000), 'The Empirical Relevance of the New Economic Geography: Testing for a Spatial Wage Structure in Germany', CESifo Working Paper, 395, Center for Economic Studies, Munich.

Brakman, S., Garretsen, H. and Schramm, M. (2002), 'New Economic Geography in Germany: Testing the Helpman-Hanson Model', HWWA Discussion paper 172.

Brakman, S., Garretsen, H. and van Marrewijk, C. (2001), *An Introduction to Geographical Economics: Trade, Location and Growth*, Cambridge University Press, Cambridge.

Brülhart, M. and Torstensson J. (1996), 'Regional Integration, Scale Economies and Industry Location in the European Union', CEPR Discussion Paper No. 1435.

Damijan, P.J. (1999), 'Interaction of Different Types of Economies of Scale, Monopolistic Competition, and Modern Patterns of Trade', Ljubljana, RCEF Working papers series, 85.

Damijan, P.J., Knell, M., Majcen, B. and Rojec, M. (2001), 'The Role of FDI, R&D Accumulation and Trade in Transferring Technology to Transition Countries: Evidence from Firm Panel Data for Eight Transition Countries', Ljubljana, IER Working papers, 10.

Damijan, P.J. and Kostevc, Č. (2002), 'The Impact of European Integration on Adjustment Pattern of Regional Wages in Transition Countries: Testing Competitive Economic Geography Models', LICOS Discussion paper, No. 118/2002.

Davis, D.R. and Weinstein, D.E. (1999), 'Economic Geography and Regional Production Structure: an Empirical Investigation', *European Economic Review*, Vol. 43, pp. 379–407.

De la Fuente, A. (2002), 'On the Courses of Convergence: a Close Look at the Spanish Regions', *European Economic Review*, 46(3), pp. 569–99.

Forslid, R., Haaland, J.I. and Midelfart Knarvik, K.H. (2002), 'A U-Shaped Europe? A Simulation Study of Industrial Location'. *Journal of International Economics*, Vol. 57, pp. 273–97.

Fujita, M., Krugman, P. and Venables, A.J. (1999), *The Spatial Economy: Cities, Regions, and International Trade*, MIT Press, Cambridge, Mass.

Hallet, M. (2000), 'Regional Specialization and Concentration in the EU', Economic papers, European Commission, 141, pp. 1–29.

Hanson, G.H. (1997), 'Increasing Returns, Trade and the Regional Structure of Wages', *The Economic Journal*, Vol. 107, pp. 113–32.

Hanson, G.H. (2000a), *Market Potential, Increasing Returns, and Geographic Concentration*, Mimeo, University of Michigan.

Hanson, G.H. (2000b), 'Scale Economies and the Geographic Concentration of Industry', NBER Working paper, 8013.

Helpman, E. (1981), 'International Trade in the Presence of Product Differentiation, Economies of Scale and Monopolistic Competition: A Chamberlin-Hecksher-Ohlin Approach', *Journal of International Economics*, Vol. 11, pp. 305–40.

Helpman, E. (1998), 'The Size of Regions', in D. Pines, E. Sadka and I. Zilcha (eds), *Topics in Public Economics*, Cambridge University Press, Cambridge.

Hummels, D. (1999a), 'Toward a Geography of Trade Costs', Mimeo, Purdue University.

Hummels, D. (1999b), 'Have International Trade Costs Declined?', Mimeo, Purdue University.

Hummels, D. (2001), 'Time as a Trade Barrier', Mimeo, Purdue University.

Krugman, P.R. (1979), 'Increasing Returns, Monopolistic Competition and International Trade', *Journal of International Economics*, Vol. 9, pp. 469–79.

Krugman, P.R. (1980), 'Scale Economies, Product Differentiation and the Pattern of Trade', *American Economic Review*, Vol. 70, pp. 950–59.

Krugman, P. (1991a), *Geography and Trade*, MIT Press, Cambridge, Mass.

Krugman, P. (1991b), 'Increasing Returns and Economic Geography', *Journal of Political Economy*, Vol. 99, pp. 483–99.

Krugman, P. (1996), *Pop Internationalism*, MIT Press, Cambridge, Mass.

Krugman, P. (1998), 'Space, the Final Frontier', *Journal of Economic Perspectives*, pp. 161–74.

Krugman, P. and Venables, A.J. (1995), 'Globalization and the Inequality of Nations', *Quarterly Journal of Economics*, Vol. 110, pp. 857–80.

Krugman P. and Venables, A.J. (1996), 'Integration, Specialization, and Adjustment', *European Economic Review*, Vol. 40, pp. 959–68.

Lancaster, K. (1980), 'Intra-Industry Trade under Perfect Monopolistic Competition', *Journal of International Economics*, Vol. 10, pp. 151–75.

Markusen, J.R. and Melvin, J.R. (1981), 'Trade, Factor Prices, and the Gains from Trade with Increasing Returns to Scale', *Canadian Journal of Economics*, Vol. 14, pp. 450–69.

Midelfart-Knarvik, K., Overman, H.G. and Venables, A.J. (2000), 'Comparative Advantage and Economic Geography', CEPR Discussion Paper, 2618.

Neary, J.P. (2001), 'Of Hype and Hyperbolas: Introducing the New Economic Geography', *Journal of Economic Literature*, Vol. 39, pp. 536–61.

Overman, H.G., Redding, S. and Venables, A.J. (2000), 'The Economic Geography of Trade, Production, and Income: A Survey of Empirics', Mimeo, London School of Economics.

Panagariya, A. (1981), 'Variable Returns to Scale and Patterns of Specialization', *American Economic Review*, Vol. 71, pp. 221–30.

Puga, D. (1999), 'The Rise and Fall of Regional Inequalities', *European Economic Review*, Vol. 43, pp. 303–34.

Puga, D. (2001), 'European Regional Policies in Light of Recent Location Theories', CEPR Discussion Paper, 2767, London.

Pratten, C. (1988), 'A Survey of the Economies of Scale', in Commission of EC: Research on the 'Cost of Non-Europe', Vol. 2: *Studies on the Economics of Integration*, Luxembourg.

UNCTAD (2001), *World Investment Report*.

Chapter 11

The Implications of European Integration and Adjustment for Border Regions in Accession Countries*

Laura Resmini

1 Introduction

In Central and Eastern Europe the process of economic change and liberalization that occurred during the 1990s has had important spatial consequences, often neglected by the literature on the effects of the enlargement, which has focussed mainly on the national level (Baldwin, Francois and Portes, 1997; Avery, Cameron, 1998). Within these spatial and socio-economic dynamics, borders and border regions[1] are likely to play a critical role for several reasons. First of all, border regions in accession countries are not the exception but the rule, accounting for almost 66 per cent of the land area and 58 per cent of the total population (EC, 2001). Secondly, the fall of the Berlin wall and the ongoing process of economic integration with the European Union (EU) have put borders in a state of flux, with changes occurring in their physical location and economic and political significance as well. Borders are no longer considered as fixed separating lines, but as 'contact' areas, bridges to new markets and cultures. Old borders have been vanishing, and a new geo-political and economic map has emerged, with a different distribution of roles and possibilities at the national and regional levels (Njikamp, 1994). Indeed, the reorientation of economic links from East to West has created new challenges and opportunities for the development of western border regions, and has raised serious concerns for regions located along the Eastern border, which is potentially more sensitive to the collapse of the CMEA and the former Soviet Union.

International trade and Foreign Direct Investment (FDI) – the two driving forces behind economic integration – have undoubtedly had a considerable impact on the economy, at both national and regional levels. The possibility to exchange goods and services internationally opens opportunities to specialize and to use economies of scale and therefore may result in the concentration of economic activities in a few locations close to international markets. Furthermore, trade occurs in a heterogeneous space, where distance and quality of infrastructure matter, so integration may have different consequences for the center and the periphery. Even more than trade, FDI affects the domestic economy through technical – transfer of technology, skills, knowledge and governance – as well as pecuniary – backward and forward linkages with domestic firms – externalities, which may generate positive spillovers to domestic economies. Since, however, FDI tends to cluster geographically in Central and Eastern Europe (Resmini, 2000), it can generate or further increase regional disparities within candidate countries.

This chapter aims at exploring and analysing, on a comparative basis, the impact of the EU's eastern enlargement on border regions in five candidate countries: Bulgaria, Estonia, Hungary, Romania and Slovenia. These countries have different development levels and geographical coordinates that make their comparative analysis interesting. Hungary and Slovenia are relatively more advanced than Estonia, Bulgaria and Romania. In addition, Estonia is a North European country sharing a border with Finland, while Hungary is a Central European country with a common border with Austria. Slovenia and Bulgaria are Southern European countries bordering, respectively, Italy and Austria, and Greece. Romania does not share any border with the EU-15. As a result, Hungary and Slovenia seem to have the advantage of geographical proximity to Western European core countries, while the others do not.

In order to achieve its overall objective, the chapter will first provide a brief overview of the main theoretical predictions about regional adjustments to trade liberalization and economic integration (section 2). Then, it will provide a definition and identification of border regions in the selected countries, as well as a descriptive analysis of their relative position within each country and with respect to the EU-15 average (section 3). Thirdly, it will develop an econometric model able to analyse the determinants of regional specialization and adjustment over time. In particular, the work will explore how the ongoing process of economic integration with the EU is affecting the location of economic activity in candidate countries and which regions are winning and losing in this process, in terms of regional growth prospects. This classification will be used to evaluate the likely distributional implications of enlargement for the accession countries under consideration.

2 Economic Integration and Border Regions

Although a systematic theory of border regions has never been developed, location theory has traditionally considered them disadvantaged areas because of international barriers to trade and the threat of military invasion (Anderson, O'Dowd, 1999). National borders negatively affect regional economies by artificially cutting up spatially complementary regions and by increasing transaction costs. Tariffs, differences in language, culture and business practices inhibit cross-border trade, while the conflict between political and economic objectives – which is at the basis of the potential political and social instability of border areas – decreases the incentive or domestic and foreign producers to locate in these regions. Moreover, it has been demonstrated that the larger the market area, the fewer the entrepreneurs who will choose a location close to the frontier, other things being equal (Hansen, 1977a).

The reversal of this unfavourable picture is that greater international economic integration – with the consequent removal of national boundaries and trade barriers – should create new prospects for growth in border regions, as happened in Europe with the realization of the Single Market in 1993[2] and in North America after the creation of NAFTA (Hanson, 1996, 1998).

From location theory's standpoint, the eastern enlargement of the EU should benefit all regions directly affected by the removal of national borders, i.e., regions directly bordering the EU, as well as those bordering other countries with interests in the enlargement process. It should have a negligible impact on internal regions and possible negative effects on regions still influenced by a frontier, such as regions bordering a third country not involved in the enlargement process, because of their peripheral position within a large market area.

However, location theory is just one field within the body of theory that attempts to explain how trade liberalization affects industry location. An answer to this question may also be found in traditional international trade theories, which emphasize international (or inter-regional) differences in factor endowments (Hecksher-Ohlin) or technologies (Ricardo), as well as in the New Trade Theories (NTTs) and in the New Economic Geography (NEG), which try to explain the spatial structure of economic activities using models with increasing returns to scale and imperfectly competitive markets (Venables, 1998; Krugman, 1998; Fujita, Krugman and Venables, 2000).

The NTTs, developed during the 1980s, stress the importance of market access as a determinant for the location of economic activities. Their most interesting prediction for the purposes of this analysis is that since firms have increasing returns to scale and trade is not costless, they will situate production in only a few locations, namely those possessing the largest market for their products.. This idea, labeled known as the "home market effect", suggests that a reduction in trade barriers will lower trade costs, thus increasing firms' incentives to relocate to regions with better access to the largest of foreign markets, such as border regions or coastal areas, given that the labor force is internationally immobile.

Although geographical advantage plays a role in NTTs, it is, however, considered exogenous, as if it were determined by physical rather than economic characteristics. However, the key determinant of geographical advantage is the interaction among different economic agents – suppliers, consumers, institutions – which, of course, is not fixed, but endogenous, as the rising and declining of economic centers over the years and across regions has suggested (Venables, 1998). According to this idea, firms locate in an economic center, which can be considered a center only because other firms locate there. This indicates the existence of a cumulative causation process in which the entry of new firms in a location makes it a more attractive location for additional firms. The functioning of this cumulative causation process depends on the presence of pecuniary (backward and forward linkages) as well as technological externalities (knowledge spillovers and learning-by-doing) between firms.[3] To the extent that such externalities are localized, production is also geographically concentrated, and the logic of increasing returns to scale implies that once a pattern of industrialization has been established, it will persist over time. In the case of trade liberalization, the presence of externalities alters firms' incentives to relocate close to foreign markets since that would mean losing the benefits of being near their suppliers, customers, sources of information or technology, or, more generally, firms from which they derive positive externalities.

The consideration of agglomeration forces makes the impact of the enlargement process on the location of economic activities in candidate countries more uncertain. The sharp increase and diversification of trade flows between the EU and the candidate countries indicates that domestic producers in candidate countries might have an incentive to relocate close to the EU border in order to exploit economies of scale and better market access. However, the presence of old industrial poles often located far from the Western border may represent an incentive for firms not to relocate.

Overall, both traditional and more recent theories of location seem to suggest that the enlargement process is likely to have an uneven impact on border and non-border regions, with the greatest impact on regions bordering the EU, because of their geographical proximity to large potential markets. The next sections will be devoted to understanding whether these theoretical predictions apply to transition countries in Central and Eastern Europe.

3   The Economic Situation of Border Regions

3.1   Definition of Border Regions

For the purposes of analysis, this study defines border regions as regions at the NUTS 3 level eligible for PHARE-CBC programmes. Within this broad category, three different sets of relatively homogeneous regions can be identified:

- borders with present EU members (hereafter BEU);
- borders with other candidate countries currently negotiating accession (hereafter BAC);
- borders with external countries (hereafter BEX).

These differ from internal regions (hereafter INT) because of their geographical position along international borders. According to this definition, the sample includes 106 regions (Table 11.1): 63 border regions – 14 bordering the EU, 21 bordering external countries and 28 bordering other candidate countries – and 43 non-border regions, located in Bulgaria, Hungary and Romania. Estonia and Slovenia, being small countries, have virtually only border regions.

Border regions display many kinds of difference and asymmetry. From a geo-economic point of view, they may have different shapes and sizes, be densely or scarcely populated, stagnate in their economic and social peripherality, or turn it to political and economic advantage (Anderson, O'Dowd, 1999). So, rather than concentrating only on internal characteristics, it is more fruitful to study a border region in terms of how it compares with other regions in its own state, as well as in other states and in the EU, the integrated economic space to which they already belong. The next section focuses on this multi-level comparative analysis. Four economic indicators have been applied in order to compare different sets of regions within and across countries. They refer to the spatial distribution and to changes in population, GDP per capita, the unemployment rate and relative employment at a sectoral level.

Table 11.1   Regions' classification in candidate countries

	BEU	BEX	BAC	INT
BG	Blagoevgrad; Kardjali; Smolyan	Bourgas; Kustendil; Pernik; Haskovo; Yambol	Vidin; Vtratza; Dobrich; Montana; Pleven; Russe; Silistra	Varna; Veliko Tarnovo; Gabrovo; Lovech; Pazardjik; Plovdiv; Razgrad; Sliven; *Sofia*; Sofia region; Stara Zagora; Targoviste; Shumen
EE	*Northern Estonia*; NE Estonia; Western Estonia		Central Estonia; Southern Estonia	
HU	Győr-Moson-Sopron; Vas	Baranya; Somogy; Bács-Kiskun	Komárom-Esztergom; Zala; Borsod-Abaúj-Zemplén; Nógrád; Hajdú-Bihar; Szabolcs-Szatmár-Bereg; Békés; Csongrád	*Budapest*; Pest; Fejér; Veszprém; Tolna; Heves; Jász-Nagykun-Szolnok
RO		Botosani; Iasi; Suceava; Vaslui; Galati; Tulcea; Caras-Severin; Maramures	Constanta; Calarasi; Giurgiu; Teleorman; Dolj; Mehedinti; Olt; Arad; Timis; Bihor; Satu Mare	Bacau; Neamt; Braila; Buzau; Vrancea; Arges; Dambovita; Ialomita; Prahova; Gorj; Valcea; Hunedoara; Bistrita-Nasaud; Cluj; Salaj; Alba; Brasov; Covasna; Harghita; Mures; Sibiu; *Mun. Bucuresti (inc.Ilfov)*
SI	Pomorska; Podravska; Koroška; Gorenjska; Goriška; Obalno-kraška	Savinjska; Spodnjeposavska; Dolenjska; Osrednjeslovenska; Notranjsko-kraška		Zasavska

BEU = regions bordering the EU-15; BEX = regions bordering other candidate countries; BAC = regions bordering with third countries; INT = non-border regions.

3.2   Comparative Analysis within and across States

Table 11.A1 considers the first three economic indicators.[4] There are striking differences between border regions in terms of socio-economic development. In 1998 border regions had a population of about 22 million inhabitants, about 50 per cent of the total population in the countries considered. The border with the EU does not seem to have had any effect on population location, since only 5.4 per cent of the total population lived there. However, the available statistics suggest that regions bordering the EU have already benefited from their location. On average, in 1995, the economic conditions in these regions were very similar to those in Eastern border regions (BEX), while those in BAC regions were closer to the level of development showed by internal regions. Proximity to the EU, however, seems to have contributed to stimulating growth: in the second half of the 1990s, GDP per capita has grown, on average, at about six per cent a year, while the unemployment rate decreased, on average, about 0.5 per cent a year. All other regions show opposite patterns for both variables. Thus, in 1998, BEU regions' GDP per capita was higher and their unemployment rate was lower than the average of other groups of border regions.[5] Consequently, one can conclude that convergence and the catching-up processes between regions bordering the EU and non-border regions have been occurring in candidate countries during the second half of the 1990s.

In evaluating the economic performance of internal regions, it is worth noting the dominant role of capital cities. Their economic impact is impressive. To give just two examples, the Tallin area (Estonia) has 95 per cent of FDI and 48 per cent of all registered firms. Budapest accounts for about 20 per cent of total population, 48 per cent of total employment in the service sectors and 52 per cent of total FDI, contributing to a GDP per capita level three times that of the worst placed county in the country.[6] The absence of other similarly dominant urban centers means that, outside the capital cities, spatial disparities in growth are more limited, as is shown by the figures reported in the bottom part of Table 11.A1. At the end of the 1990s, BEX regions were, on average, the poorest ones. Their geographical location at the extreme periphery of Europe, and the poor economic conditions in the countries they border – Russia, the Ukraine, Belarus, Serbia and Croatia – partially explain their overall economic weakness. BAC regions did not show significant changes in their economic conditions during the second half of the nineties, becoming more similar to BEX regions.

A more comprehensive analysis reveals different pictures across countries. In 1998, border regions in Bulgaria had a population of 3.5 million inhabitants, or 42 per cent of the country's population. These figures are 5.5 and 54 per cent for Hungary and 9.9 and 44 per cent for Romania. Since the border area is so large in Slovenia and Estonia, these figures are not significant for both countries. As far as GDP per capita and unemployment are concerned, border regions show different levels of development across countries. In Bulgaria, at the beginning of the period, regional disparities did not seem particularly large, with BEX and INT (Sofia included) regions above (below) and BEU and BAC regions below (above) the national average, in terms of GDP per capita (unemployment rate). However, BEU and BAC regions experienced GDP per capita growth rates above the national average during the second half of the 1990s, thus reducing the disparities with the other groups of regions. These patterns remain unchanged when Sofia is not included in the calculation, though in this case the rate of growth of GDP per capita in non-border regions is substantially higher than before, indicating that Sofia suffered more during economic restructuring during the transition than other internal regions did, thus reducing the regional disparities within the country. The unemployment rate has increased over time in all regions but those bordering with the EU, with the highest increases in BAC and INT regions.

In the second half of the 1990s, economic development has been positive in Hungarian BEU regions and in the internal ones (Budapest included), which were more similar in 1998 than at the beginning of the transition. BEX regions show deterioration in their relative position within the country, becoming more and more similar to BAC regions, which stagnated during the second half of the period. The dominant role of Budapest is evident from the comparison of the performance of internal regions with and without the Budapest district.

In Romania, GDP per capita in internal regions (Bucharest included) was, in 1998, more than double that in border regions, which show the best (BAC regions) and the worst (BEX regions) positions in term of the unemployment rate. Unlike in the other countries in the sample, regional differences among border and non-border regions seem to have increased over the period, due to the bad performance of border regions when taken as a whole. However, when the district of Ilfov, which includes Bucharest, is excluded from calculations, regional disparities become less evident.

3.3   Comparison at the EU Level

This section focuses on the position of border and non-border regions in candidate countries relative to the present EU average. The discussion is based on a transition matrix (Puga, 2001; Overman, Puga, 1999) that tracks changes over time in the relative position of regions within a given distribution. The transition matrix in Figure 11.1 reports changes between the 1992 and the 1999 distributions of GDP per capita relative to the EU average.[7] The transition matrix gives us several pieces of information. The first column shows the classes that divide up the distribution of relative regional income levels. The second column gives the number of regions that begin their transition in that range of the distribution and their sub-division among types of regions. Rows refer to 1992 distribution and columns to the distribution at the end of the period. The main diagonal gives the most important piece of information: it shows the fraction of regions that were in the same range of the distribution in 1992 and in 1999. The top row of the matrix indicates that, in 1992, only four regions (one for each type, all belonging to Bulgaria) had a GDP per capita below 0.05 times the present EU average.[8] Half of them remained in the same range in 1999, while the other 50 per cent saw their relative income increase between 0.05 and 0.1 times the present EU average. Both of them are border regions: Blagoevgrad, bordering with Greece, and Montana, on the Northern border with Romania. The number of regions that experienced little relative change is very high for all ranges of the distribution, although regions with the highest 1992 relative GDP per capita (first two rows from the bottom) showed more mobility: most of them, however, saw their relative income fall. Considering the different types of regions, only one non-border region (Fejér, Hungary) improved its relative GDP per capita, while BEX and BAC regions saw their relative per capita income decrease. BEU regions (all located in Hungary) remained in the same range.

It is interesting to compare the distribution of GDP per capita with unemployment rates. Reading the corresponding transition matrix (Figure 11.2) along the main diagonal, it shows that of the 12 regions that had an unemployment rate in 1992 below 0.75 times the European average, none remained in that range in 1999. All of them but one (an internal Hungarian region, Budapest) saw their relative unemployment rate increase. Jumping to the bottom, we see strong persistence in the regions with the highest unemployment rates. However, 40 per cent of BEU regions and 12 per cent of BEX ones saw their relative unemployment rates fall into an inferior range, as well as 20 per cent of non-border regions, while BAC regions did not seem to have been able to decrease their unemployment rates over the 1990s.

The contrast between changes in relative GDP per capita and changes in relative unemployment rates can be seen more clearly if we compare the two matrices. This comparison shows that, while regions exhibited strong persistence in their relative income per capita levels, they have experienced

1995	n.	1998 [0–0.05)	[0.05–0.10)	[0.10–0.15)	[0.15–0.20)	[0.20–)
[0–0.05)	**4**	**0.50**	**0.5**			
BEU	*1*		*1.00*			
BEX	*1*	*1.00*				
BAC	*1*		*1.00*			
INT	*1*	*1.00*				
[0.05–0.10)	**32**		**1.00**			
BEU	*2*		*1.00*			
BEX	*5*		*1.00*			
BAC	*9*		*1.00*			
INT	*16*		*1.00*			
[0.10–0.15)	**22**			**0.91**	**0.09**	
BEU						
BEX	*4*			*1*		
BAC	*6*			*1*		
INT	*12*			*0.83*	*0.17*	
[0.15–0.20)	24			**0.13**	**0.83**	**0.04**
BEU						
BEX	*6*			*0.17*	*0.83*	
BAC	*8*			*0.25*	*0.75*	
INT	*10*				*0.90*	*0.01*
[0.20–)	**7**				**0.14**	**0.86**
BEU	*2*					*1.00*
BEX						
BAC	*2*				*0.5*	*0.50*
INT	*3*					*1.00*

	[0–0.05)	[0.05–0.10)	[0.10–0.15)	[0.15–0.20)	[0.20–)
N.	**3**	**33**	**24**	**23**	**7**
BEU		3	1		2
BEX	1	5	5	5	
BAC		10	8	7	*1*
INT	2	15	10	11	4

Bulgaria, Hungary and Romania, 1995–99.

Figure 11.1 Transition matrix (GDP per capita)

		n	1999				
			[0–0.6)	[0.60–0.75)	[0.75–1)	[1–1.30)	[1.30–)
	[0–0.6)	4	0.00	0.25	0.25	0.25	0.25
	BEU						
	BEX						
	BAC	*1*			*1.00*		
	INT	*3*		*0.33*		*0.33*	*0.33*
	[0.60–0.75)	8	0.13	0.00	0.13	0.63	0.13
	BEU						
	BEX						
	BAC	*3*		*0*	*0.33*	*0.66*	
	INT	*5*	*0.20*	*0.00*		*0.60*	*0.20*
	[0.75–1)	17	0.06	0.12	0.18	0.24	0.41
	BEU						
1992	*BEX*	*1*				*1.00*	
	BAC	*6*		*0.33*	*0.17*	*0.33*	*0.17*
	INT	*10*	*0.1*		*0.20*	*0.1*	*0.6*
	[1–1.30)	18	0.00	0.00	0.06	0.17	0.78
	BEU						
	BEX	*7*				*0.29*	*0.71*
	BAC	*3*				*0.33*	*0.67*
	INT	*8*			*0.12*		*0.88*
	[1.30–)	42	0.00	0.02	0.02	0.1	0.86
	BEU	*5*		*0.2*	*0.2*		*0.6*
	BEX	*8*				*0.12*	*0.88*
	BAC	*13*					*100*
	INT	*16*				*0.19*	*0.81*
			[0–0.6)	[0.60–0.75)	[0.75–1)	[1–1.30)	[1.30–)
	N.		2	4	7	17	59
	BEU			1	1		3
	BEX					4	12
	BAC			2	3	5	16
	INT		2	1	3	8	28

Bulgaria, Hungary and Romania, 1992–99.

Figure 11.2   Transition matrix (unemployment)

a polarization of regional unemployment rates towards the upper extreme of the distribution. As a result, in 1999 there were more regions with very high unemployment rates and fewer regions with very low relative unemployment rates. This polarization does not seem to have a geographical component since it involves both border and non-border regions.

This simple exercise allows us to conclude that the transition towards a market economy and economic integration with the EU do not seem to have been a positive contribution to regional convergence in Europe.

3.4   Regional Employment in Border Regions

In order to identify regional patterns of specialization, it is useful to analyse the employment structure and its changes at a regional level. Table 11.2 shows regional shares in national employment by groups of economic activity for 1992 and 1999, while the average annual relative employment growth rates are summarized in Table 11.3.[9]

Although the time period is too short to highlight clear patterns of change, some interesting features emerge. The first is that employment adjustments seem to be *country-* and *sector-* rather than *region-*specific. Economic activities are spread between border and non-border regions relatively more evenly in Romania than in Bulgaria and Hungary. At the sectoral level, it is worth noting the almost overall geographic concentration of natural resource-based activities – such as agriculture and mining and quarrying – in border regions. In 1992, 53 per cent of employment in agriculture and 61 per cent of employment in mining and quarrying was concentrated in border regions in Bulgaria. These percentages are respectively 66 and 65 per cent in Hungary and 53 and 29 per cent in Romania. Relative employment remained more or less unchanged over the 1990s in all countries, with the exception of Bulgaria, whose mining and quarrying sector experienced a dramatic change in favour of internal regions. Most of the adjustment occurred in BEU regions. Services are mainly concentrated in internal regions, which include the capital city.[10]

As far as the manufacturing sector is concerned, it is worth noting that relocation activity was very intensive, and mainly in favour of border regions, though marked differences across countries do exist. In Bulgaria, border regions reinforced their specialization only in textiles and clothing production, while other sectors relocated mainly in internal regions. Regions bordering the EU were the only ones to benefit from increased specialization in textiles and clothing. Data also indicate a relocation of furniture and other manufacturing products from BAC to BEU regions. In Hungary, relocation activity within the manufacturing sector was very intense. Overall, it occurred in favour of border regions, and especially regions bordering the EU. Negative adjustments, i.e., a decrease in relative employment, happened only for furniture and other manufacturing, n.e.c. in BAC regions. In Romanian border regions, relative employment increased mainly in wood and paper products and in machinery, equipment and motor vehicles. Most of this adjustment, however, is within border regions, from BAC to BEX regions. In Estonia, it is interesting to note that adjustments in relative employment occurred from BEU to BAC regions in all sectors but machinery, equipment and motor vehicles, whose level of agglomeration in BEU regions increased over time.

Table 11.3 shows average annual relative employment growth rates by region and economic activity for the period 1992–99. Again, the data indicate that sector-specific effects are stronger than region-specific effects. Rates of growth, in fact, are more homogenous across regions than across sectors and countries, with a few remarkable exceptions. Border regions, taken as a whole, perform better than internal ones in Hungary and in Romania but not in Bulgaria. In Hungary, relative employment growth rates in BEU and, to a lesser extent, BEX regions have a positive sign in several manufacturing sectors, while the country trend is negative. In Romania, differences in relative employment growth rates between border and non-border regions are less pronounced than in Hungary and both follow the same negative trend. In Bulgaria, relative employment growth rates in the manufacturing sector are negative in all regions and larger in border than in non-border regions, with the exception of the textiles and clothing sector, which shows a positive relative employment growth rate in the BEU regions.

4   The Econometric Model

In this section more formal empirical techniques are used to study how relative employment at a regional level in candidate countries responds to economic integration. Estimation has been undertaken using data for 94 regions and seven manufacturing sectors in Bulgaria, Estonia, Hungary and Romania during the period 1992–99.[11] This data set has the advantage of relatively straightforward geography with a clear set of border and internal regions, and of covering a period of increasing economic integration with the EU.

In order to identify region-specific factors able to condition adjustments to trade liberalization and economic integration, I have first studied the determinants of industry location in different types of border and non-border

Table 11.2 Regional shares of national employment by groups of economic activities, 1992 and 1999

| | Regional share of national employment, 1992 | | | | | Regional share of national employment, 1999 | | | | |
| | Total | Border regions | | | Internal regions | Total | Border regions | | | Internal regions |
		BEU	BEX	BAC			BEU	BEX	BAC	
Bulgaria										
Agriculture	0.53	0.11	0.16	0.25	0.47	0.52	0.12	0.16	0.25	0.48
Mining and quarrying	0.61	0.22	0.32	0.07	0.39	0.52	0.12	0.33	0.08	0.48
Manufacturing	0.39	0.08	0.13	0.18	0.61	0.39	0.09	0.14	0.17	0.61
Food, beverages and tobacco	0.44	0.07	0.15	0.22	0.56	0.40	0.07	0.14	0.18	0.60
Textiles, clothing and leather	0.50	0.15	0.13	0.22	0.50	0.54	0.20	0.13	0.21	0.46
Wood and paper products	0.32	0.10	0.10	0.12	0.68	0.25	0.08	0.08	0.09	0.75
Fuel and chemicals, rubber and plastic	0.49	0.04	0.27	0.19	0.51	0.47	0.04	0.26	0.17	0.53
Non-metallic mineral products	0.33	0.03	0.12	0.19	0.67	0.30	0.02	0.11	0.18	0.70
Metallurgy, machinery and equip., motor vehicles	0.32	0.05	0.11	0.16	0.68	0.29	0.04	0.11	0.14	0.71
Furniture and other manufacturing products	0.40	0.09	0.10	0.20	0.60	0.37	0.11	0.09	0.16	0.65
Energy	0.45	0.04	0.13	0.27	0.55	0.45	0.04	0.15	0.26	0.55
Construction	0.35	0.06	0.14	0.15	0.65	0.34	0.06	0.14	0.14	0.66
Services	0.39	0.07	0.13	0.17	0.61	0.35	0.07	0.13	0.15	0.65
Hungary										
Agriculture	0.66	0.09	0.19	0.38	0.34	0.67	0.09	0.20	0.38	0.33
Mining and quarrying	0.65	0.00	0.17	0.48	0.35	0.62	0.01	0.07	0.54	0.38
Manufacturing	0.47	0.08	0.09	0.29	0.53	0.51	0.12	0.10	0.29	0.49

Table 11.2 cont'd

	Regional share of national employment, 1992					Regional share of national employment, 1999				
		Border regions			Internal		Border regions			Internal
	Total	BEU	BEX	BAC	regions	Total	BEU	BEX	BAC	regions
Food, beverages and tobacco	0.62	0.08	0.15	0.39	0.38	0.63	0.09	0.14	0.40	0.37
Textiles, clothing and leather	0.56	0.15	0.12	0.29	0.44	0.64	0.18	0.14	0.33	0.36
Wood and paper products	0.45	0.08	0.11	0.26	0.55	0.45	0.08	0.12	0.26	0.55
Fuel and chemicals, rubber and plastic	0.32	0.04	0.03	0.25	0.68	0.38	0.08	0.04	0.25	0.62
Non-metallic mineral products	0.47	0.03	0.08	0.36	0.53	0.58	0.08	0.07	0.44	0.42
Metallurgy, machinery and equip., motor vehicles	0.38	0.08	0.06	0.24	0.62	0.43	0.12	0.09	0.22	0.57
Furniture and other manufacturing products	0.58	0.07	0.12	0.39	0.42	0.53	0.12	0.10	0.31	0.47
Energy	0.61	0.10	0.14	0.38	0.39	0.60	0.07	0.13	0.40	0.40
Construction	0.46	0.08	0.10	0.28	0.54	0.46	0.07	0.11	0.27	0.54
Services	0.44	0.06	0.10	0.27	0.56	0.37	0.05	0.09	0.23	0.63
Estonia										
Agriculture	1.00	0.56	—	0.44	—	1.00	0.47	—	0.53	—
Mining and quarrying	NA	NA	—	NA	—	NA	NA	—	NA	—
Manufacturing	1.00	0.79	—	0.21	—	1.00	0.73	—	0.27	—
Food, beverages and tobacco	1.00	0.70	—	0.30	—	1.00	0.72	—	0.28	—
Textiles, clothing and leather	1.00	0.81	—	0.19	—	1.00	0.80	—	0.20	—
Wood and paper products	1.00	0.73	—	0.27	—	1.00	0.52	—	0.48	—

Table 11.2 cont'd

| | Regional share of national employment, 1992 | | | | | Regional share of national employment, 1999 | | | | |
| | | Border regions | | | Internal | | Border regions | | | Internal |
	Total	BEU	BEX	BAC	regions	Total	BEU	BEX	BAC	regions
Fuel and chemicals, rubber and plastic	1.00	0.95	—	0.05	—	1.00	0.78	—	0.22	—
Non-metallic mineral products	1.00	0.87	—	0.13	—	1.00	0.75	—	0.25	—
Metallurgy, machinery and equip., motor vehicles	1.00	0.77	—	0.23	—	1.00	0.85	—	0.15	—
Furniture and other manufacturing products	1.00	0.72	—	0.28	—	1.00	0.67	—	0.33	—
Energy	NA	NA	—	NA	—	NA	NA	—	NA	—
Construction	NA	NA	—	NA	—	NA	NA	—	NA	—
Services	1.00	0.74	—	0.26	—	1.00	0.73	—	0.27	—
Slovenia (1997 and 1999)										
Agriculture	0.99	0.53	0.47	—	0.01	1.00	0.54	0.46	—	0.00
Mining and quarrying	0.80	0.13	0.67	—	0.20	0.80	0.12	0.68	—	0.20
Manufacturing	0.98	0.48	0.50	—	0.02	0.98	0.48	0.49	—	0.02
Food, beverages and tobacco	0.99	0.51	0.48	—	0.01	0.99	0.52	0.47	—	0.01
Textiles, clothing and leather	0.98	0.53	0.45	—	0.02	0.98	0.53	0.45	—	0.02
Wood and paper products	0.99	0.38	0.61	—	0.01	0.99	0.38	0.61	—	0.01
Fuel and chemicals, rubber and plastic	0.99	0.42	0.56	—	0.01	0.99	0.45	0.54	—	0.01
Non-metallic mineral products	0.89	0.42	0.47	—	0.11	0.89	0.44	0.45	—	0.11

Table 11.2 cont'd

| | Regional share of national employment, 1992 | | | | | Regional share of national employment, 1999 | | | | |
| | | Border regions | | | | | Border regions | | | |
	Total	BEU	BEX	BAC	Internal regions	Total	BEU	BEX	BAC	Internal regions
Metallurgy, machinery and equip., motor vehicles	0.97	0.51	0.46	–	0.03	0.97	0.51	0.46	–	0.03
Furniture and other manufacturing products	0.98	0.40	0.58	–	0.02	0.98	0.39	0.59	–	0.02
Energy	0.95	0.41	0.54	–	0.05	0.94	0.41	0.54	–	0.06
Construction	0.98	0.45	0.53	–	0.02	0.98	0.44	0.54	–	0.02
Services	0.98	0.44	0.54	–	0.02	0.98	0.44	0.54	–	0.02
Romania										
Agriculture	0.53	–	0.22	0.31	0.47	0.53	–	0.22	0.30	0.47
Mining and quarrying	0.29	–	0.15	0.14	0.71	0.31	–	0.13	0.17	0.69
Manufacturing										
Food, beverages and tobacco	0.44	–	0.16	0.28	0.56	0.39	–	0.15	0.25	0.61
Textiles, clothing and leather	0.43	–	0.20	0.23	0.57	0.44	–	0.19	0.25	0.56
Wood and paper products	0.31	–	0.16	0.15	0.69	0.36	–	0.24	0.12	0.64
Fuel and chemicals, rubber and plastic	0.24	–	0.09	0.15	0.76	0.25	–	0.09	0.17	0.75
Non-metallic mineral products	0.28	–	0.12	0.16	0.72	0.25	–	0.12	0.13	0.75
Metallurgy, machinery and equip., motor vehicles	0.33	–	0.15	0.18	0.67	0.37	–	0.17	0.19	0.63
Furniture and other manufacturing products	NA	–	NA	NA	NA	NA	–	NA	NA	NA

Table 11.2 cont'd

| | Regional share of national employment, 1992 | | | | | Regional share of national employment, 1999 | | | | |
| | *Total* | *Border regions* | | | *Internal regions* | *Total* | *Border regions* | | | *Internal regions* |
		BEU	*BEX*	*BAC*			*BEU*	*BEX*	*BAC*	
Energy	0.38	–	0.14	0.24	0.62	0.39	–	0.15	0.24	0.61
Construction	0.37	–	0.14	0.24	0.63	0.39	–	0.15	0.24	0.61
Services	0.40	–	0.15	0.24	0.60	0.41	–	0.17	0.25	0.59

Table 11.3 Annual average employment growth in region by groups of economic activities, 1992–99

	BEU	BEX	BAC	BORDER	INT	Country average
Bulgaria						
Agriculture	0.026	0.013	0.018	0.018	0.022	0.02
Mining and quarrying	−0.156	−0.071	−0.063	−0.095	−0.045	−0.074
Manufacturing	−0.038	−0.052	−0.068	−0.056	−0.053	−0.054
Food, beverages and tobacco	−0.009	−0.035	−0.051	−0.039	−0.018	−0.027
Textiles, clothing and leather	0.01	−0.029	−0.033	−0.018	−0.041	−0.029
Wood and paper products	−0.103	−0.115	−0.12	−0.113	−0.065	−0.079
Fuel and chemicals, rubber and plastic	−0.036	−0.038	−0.053	−0.044	−0.029	−0.036
Non–metallic mineral products	−0.09	−0.07	−0.065	−0.069	−0.051	−0.057
Metallurgy, machinery and equip., motor vehicles	−0.102	−0.061	−0.089	−0.081	−0.063	−0.068
Furniture and other manufacturing products	−0.121	−0.157	−0.169	−0.154	−0.136	−0.143
Energy	0.01	0.013	−0.006	0.001	−0.001	0
Construction	−0.056	−0.061	−0.07	−0.064	−0.059	−0.061
Services	0.006	0.004	−0.007	−0.001	0.021	0.013
Hungary						
Agriculture	−0.108	−0.1	−0.103	−0.103	−0.108	−0.105
Mining and quarrying	0.083	−0.305	−0.202	−0.22	−0.207	−0.215
Manufacturing	0.027	−0.005	−0.023	−0.009	−0.03	−0.02
Food, beverages and tobacco	−0.027	−0.059	−0.046	−0.046	−0.052	−0.049
Textiles, clothing and leather	0.003	−0.002	−0.001	0	−0.048	−0.019
Wood and paper products	−0.017	−0.015	−0.02	−0.018	−0.018	−0.018
Fuel and chemicals, rubber and plastic	0.099	0.015	−0.02	0.002	−0.036	−0.023
Non–metallic mineral products	0.105	−0.065	−0.005	−0.004	−0.065	−0.034

Table 11.3 cont'd

	BEU	BEX	BAC	BORDER	INT	Country average
Metallurgy, machinery and equip., motor vehicles	0.058	0.066	−0.018	0.015	−0.016	−0.003
Furniture and other manufacturing products	0.067	−0.039	−0.043	−0.025	0.002	−0.013
Energy	−0.065	−0.026	−0.017	−0.026	−0.022	−0.024
Construction	−0.044	−0.028	−0.045	−0.041	−0.039	−0.04
Services	−0.027	−0.034	−0.028	−0.03	0.011	−0.006
Estonia						
Agriculture	−0.154	–	−0.11	−0.133	–	−0.133
Mining and quarrying	NA	–	NA	NA	–	NA
Manufacturing	−0.059	–	−0.014	−0.048	–	−0.048
Food, beverages and tobacco	−0.023	–	−0.038	−0.027	–	−0.027
Textiles, clothing and leather	−0.065	–	−0.05	−0.062	–	−0.062
Wood and paper products	0.028	–	0.169	0.078	–	0.078
Fuel and chemicals, rubber and plastic	−0.139	–	0.086	−0.114	–	−0.114
Non–metallic mineral products	−0.139	–	−0.032	−0.12	–	−0.12
Metallurgy, machinery and equip., motor vehicles	−0.068	–	−0.131	−0.08	–	−0.08
Furniture and other manufacturing products	−0.042	–	−0.011	−0.033	–	−0.033
Energy	NA	–	NA	NA	–	NA
Construction	NA	–	NA	NA	–	NA
Services	0.003	–	0.011	0.005	–	0.005
*Slovenia**						
Agriculture	−0.073	−0.089	–	−0.081	−0.159	−0.081
mining and quarrying	−0.078	−0.036	–	−0.043	−0.056	−0.046
Manufacturing	−0.016	−0.022	–	−0.019	−0.025	−0.019
Food, beverages and tobacco	−0.005	−0.027	–	−0.015	−0.038	−0.016

Table 11.3 cont'd

	BEU	BEX	BAC	BORDER	INT	Country average
Textiles, clothing and leather	−0.066	−0.059	—	−0.063	−0.134	−0.064
Wood and paper products	−0.019	−0.017	—	−0.018	−0.038	−0.018
Fuel and chemicals, rubber and plastic	0.03	−0.009	—	0.008	−0.104	0.006
Non−metallic mineral products	−0.002	−0.052	—	−0.028	−0.001	−0.025
Metallurgy, machinery and equip., motor vehicles	−0.002	−0.005	—	−0.003	0.016	−0.003
Furniture and other manufacturing products	−0.048	−0.029	—	−0.037	−0.006	−0.036
Energy	−0.034	−0.037	—	−0.036	−0.022	−0.035
Construction	0.01	0.035	—	0.024	0.031	0.024
Services	0.022	0.022	—	0.022	0.04	0.022
Romania						
Agriculture	—	0.001	−0.001	0	0.002	0.001
Mining and quarrying	—	−0.093	−0.052	−0.072	−0.08	−0.077
Manufacturing	—	−0.06	−0.065	−0.063	−0.075	−0.071
Food, beverages and tobacco	—	−0.047	−0.047	−0.047	−0.018	−0.03
Textiles, clothing and leather	—	−0.059	−0.045	−0.052	−0.057	−0.055
Wood and paper products	—	0.029	−0.065	−0.018	−0.039	−0.029
Fuel and chemicals, rubber and plastic	—	−0.085	−0.075	−0.08	−0.087	−0.085
Non−metallic mineral products	—	−0.071	−0.098	−0.085	−0.064	−0.07
Metallurgy, machinery and equip., motor vehicles	—	−0.075	−0.082	−0.078	−0.101	−0.093
Furniture and other manufacturing products	—	NA	NA	NA	NA	NA
Energy	—	0.022	0.012	0.017	0.013	0.014
Construction	—	−0.062	−0.071	−0.066	−0.078	−0.074
Services	—	−0.016	−0.027	−0.022	−0.032	−0.029

*1997–99

regions, and then verified in which locations industrial employment has grown faster (Hanson, 1998).[12]

To reach these simple goals, I have assumed the following general expression for labor demand in industry j located in region i at time t:

$$E_{ijt} = \alpha_{ijt} + \beta W_{ijt} \, \gamma X_{ijt} \, \varepsilon_{ijt} \qquad (1)$$

where E_{ijt} denotes employment, W_{ijt} the wage, and X_{ijt} a vector of variables able to affect the location of economic activities at the regional and sectoral level, while ε_{ijt} is an i.i.d labor demand shock that has mean zero and constant variance.

Following the most recent developments in the literature, I assume that both comparative advantage and economic geography factors might determine the location of economic activities both at a national and a sub-national level (Overman, Redding, Venables, 2001). This implies that vector X in eq. (1) should include at least two types of variables:

1) *geography* variables, such as the distance between economic agents and agglomeration economies. Distance is directly related to transaction costs, because of the transport costs of shipping goods, the costs of contracting at a distance, and the costs of acquiring information about distant economies. Intuitively, this implies that economic activities will concentrate close to large markets to minimize transport costs. Agglomeration economies (i.e., opportunities to create a network with other firms operating in the same sector or in a different industrial branch) may explain why firms locate close to each other. They might reinforce cumulative causation processes of location or cause firms to refrain from relocating elsewhere;
2) *comparative advantage* variables, deriving from natural, i.e., exogenous, factors such as proximity, regional accessibility, and the endowment of natural resources, a well as characteristics of the local economic environment, such as the structure of the labor force, the educational level, the availability of services related to production activities, etc.

Eventually, the choice of the variables to include in the empirical analysis has to take into account two further elements: the peculiar experience of transition countries and the availability of reliable figures for a sufficiently long time series.[13] Concerning the former, several empirical studies have shown the key role played by *foreign direct investment* (FDI) in transition countries.[14] FDI, even more than trade, has driven the integration process with the EU

(Döhrn, 2001), and has contributed to the economic restructuring process, bringing into the area financial capital as well as new technology, skills and managerial know-how, which in turn have generated positive spillovers to the domestic economy (Konings, 1999; Damijan, Majcen, 2000; Djankov, Hoeckman, 2000). Finally, FDI may also generate agglomeration processes for domestic firms through linkages with local suppliers (Altomonte, Resmini, 2001). These consideration and data constraints yield the equation that will be estimated:

$$
\ln\left(\frac{E_{ijt}}{E_{ijt}}\right) = \alpha_{ijt} + \beta_1 \ln(\frac{W_{it-1}}{W_{t-1}}) + \beta_2 \ln\left(\frac{DIST_i}{\sum_i \omega_{ijt} DIST_i}\right) + \beta_3 \ln(FDI_i)
$$
$$
+ \beta_4 \ln(\frac{E_{ist}}{E_{st}}) + \beta_5 \ln(ROAD_{it}) + \beta_6 \ln(STUD_{it}) + \varepsilon_{ijt}
$$

$$(2)$$

where i indicates regions, j industries, t time and s the service sector.

The dependent variable is regional employment in sector j, measured relative to national employment in order to control for national trends and country size. The first term on the right side of eq (2) is the average regional wage relative to the national wage. In order to avoid introducing simultaneity into the regression, relative wages have been lagged by one period. Relative employment might be decreasing or increasing with relative wages, according to the conditions of local labor markets. The second term is a proxy for geographical distance, which I measure as road distance from region i to the capital city relative to the industry-weighted average distance from the capital. The distance variable should be uncorrelated with relative employment if trade liberalization and transition have reoriented core markets towards foreign markets; otherwise it should be negatively correlated with relative employment, since transport costs increase with distance. The third term in eq. (2) captures the role of FDI in developing regional economies. I measured FDI as the number of foreign firms in region i at time t per 100,000 inhabitants, in order to take regional size effects into account. To the extent that FDI plays a positive role in promoting local development through spillovers and linkages, I expect relative employment to be increasing with FDI. However, since foreign firms have been heavily involved in restructuring activities, mainly in the early transition period, the impact on relative employment might be negative. The fourth term in eq. (2) measures relative employment in the service sector. Since services are supposed to give a positive contribution to economic activity, I expect that it positively affects the location of economic activities. The fifth term in eq. (2) is a proxy for the region's accessibility, which I measure as road density. I expect relative employment to be higher where the endowment of infrastructure is higher. Finally, the sixth term of eq. (2) indicates the size of the skilled labor force, measured indirectly through the number of third level students per 100,000 inhabitants at a regional level. Again, normalization is needed to take into account effects related to differences in region size.

Concerning the error term, I control for the possibility that there are idiosyncratic components within economic activity location at a regional level by allowing it to have the following structure:

$$\varepsilon_{ijt} = \tau_i + \kappa_j + \eta_i + \mu_{ijt} \tag{3}$$

where τ_i is a fixed region-type effect,[15] κ_j is a fixed industry effect, η_t is a fixed year effect and μ_{ijt} is an i.i.d. random variable with mean zero and variance σ^2. I choose fixed effects rather than random effects estimation since relative employment is a consequence of both regional and industrial characteristics. From a technical point of view, this indicates that μ_{ijt} cannot be considered uncorrelated across regions and industries. Thus, a fixed effects estimation is more appropriate (Baltagi, 2001). Given the size of the sample, using dummy variables to control for fixed effects does not substantially reduce the degrees of freedom of the regression. The relatively large number of observations also allows for the estimation of a variable coefficient model, which aims at evaluating potential differences in the explanatory power of the exogenous variables in each group of regions.

In order to study in which location relative employment has grown faster, I assume that the average growth rate of relative employment over the period can be expressed as a function of the initial conditions of the relative employment of industry j in region i, and other regions' characteristics as well. This specification allows us to avoid introducing simultaneity into the regression, which has the following structure:

$$\ln\left(\frac{E_{ijT}/E_{jT}}{E_{ijt}/E_{jt}}\right) = \alpha_{ij} + \beta_1 \ln(\frac{W_{it-1}}{W_{t-1}}) + \beta_2 \ln\left(\frac{DIST_i}{\sum_i \omega_{ijt} DIST_i}\right) + \beta_3 \ln(FDI_{it})$$

$$+ \beta_4 \ln(\frac{E_{ist}}{E_{st}}) + \beta_5 \ln(ROAD_{it}) + \beta_6 \ln\left(\frac{E_{ijt}}{E_{jt}}\right) + \beta_7 \ln(STUD_{it}) + \varepsilon_{ijt}$$

$$\tag{4}$$

where T indicates the final period (1999) and t the initial period (1993).

Eq. (4) has been estimated twice, first without controlling for fixed effects and then including dummy variables for region types (different types of borders and non-border regions) and industries. The equation has been estimated by OLS. Since there are two potential sources of heteroscedasticity (across regions and across industries), I use White's (1980) correction in order to obtain consistent standard errors.

5   Estimation Results

Table 11.4 gives estimation results on relative employment (equation 2). Column (1) presents estimation results for pooling all observations across sectors, years and regions, while in the following columns the hypotheses of common intercepts across regions, sectors and years have been progressively relaxed.

All control variables significantly affect the location of the manufacturing activities, with the exception of the relative wage variable, which is never significant. Relative distance to the capital is negatively related to relative employment in all regressions. This suggests that distance from the capital reduces regional labor demand. Despite trade liberalization and economic integration with the EU, the domestic market still determines the location of economic activities within a country. The results also show that relative employment is positively correlated with the infrastructure variable, the FDI variable, the number of students and relative employment in the service sectors. The largest quantitative effects are those related to the service sector and the road variable.

These results hold also when controlling for fixed effects. Relative employment is different across regions and sectors, while the location of economic activities is only weakly affected by the passage of time.

The estimated coefficients are averages for all regions included in the sample. However, the location of economic activities may respond differently to the explanatory variables according to the geographical position of each region with respect to the borders. To determine the individual influence of each explanatory variable, I re-estimate equation (2), allowing for separate slope parameters in each of the four groups of regions previously identified. The resulting coefficients are shown in Table 11.A2. The most striking changes from the previous results concern internal regions, whose capacity to attract economic activities seems to rely only on the presence of foreign firms, an educated labor force and services.

Table 11.4 Regression results on regional industry relative employment, 1992–99

Variable	Pool (1)	(2)	FE (3)	(4)
Rel. wage	0.044	0.0043	0.043	0.038
	(0.033)	(0.038)	(0.037)	(0.038)
Rel. distance	−0.080	−0.062	−0.051	−0.055
	(0.013)***	(0.015)***	(0.015)***	(0.015)***
FDI	0.085	0.089	0.085	0.092
	(0.012)***	(0.012)***	(0.012)***	(0.013)***
Roads	0.225	0.226	0.229	0.235
	(0.026)***	(0.031)***	(0.029)***	(0.029)***
Services	0.656	0.676	0.690	0.684
	(0.035)***	(0.027)***	(0.026)***	(0.026)***
Students	0.054	0.048	0.046	0.045
	(0.011)***	(0.011)***	(0.011)***	(0.011)***
Constant	−2.787	−2.699	−2.409	−2.93
	(0.248)***	(0.205)***	(0.203)***	(0.249)***
Region dummy	–	$F_{(3,3814)} = 5.91***$	$F_{(3,3807)} = 6.79***$	$F_{(3,3800)} = 6.44***$
Industry dummy	–	–	$F_{(6,3807)} = 33.03***$	$F_{(6,3800)} = 32.90***$
Year dummy	–	–	–	$F_{(6,3800)} = 2.04**$
No. of obs	3824	3824	3824	3824
R2	0.40	0.38	0.41	0.42
Root MSE	0.83	0.82	0.80	0.80

Robust standard errors are in parentheses.

*** indicates statistical significance at 0.01 level; ** indicates statistical significance at 0.05 level.

Differences across border regions are less marked, but perhaps more interesting. Relative employment in regions bordering external (BEX) and other candidate countries (BAC) is lower when the functional distance from the capital is higher, indicating a strong dependence of these peripheral regions on domestic markets. In BEU regions, instead, the interaction between relative employment and distance is still significant, but positive. This result indicates that bordering advanced countries – as the EU may be in comparison with transition countries – may mitigate the disadvantage of being in a peripheral position. FDI contributes positively to relative employment in all regions, except those bordering the EU, indicating that in BEU regions, foreign firms have a dominant role in the economic system. Road density does not affect the location of economic activities in BEU regions, suggesting that in these regions, economic links with foreign markets are stronger than those with internal markets, thus reducing the importance of a good system of infrastructure connecting regions *within* a country. The presence of a skilled labor force positively affects relative employment in all regions, but its level of significance is very weak in BAC regions (at 10 per cent), while manufacturing activities in BEX regions do not seem to be affected by the location of tertiary activities within the regions. Finally, it is worth noticing that, from a quantitative point of view, the service variable exerts the strongest impact in BEU regions, while the road variable coefficient takes its highest value in BEX regions, indicating that external regions need to have good accessibility in order not to be penalized by their peripheral location.

Overall, these results indicate that economic integration and trade liberalization with the EU have had different impacts on the location of economic activities in border and non-border regions. Moreover, they also confirm that border regions cannot be treated as a homogenous set of regions. The location of economic activities in border regions responds differently to the explanatory variables according to their geographical location.

5.1   Prospects for Growth

Table 11.5 gives the estimation results for eq. (4), i.e., relative employment growth over the period 1993–99. Among the control variables, only initial level of relative employment and FDI seem to be able to generate some relocation activities. In particular, relative employment growth is higher where the initial level of relative employment is lower, a sign of convergence across regions, and where the initial levels of FDI are higher. There is no evidence that relative distance is related to relative employment growth. The coefficient of

Table 11.5   Regression results: regional industry relative employment growth over the period 1993–99

Variable	(1)	(2)	(3)
Relative employment	−0.219	−0.219	−0.215
	(0.044)***	(0.044)***	(0.043)***
Relative wage	0.153	0.214	0.211
	(0.208)	(0.217)	(0.218)
Relative distance	−0.001	−0.008	−0.009
	(0.017)	(0.018)	(0.018)
FDI	0.080	0.071	0.071
	(0.024)***	(0.025)***	(0.025)***
Road	0.044	0.026	0.026
	(0.050)	(0.053)	(0.054)
Services	0.071	0.055	0.051
	(0.047)	(0.047)	(0.048)
Student	−0.012	−0.007	−0.007
	(0.022)	(0.023)	(0.024)
Constant	−0.879	−0.879	−0.884
	(0.327)***	(0.350)**	(0.372)**
Region dummies	–	$F_{(3,468)}=3.48$**	$F_{(3,461)}=3.41$**
industry dummies	–	–	$F_{(7,461)}=0.59$
No. of obs	479	479	479
R^2	0.17	0.17	0.18
Root MSE	0.402	0.404	0.404

Robust standard errors are in parenthesis.
*** indicates statistical significance at 0.01 level; ** indicates statistical significance at 0.05 level; * indicates statistical significance at 0.10 level.

the variable is negative, but statistically insignificant in all regressions. These effects are common to all sectors, and the hypothesis of heterogeneity among regions is supported by data only at a 0.05 level of significance. Consequently, economic integration and trade liberalization are likely to affect economic growth across regions only weakly, depending on their location within the country or along the borders.

In order to better understand how the enlargement process will affect regions' prospects for growth, I construct predicted growth rates using the estimated coefficients of model (3) in Table 11.5. The results are given in Table 11.6, which points out several striking features. On average, border regions have better prospects for growth than internal ones, which tend to stagnate. Among border regions, BEU ones show the highest rate of growth in relative employment, followed by BEX regions. Regions bordering other candidate countries enjoy positive rates of growth, but these are much lower than those of other border regions, and are more similar to those of internal regions. With respect to countries, all Hungarian regions are above the average in their respective categories, with the exception of BAC regions, while Bulgarian and Romanian regions show growth rates below the average, with the exception of BEX and BAC regions, respectively, indicating that regional adjustments are not independent of individual country effects.

Table 11.6   **Predicted growth rates over the period by groups of regions and country (per cent)**

	BEU	BEX	BAC	INT	INT*
Group average	10.7	10.3	3.6	−1.2	−0.6
Within country:					
Bulgaria	9.6	10.8	−4.0	−1.9	−1.9
Estonia	−7.8	–	NA	–	–
Hungary	18.3	13.3	1.7	3.0	5.4
Romania	–	8.6	11.6	−2.9	−2.8

* Without capital city districts.
NA = not estimated.

6   Concluding Remarks

This chapter provides an initial rigorous framework in which regional adjustments in Central and Eastern Europe may be assessed and understood. The need for an in-depth analysis of the impact of the enlargement process on candidate countries at the regional level has often been highlighted, but the lack of consistent and reliable statistics that are homogenous across countries and regions has made this analysis difficult and has limited it to qualitative insights about the spatial effects generated by the strengthening of economic integration with the EU. This chapter aims to fill this gap. It presents empirical evidence that the location and growth of economic activities in candidate countries may be conditioned by region-specific effects. The analysis provides interesting results which, interpreted cautiously, can be summarized as follows.

Border regions do not represent a homogeneous set of regions, since the economic performance of frontier areas is affected not only by their relative position within a country with respect to its economic center – which often coincides with the capital city in transition countries – but also by the economic conditions of the neighbouring foreign countries. For these reasons, border areas are more sensitive to regional accessibility and distance from the capital city than internal regions, though interesting differences can be identified within each group of homogeneous border regions.

BEU regions seem to take advantage of their location since it has stimulated the catching-up process: economic activity is attracted by a skilled labor force and a well-developed service sector, while FDI increases productivity and efficiency while reducing relative employment. Their peripheral location with respect to their respective capitals does not seem to be a problem, since economic activity is not affected by the regions' accessibility (measured in a national dimension). In conclusion, BEU regions seem to have many of the characteristics of what has been defined as an 'active contact space' (Nijkamp, 1998; Van Geenhuizen, Ratti, 2001), and the analysis of the prospects for growth further reinforces this conclusion.

BEX regions have caused concern among both economists and policymakers. It was thought that their very peripheral position, not only within their respective countries but also with respect to the EU, and their proximity to economically weak countries would present a serious obstacle to their economic development. However, this chapter does not confirm this pessimistic picture. FDI and infrastructure connections with the capital city are able to attract economic activities to these regions, and also to overcome the negative effect generated by the distance variable.

BAC regions do not present serious concerns. Manufacturing activity is penalized by their distance from the capital city, but takes advantage of infrastructure, FDI, and the presence of service activities. A skilled labor force does not seem to have any effect on the location of manufacturing within this group of regions, indicating the prevalence of traditional, labor-intensive activities.

Manufacturing activities in internal regions seem to be attracted only by a well-developed service sector (as it usually is in the capital city, which belongs to this group of regions with the exception of Tallin, Estonia) and to a lesser extent by FDI and a skilled labor force.

Concerning growth rates, two interesting results deserve particular attention. First, employment growth at the regional level depends negatively on the initial level of employment in each sector and positively on FDI. Region-specific effects are weakly supported by the data. Overall, these results suggest that a convergence process is working within countries, but not with respect to the EU average, as is shown by the transition matrices computed in section 3.

Finally growth prospects seem to confirm the better position of border regions relative to internal ones. The former are, on average, expected to grow, while the latter show stagnation or a small decline, other things being equal. Among border regions, BEX and BEU show the highest predicted growth rates. It is, however, worth noticing that prospects for growth are country-specific.

Many of these results are in the range of what one might have expected and therefore allow for some confidence in the reliability of the data and methodology. Altogether, they suggest a less dramatic view of the spatial effects of the enlargement process in candidate countries. However, the time period considered is too short and eventful, which makes the availability of more detailed and longer time series data desirable and necessary to completely understand the consequences of the enlargement process.

Notes

*   This research was undertaken with support from the European Community's Phare ACE Programme 1998. The content of the publication is the sole responsibility of the authors and it in no way represents the views of the Commission or its services.

1   Borders are defined as 'external state boundaries' (Anderson, O'Dowd, 1999), while border regions are 'sub-national areas, whose economic and social life is directly and significantly affected by proximity to an international frontier' (Hansen, 1997a, 1997b, p. 1).

2   In this case, however, it is hard to see some of the advantages for border regions, since regions affected by trade liberalization can no longer be considered border areas, as pointed out by Hansen (1977a).

3   This idea is not new in economics. It can be found in the pioneering works of Myrdal (1957), Hirschman (1958) and Pred (1966). Only its formal analysis can be ascribed to NEG. See Fujita, Krugman and Venables (2000) for a survey of the links between old and new agglomeration theories.

4   The aggregate analysis considers mainly Bulgaria, Hungary and Romania. Estonia and Slovenia have been studied separately for two reasons. First, figures cover a period of time shorter than those for the other three countries. Secondly, the large proportion of 'border area' in these countries makes any comparison between border and non-border regions worthless. For an in-depth analysis of the relative position of border regions in each country see Bosco, Resmini (2001).

5   Non-border regions as a whole perform better than BEU regions only when capital cities are considered.

6   On the dominant role of capital city regions, see also Weise, Butcher, Downs et al. (2001).

7   Estonia and Slovenia are excluded from this exercise since the data cover a different time period.
8   Considering the EU-27 instead of the EU-15 would make these figures less dramatic since the EU average would be lower than the present one. See, for example, EC (2001).
9   Groups of economic activities include agriculture, mining and quarrying, manufacturing, energy, construction and services. The manufacturing sector has been further split up into seven sub-sectors. A more disaggregated analysis was not possible because manufacturing activity's classification varies across countries.
10   This is true for all countries but Estonia and Slovenia. See Table 11.1.
11   Slovenia is not included in the analysis because its figures cover a shorter period of time (1997–99). Concerning economic activity, I omit agriculture, mining and quarrying (whose location is mainly natural resource driven), services (given the impossibility of distinguishing between tradable and non tradable services) and metallurgy, machinery and equipment and transportation vehicles (a composite sector made up of industries very different from each other, created only to harmonize data across countries).
12   From a theoretical standpoint, the location of economic activities is endogenous, since it can generate cumulative causation agglomeration (Fujita, Krugman and Venables, 2000).
13   The latter aspect is, needless to say, more serious than the former, especially because I am working at a regional level.
14   See UN/ECE (2001) for a comprehensive survey on the role of FDI in transition countries.
15   Given the objective of the chapter, region-fixed effects have been considered as constant within the groups of homogeneous regions previously identified (see Table 11.1).

References

Altomonte, C. and Resmini, L. (2002), 'Multinational Corporations as a Catalyst for Local Industrial Development: the Case of Poland', *Scienze Regionali. Italian Journal of Regional Science*, No. 2, pp. 29–57.
Anderson, J. and O'Dowd, L. (1999), 'Borders, Border Regions and Territoriality: Contradictory Meanings, Changing Significance', *Regional Studies*, Vol. 33(7), pp. 593–604.
Avery, G. and Cameron, F. (1998), *The Enlargement of the European Union*, Sheffield Academic Press, Sheffield.
Baldwin, R., Francois, J. and Portes, R. (1997), 'EU Enlargement – Small Costs for the West, Big Gains for the East', *Economic Policy*, April.
Baltagi, B.H. (2001), *Econometrics Analysis of Panel Data*, 2nd edn, John Wiley & Sons, Ltd., Chichester.
Bosco, M.G. and Resmini, L. (2001), 'Border Regions in Candidate Countries: Comparative Facts and Figures', PHARE ACE Project n. P98–1117–R.
Damijan, J. and Majcen, B. (2000), 'Transfer of Technology through FDI, Spillover Effects and Recovery of Slovenian Manufacturing Firms', Mimeo, University of Ljubliana.
Djankov, S. and Hoeckman, B. (1998), 'Avenues of Technology Transfer: Foreign Investment and Productivity Change in the Czech Republic', CEPR Discussion paper 1883.
Döhrn, R. et al. (2001), 'The Impact of Trade and FDI on Cohesion', background paper for the Second Report on Cohesion, RWI, Essen, April.

EC (2001), *Unity, Solidarity, Diversity for Europe, its People and its Territory*, Second Report on Economic and Social Cohesion, vols 1 and 2, EC, Brussels.

Fujita, M., Krugman, P. and Venables, A.J. (2000), *The Spatial Economy. Cities, Regions and International Trade*, MIT Press, Cambridge, Mass.

Van Geenhuizen, M. and Ratti, R. (eds) (2001), *Gaining Advantage from Open Borders. An active Space for Regional Development*, Ashgate, Aldershot.

Hansen, N. (1977a), 'Border Regions: a Critique of Spatial Theory and a European Case Studies', *Annals of Regional Science*, Vol. 11, pp. 1–12.

Hansen, N. (1977b), 'The Economic Development of Border Regions', *Growth and Change*, Vol. 8, pp. 2–8.

Hanson, G. (1996), 'Economic Integration, Intra-industry Trade, and Frontier Regions', *European Economic Review*, Vol. 40, pp. 941–9.

Hanson, G. (1998), 'Regional Adjustment to Trade Liberalisation', *Regional Science and Urban Economics*, Vol. 28, pp. 419–44.

Hirschman, A. (1958), *The Strategy of Economic Development*, Yale University Press, New Haven, CT.

Konings, J. (1999), 'The Effects of Direct Foreign Investment on Domestic Firms: Evidence from Firm Level Panel Data in Emerging Economies', LICOS Discussion Paper No. 86, Leuven.

Krugman, P. (1998), 'What's New about the New Economic Geography?', *Oxford Review of Economic Policy*, Vol. 14(2), pp. 7–17.

Myrdal, G. (1957), *Economic Theory and Under-Developed Regions*, Duckworth, London.

Nijkamp, P. (1998), 'Moving Frontiers: a Local-Global Perspective', Vrije Universiteit of Amsterdam, Faculty of Business Administration and Econometrics, Research Memorandum no. 22.

Nijkamp, P. (1994), *New Borders and Old Barriers in Spatial Development*, Avebury, Aldershot.

Overman, H.G. and Puga, D. (1999), 'Unemployment Clusters across European Regions and Countries', CEPR discussion paper 2250.

Overman, H.G., Redding, S. and Venables, A. (2001), 'The Economic of Geography of Trade, Production and Income: a Survey of Empirics', CEPR discussion paper.

Pred, A.R. (1966), *The Spatial Dynamics of US Urban-industrial Growth, 1800–1914*, MIT Press, Cambridge, Mass.

Puga, D. (2001), 'European Regional Policies in Light of Recent Location Theories', CEPR Discussion paper 2767.

Resmini, L. (2000), 'The Determinants of Foreign Direct Investment in the CEECs: New Evidence from sectoral patterns', *The Economics of Transition*, Vol. 8(3), pp. 665–89.

UN/ECE (2001), 'Economic Growth and Foreign Direct Investment in the Transition Economies', ch. 5, *Economic Survey of Europe*, No.1, United Nations, Geneva, pp. 185–225.

Venables, A. (1998), 'The assessment: Trade and Location', *Oxford Review of Economic Policy*, Vol. 14(2), pp. 1–6.

Weise, C., Bachtler, J., Downes, R. et al. (2001), 'The Impact of EU Enlargement on Cohesion', background paper for the Second Report on Cohesion, DIW and EPRC, Berlin and Glasgow, March.

White, H. (1980), 'A Heterosckedasticity-consistent Covariace Matrix Estimator and a Direct Test for heteroskedasticity', *Econometrica*, vol. 48, pp. 817–38.

Appendix

Table 11.A1 Border regions: comparative facts and figures

a) With capital cities

		BEU			BEX			BAC			INT			COUNTRY		
		95	98	Var	95	98	Var	95	98	Var	95	98	Var	95	98	Var
BG	GDP	1058.43	1270.86	6.29	1301.62	1435.34	3.31	1153.02	1343.45	5.23	1201.94	1309.96	2.91	1194.9	1331.42	3.67
	POP	721.43	710.93	-0.49	1240.97	1270.86	0.8	1619.67	1573.41	-0.96	4802.65	4735.8	-0.47	8384.72	8291	-0.37
	UN	16.83	15.18	-3.38	11.66	12.72	2.95	15.2	16.69	3.16	11.85	13.24	3.77	13.11	14.09	2.43
HU	GDP	4706.86	5499.42	5.32	3418.02	3385.42	-0.32	3373.11	3382.05	0.09	5308.43	5797.71	2.98	4542.09	4640.26	0.72
	POP	698.10	698.10	0	1291.87	1275.98	-0.41	3545.09	3232.26	-3.03	4710.62	4660.04	-0.36	10245.68	9866.38	-1.25
	UN	10.77	7.9	-9.81	18.46	19.71	2.21	22.24	22.7	0.68	17.76	15.89	-3.64	16.5	14.1	-5.1
RO	GDP	–	–	–	3104.52	3066.41	-0.41	3403.42	3285.51	-1.17	6244.8	6749.58	2.62	4944.34	5188.49	1.62
	POP	–	–	–	4268.9	4268.43	0	5662.79	5602.55	-0.36	12749.26	12631.83	-0.31	22680.95	22502.8	-0.26
	UN	–	–	–	12.06	12.98	2.48	7.83	7.97	0.59	10	11.73	5.46	9.51	10.4	3.03
Total	GDP	2852.67	3375.04	5.77	2835.13	2834.43	-0.01	3056.87	3031.11	-0.28	4958.79	5378.73	2.75	–	–	–
	POP	1419.53	1409.04	-0.25	6801.74	6754.64	-0.23	10827.55	10408.21	-1.31	22262.523	22067.99	-0.29	–	–	–
	UN	14.41	12.35	-5	12.94	13.94	2.53	14.46	14.58	0.26	11.87	12.89	2.79	–	–	–
EE*	GDP*	3433.43	3733.66	4.28	–	–	–	1996.45	1976.57	-0.5	–	–	–	2981.42	3180.23	3.28
	POP*	1002.22	990.27	-0.6	–	–	–	459.91	455.31	-0.5	–	–	–	1462.13	1445.58	-0.57
	UN	5.09	4.8	-1.45	–	–	–	6.34	5.47	-4.79	–	–	–	5.04	4.75	-1.96
SLO**	GDP	6455.73	7519.55	7.93	5578.19	6397.68	7.09	–	–	–	20152.24	22975.57	6.78	6339.2	7318.09	7.44
	POP	940.32	938.27	-0.11	1000	999.96	0	–	–	–	47.16	46.71	-0.48	1987.5	1984.94	-0.06
	UN	13.96	13.31	-2.35	13.6	12.85	-2.8	–	–	–	17.74	18.48	2.08	15.1	14.88	-0.73

b) Without capital cities

		BEU			BEX			BAC			INT			COUNTRY		
		95	98	Var	95	98	Var	95	98	Var	95	98	Var	95	98	Var
BG	GDP	1058.43	1270.86	6.29	1301.62	1435.34	3.31	1153.02	1343.45	5.23	1167.69	1293.83	3.48	1176.54	1326.97	4.09
	POP	721.43	710.93	-0.37	1240.97	1270.86	0.8	1619.67	1573.41	-0.96	3609.91	3536.09	-0.69	7191.98	7091.29	-0.47
	UN	16.83	15.18	-3.38	11.66	12.72	2.95	15.2	16.69	3.16	12.48	14.01	3.92	13.4	14.47	2.59
HU	GDP	4706.86	5499.42	5.32	3418.02	3385.42	-0.32	3373.11	3382.05	0.09	3542.26	3927.2	3.5	3548.62	3747.18	1.83
	POP	698.1	698.1	0	1291.87	1275.98	-0.41	3545.09	3232.26	-3.03	2780.6	2798.66	0.22	8315.66	8005	-1.26
	UN	10.77	8.11	-9.02	18.46	19.71	2.21	22.24	21.82	-0.63	19.55	17.95	-2.81	19.74	19.18	-0.95
RO	GDP	-	-	-	3104.52	3066.41	-0.41	3403.42	3285.51	-1.17	3246.61	3144.46	-1.06	3260.44	3167.08	-0.96
	POP	-	-	-	4268.9	4268.43	0	5662.79	5602.55	-0.36	10416.64	10338.89	-0.25	20348.33	20209.86	-0.23
	UN	-	-	-	12.06	12.98	2.48	7.83	7.97	0.59	10.24	12.05	5.58	9.9	10.99	3.54
Total	*GDP*	*2852.67*	*3375.04*	*5.77*	*2835.13*	*2834.43*	*-0.01*	*3056.87*	*3031.11*	*-0.28*	*2929.93*	*2883.37*	*-0.53*	-	-	-
	POP	*1419.53*	*1409.04*	*-0.25*	*6801.74*	*6754.64*	*-0.23*	*10827.55*	*10408.21*	*-1.31*	*16807.15*	*16673.63*	*-0.27*	-	-	-
	UN	*14.41*	*12.35*	*-5*	*12.94*	*13.94*	*2.53*	*14.46*	*14.58*	*0.26*	*13.01*	*13.56*	*1.39*	-	-	-

GDP = GDP per capita; POP= population; UN= unemployment rate.
* 1996–98; ** Slovenia: GDP, POP 1995–97; UN 1997–98.

Table 11.A2 Regression results on regional industry relative employment (1992–99): variable coefficient model

Variable	(1)	(2)	BEU (1)	BEU (2)	BEX (1)	BEX (2)	BAC (1)	BAC (2)	INT (1)	INT (2)	Wald (1)	Wald (2)
Rel. wage			0.079	0.082	0.083	-0.143	0.033	-0.05	-0.232	-0.276	$F_{(3,3796)} = 1.53$	$F_{(3,3740)} = 1.27$
			(0.058)	(0.067)	(0.287)	(0.292)	(0.053)	(0.052)	(0.189)	(0.184)		
Rel. distance			0.295	0.227	-0.321	-0.184	-0.342	-0.289	-0.009	0.013	$F_{(3,3796)} = 25.77^{***}$	$F_{(3,3740)} = 17.01^{***}$
			(0.067)***	(0.068)***	(0.101)***	(0.103)*	(0.047)***	(0.046)***	(0.021)	(0.020)		
FDI			-0.259	-0.321	0.217	0.215	0.15	0.167	0.048	0.049	$F_{(3,3796)} = 17.84^{***}$	$F_{(3,3740)} = 17.09^{***}$
			(0.074)***	(0.092)***	(0.028)***	(0.028)***	(0.027)***	(0.027)***	(0.017)***	(0.017)***		
Road			0.147	0.129	1.204	1.230	0.167	0.187	0.026	0.019	$F_{(3,3796)} = 10.86^{***}$	$F_{(3,3740)} = 12.37^{**}$
			(0.182)	(0.227)	(0.161)***	(0.183)***	(0.058)***	(0.056)***	(0.079)	(0.076)		
Student			0.310	0.316	0.078	0.081	0.037	0.037	0.047	0.045	$F_{(3,3796)} = 4.14^{***}$	$F_{(3,3740)} = 4.01$
			(0.076)***	(0.080)***	(0.035)**	(0.034)**	(0.020)*	(0.020)	(0.019)**	(0.018)**		
Services			1.180	1.119	0.135	0.168	0.750	0.749	0.820	0.825	$F_{(3,3796)} = 22.15^{***}$	$F_{(3,3740)} = 18.77^{**}$
			(0.166)***	(0.195)***	(0.076)	(0.077)	(0.062)***	(0.060)***	(0.058)***	(0.057)***		
Industry dummy			No	Yes	No	Yes	No	Yes	No	Yes		
No. of obs	3824	3824	330		505		1208		1781			
R2	0.42	0.48										
Root MSE	0.79	0.76										

Robust standard errors are in parentheses. Regional fixed effects are not reported.

*** indicates statistical significance at 0.01 level; ** indicates statistical significance at 0.05 level; * indicates statistical significance at 0.1 level.

Last two columns give the Wald test statistics for the nul hypothesis of equal slope coefficients among groups of regions.

Chapter 12

The Emerging Economic Geography in EU Accession Countries: Concluding Remarks and Policy Implications

Iulia Traistaru, Peter Nijkamp and Laura Resmini

1 Introduction

This book provides evidence of the increased economic integration of five accession countries with the European Union and analyses its spatial implications for the five countries: Bulgaria, Estonia, Hungary, Romania and Slovenia. In particular, the following questions have been addressed: Has relocation of manufacturing activity taken place? How specialized/diversified are regions? How concentrated/dispersed are industries? Have patterns of regional specialization and concentration of manufacturing changed over the period 1990–99? What is the relationship between specialization and economic performance? How has increased trade between the European Union and these countries affected regional nominal and relative wages? What types of regions are potential winners and what types of regions are potential losers in the process of economic integration with the European Union? What policy implications could be concluded from the analysis of regional specialization and location of industrial activity in accession countries?

The research approach taken here combines five country analyses and three comparative analyses focused on patterns of regional specialization and concentration of industrial activity, adaptation processes in border regions and the impact of economic integration on the adjustment pattern of regional relative wages. The comparative analyses are particularly interesting because the five countries included in this study differ with respect to size, progress toward accession and geographical position. Romania is relatively large, while Hungary and Bulgaria are small and Estonia and Slovenia are very small. Estonia, Hungary and Slovenia will become EU members in 2004, while Bulgaria and Romania need to make further progress in fulfilling the accession criteria. Hungary and Slovenia are closer to the European core, Estonia is part of Northern Europe, and Bulgaria and Romania have a peripheral position in the southeastern part of Europe. Romania has no border with any current EU member state. Finally, Hungary is land-locked, while the other four countries have coastal areas.

This chapter summarizes the main findings of the papers included in this book and, on this basis, explores lessons that can be learned from them and policy implications.

2 Relocation of Manufacturing Activity

Over the last decade, in the five accession countries included in this study, there has been a relocation of manufacturing activities across regions. The structural changes in regional manufacturing employment are country-specific.

In Bulgaria, most industries seem to be moving away from border regions, in particular from regions bordering accession countries and regions bordering countries that are outside the EU enlargement area. Regions bordering the EU have gained employment in *textiles, clothing and leather* as well as *furniture and other manufacturing*, while non-border regions have increased their employment shares in the branches of *food, beverages and tobacco*, *wood and paper products* and *furniture and other manufacturing*, mainly at the expense of regions bordering accession countries.

In Hungary, there is a clear pattern of relocation of manufacturing to regions bordering the EU and regions bordering other accession countries at the expense of non-border regions and regions bordering countries outside the EU enlargement area. Regions bordering the EU have increased their employment shares, particularly in resource-intensive industries (*non-metallic mineral products, fuel, chemicals, rubber and plastics*, and *furniture*) and high-technology industries (*machinery, equipment and motor vehicles*), while regions bordering accession countries have gained employment in resource-intensive industries (*non-metallic mineral products*) and labour-intensive industries (*textiles, clothing and leather*).

In Romania, the pattern of manufacturing relocation is less clear. Non-border regions have gained employment in labour-intensive (*food, beverages and tobacco*) and resource-intensive (*non-metallic mineral products*) industries at the expense of regions bordering accession countries. Regions bordering countries outside the EU enlargement area have attracted, in particular, the *manufacturing of wood and paper products* (resource intensive) as well as *machinery, equipment and motor vehicles.* Regions bordering accession countries have gained employment in the manufacturing of *textiles, clothing and leather.*

In Estonia, manufacturing has moved away from regions bordering the EU to regions bordering accession countries, with the exception of *machinery, equipment and motor vehicles* and food, *beverages and tobacco.* The biggest shifts in regional employment shares have taken place for resource-intensive industries.

In Slovenia, there has been very little change in regional employment shares. The evidence suggests a possible tendency towards the relocation of manufacturing activities to regions bordering the EU at the expense of those bordering countries outside the EU enlargement area. In particular, regional employment shares in regions bordering the EU have increased in the cases of resource-intensive industries (*fuel, chemicals, rubber and plastics* as well as *non-metallic mineral products*).

3   Patterns of Regional Specialization and Concentration of Manufacturing

In the five accession countries included in this study, patterns of regional manufacturing specialization are found to differ depending on the geographical position of the regions and their proximity to the EU markets. For countries close to the EU accession area (Estonia, Hungary, Slovenia), regions bordering the EU are less specialized than the national average. In Bulgaria, which lags behind with the accession, regions bordering the EU are more specialized than other regions. Regions bordering other accession countries are found to be more specialized than the national average in Hungary, while in Estonia, Bulgaria and Romania, these regions are less specialized. Regions bordering countries outside the EU enlargement area are more specialized for all countries, with the exception of Romania. Non-border regions are less specialized in Bulgaria and Hungary and more specialized in Romania and Slovenia.

With a few exceptions, high specialization is associated with inferior economic performance, while regions with low specialization perform better than the national averages. High specialization is associated with superior economic performance in regions bordering countries outside the EU enlargement area in Romania and non-border regions in Bulgaria.

Overall, our findings suggest that highly specialized regions show economic performance that is inferior to national averages, while less specialized regions have better economic performance. Over the period 1990–99, regional specialization has increased significantly in Bulgaria and Romania, has decreased in Estonia and has not changed significantly in Hungary and Slovenia.

Manufacturing concentration patterns differ depending on industry characteristics such as economies of scale, technology intensity and the level of wages. Highly concentrated industries are those with large economies of scale, high technology and high wages. Industries with low technology and low wages are dispersed. Geographical concentration of manufacturing has increased significantly in Bulgaria and has not changed significantly in the remaining countries analysed here.

In Bulgaria, in the period 1990–99, absolute and relative specialization increased in 21 regions out of 28. Most manufacturing industries have become more concentrated. The most concentrated industry is *fuels* and the least concentrated is *food, beverages, and tobacco*. There has been a relative decrease in industrial employment in the big cities (Sofia, Varna, Plovdiv, Gabrovo, Stara Zagora) and traditional industrial centres (southwest, south central and north central).

In Estonia, over the period 1990–99, absolute and relative specialization decreased in four out of five regions. Absolute concentration increased in four out of 13 manufacturing branches, while relative concentration increased in six manufacturing branches. The most concentrated industry is *motor vehicles and transport equipment* and the least concentrated is *food, beverages and tobacco*.

In Hungary, in the period 1992–98, six out of 20 regions experienced increasing specialization in absolute terms, but relative specialization increased in only five regions. Western regions are more specialized than the rest of the country. Geographic concentration has increased in two out of eight industries. The most concentrated industry is *chemicals* and the least concentrated is *machinery and equipment*. Western and central regions are more dynamic and are catching up with the European averages, while increasing polarization is the trend in eastern and southern regions.

In Romania, in the period 1991–99, absolute specialization increased in 25 out of 41 regions, while relative specialization increased in 29 regions. The highest regional specialization is found in Bucharest, northeast, southwest and south and the lowest in the centre, west and northwest. Most manufacturing industries (ten out of 13) have become more concentrated. The most concentrated industry is *electrical machinery* and the least concentrated is *food, beverages and tobacco*. The evidence indicates that the share of manufacturing employment in the capital city is decreasing and the share of manufacturing employment in the centre region is increasing.

In Slovenia, in the period 1994–98, absolute and relative specialization increased in seven out of 12 regions. The lowest level of regional specialization is observed in the largest regions (Osrednjeslovenska and Podravska), while specialization is highest in the smallest regions (Spodnjeposavska and Zasavska) and in Pomurska region. There is evidence of the relocation of manufacturing activity between regions. Regions with initially large shares of manufacturing employment have benefited more from economic integration with the European Union, while regions with small shares of manufacturing employment have not gained in the relocation process.

4 Determinants of Industrial Location Patterns

Patterns of location for industrial activity are determined by an interaction between regional and industry characteristics. Relevant regional characteristics include size, factor endowments and geographical position relative to industrial centres. Industry characteristics that might affect location include economies of scale, technology and wages, as well as factor intensities.

The results of the estimated econometric model of determinants of industrial location patterns in the five accession countries suggest that both factor endowments and proximity to capital regions and EU markets explain the emerging economic geography in these countries. On average and other things being equal, industries are attracted by large markets. Industries with large economies of scale tend to locate close to capital region and regions bordering the European Union. Labour-intensive industries locate in regions endowed with large labour forces, while research-oriented industries are attracted by regions endowed with researchers.

5 Economic Integration and Adjustment in Border Regions

The analysis of determinants of location and growth of manufacturing activities in border regions suggests that regions bordering the European Union have taken advantage of their location since the beginning of the transition process. High wages, a skilled labour force and a well developed service sector have contributed to increasing employment in manufacturing activities relative to national averages. Foreign direct investment (FDI), however, has a negative impact on relative employment, probably due to the severe restructuring processes adopted by foreign firms to make the former state-owned enterprises competitive.

The location of manufacturing activities in other border regions has been positively affected not only by FDI, but also by regional accessibility – measured by road density – and the presence of a skilled labour force and service sector, while distance from the capital city, the core of economic activity in several candidate countries, exerts a negative effect on relative employment. Regional differences seem to be less marked when one considers relative employment growth, which seems to have followed a pattern of convergence within regions and to have been led mainly by FDI. Region specific effects are only weakly supported by the data.

Growth prospects seem to confirm the better position of border regions relative to internal ones. The former are, on average, expected to grow, while the latter show stagnation or a small decline, other things being equal. Among border regions, regions bordering the European Union and external countries (non-European Union, non-accession countries) show the highest predicted growth rates. It is, however, worth noting that prospects for growth are country-specific.

6 Economic Integration and Regional Relative Wages

Previous theoretical and empirical studies have indicated that regional nominal and relative wages are decreasing with transport cost (often proxied by distance) to industry centres. In a closed economy, the industrial centre is, in most cases, the capital region, which concentrates human and capital resources in the process of industrialization. Trade liberalization makes access to foreign markets more important and as a consequence some industrial activity is likely to relocate to border regions. As a consequence of a larger labour demand, wages in border regions are expected to rise and the importance of their distance from the capital region for explaining regional wage differentials is expected to decline.

The five country studies included in this book found empirical support for the increasing importance of geographical proximity to EU markets as a determinant of regional relative wages. Regional relative wages decrease with the distance from the capital cities, which remain the main industrial centres in the five countries. The effect of the distance to the capital regions on regional relative wages is, however, smaller for border regions and has become less important with the establishment of the European Agreements.

The analysis of the adjustment of regional wages in response to falling trade barriers between the EU and accession countries suggests that there is a tendency towards income polarization in Estonia and Romania, while declining regional relative wages are found in Bulgaria and Hungary. In Slovenia both catching-up and polarization tendencies are found.

7 Policy Implications

Increasing economic integration with the European Union during the 1990s has resulted in changes in the regional production structures in Bulgaria, Estonia, Hungary, Romania and Slovenia. This research has found that average regional specialization has increased in Bulgaria and Romania, decreased in Estonia and has not changed significantly in Hungary and Slovenia. Highly specialized regions are found to have economic performance below the national averages, while diversified regions have better economic performance. While capital regions are the main industrial centres in all five accession countries, we find evidence of the relocation of industrial activity to border regions near European core markets. The main factors driving industrial location in accession countries seem to be factor endowments and proximity to European markets. Border regions appear to be the winners in the integration process with positive and high growth rates, while internal regions are predicted to stagnate or decline. Regions bordering the EU and regions bordering external countries (non-European Union, non-accession countries) seem to have the best prospects for growth.

A differentiated policy is necessary, on the one hand, to encourage the specialization of regions in industries with growth potential and to reduce income polarization, on the other hand. In the first case, policy should concentrate on further restructuring and internationalizing the regional production structure, attracting foreign direct investment and enhancing innovative and technological potential. In the second case, support for local entrepreneurship and human resources development, upgrading of infrastructures aimed at capacity-building for dealing with structural weaknesses in performance, in particular, and expansion of the limited marketing horizons, in general, are suggested.

It is known that the determinants of industrial location patterns discriminate between large, small and medium-size firms. Especially in the accession countries with less access to international markets, the creation of a policy favouring the emergence of small and medium-size companies may be a wise strategy.

Index

Printed and bound by CPI Group (UK) Ltd, Croydon, CR0 4YY

22/10/2024

01777625-0018